Mathematics – Key Technology for the Future

Willi Jäger · Hans-Joachim Krebs
Editors

Mathematics – Key Technology for the Future

Joint Projects Between Universities and Industry 2004–2007

Willi Jäger
IWR, Universität Heidelberg
Im Neuenheimer Feld 368
69120 Heidelberg
Germany
jaeger@iwr.uni-heidelberg.de

Hans-Joachim Krebs
Projektträger Jülich
Forschungszentrum Jülich GmbH
52425 Jülich
Germany
h.-j.krebs@fz-juelich.de

ISBN 978-3-540-77202-6   e-ISBN 978-3-540-77203-3

DOI 10.1007/978-3-540-77203-3

Library of Congress Control Number: 2008923103

Mathematics Subject Classification (2000): 34-XX, 35-XX, 45-XX, 46-XX, 49-XX, 60-XX, 62-XX, 65-XX, 74-XX, 76-XX, 78-XX, 80-XX, 90-XX, 91-XX, 92-XX, 93-XX, 94-XX

© 2008 Springer-Verlag Berlin Heidelberg

This work is subject to copyright. All rights are reserved, whether the whole or part of the material is concerned, specifically the rights of translation, reprinting, reuse of illustrations, recitation, broadcasting, reproduction on microfilm or in any other way, and storage in data banks. Duplication of this publication or parts thereof is permitted only under the provisions of the German Copyright Law of September 9, 1965, in its current version, and permission for use must always be obtained from Springer. Violations are liable to prosecution under the German Copyright Law.

The use of general descriptive names, registered names, trademarks, etc. in this publication does not imply, even in the absence of a specific statement, that such names are exempt from the relevant protective laws and regulations and therefore free for general use.

Cover design: WMXDesign, Heidelberg
Typesetting and production: le-tex publishing services oHG, Leipzig, Germany

Printed on acid-free paper

9 8 7 6 5 4 3 2 1

springer.com

Solving real problems requires teams – mathematicians, engineers, scientists and users from various fields, and – last but not least – it requires money!

Preface

In 1993, the Federal Ministry of Research and Education of the Federal Republic of Germany (BMBF) started the first of now five periods of funding mathematics for industry and services. To date, its efforts have supported approximately 280 projects investigating complex problems in industry and services. Whereas standard problems can be solved using standard mathematical methods and software off the shelf, complex problems arising from e.g. industrial research and developments require advanced and innovative mathematical approaches and methods. Therefore, the BMBF funding programme focuses on the transfer of the latest developments in mathematical research to industrial applications. This initiative has proved to be highly successful in promoting mathematical modelling, simulation and optimization in science and technology.

Substantial contributions to the solution of complex problems have been made in several areas of industry and services.

Results from the first funding period were published in "Mathematik – Schlüsseltechnologie für die Zukunft, Verbundprojekte zwischen Universität und Industrie" (K.-H. Hoffmann, W. Jäger, T. Lohmann, H. Schunck (Editors), Springer 1996). The second publication "Mathematics – Key Technology for the Future, Joint Projects between Universities and Industry" (W. Jäger, H.-J. Krebs (Editors), Springer 2003) covered the period 1997 to 2000. Both books were out of print shortly after publication.

This volume presents the results from the BMBF's fourth funding period (2004 to 2007) and contains a similar spectrum of industrial and mathematical problems as described in the previous publications, but with one additional new topic in the funding programme: risk management in finance and insurance.

Other topics covered are mathematical modelling and numerical simulation in microelectronics, thin films, biochemical reactions and transport, computer-aided medicine, transport, traffic and energy.

As in the preceding funding periods, novel mathematical theories and methods as well as a close cooperation with industrial partners are essential components of the funded projects. In order to strengthen and document this cooperation with industry, the projects starting in the 5th period are based on consortium agreements between universities, industry and services, though there is no direct funding for the industrial partners.

The BMBF has declared the year 2008 as "The Year of Mathematics" in Germany. This book is an excellent contribution to this special year, demonstrating the usefulness of and prospects for mathematics in technology and economics. It is widely acknowledged that the support provided by the Ministry has been substantial in advancing the transfer of mathematics to industry and services and in opening academic mathematical research for challenges arising in industrial and economic applications.

In all areas of society, global aspects are becoming more and more important. Therefore it is extremely important to promote mathematics and its transfer as a key technology to sciences and industry both on the national and on the international level. Here, we take the opportunity to acknowledge the assistance of the BMBF, providing not only necessary national funding, but also emphasizing the importance of mathematics and the necessity of promoting its applications in national and international cooperation. By initiating an OECD Global Science Forum on "Mathematics for Industry" and by dedicating the scientific year 2008 to mathematics, the BMBF has made a clear statement. Special thanks for their support go to Dr. H.-F. Wagner, Dr. R. Koepke and Professor Dr. J. Richter from the BMBF. Last but not least, we would also like to acknowledge the valued support of Dr. Stefan Michalowski, representing the OECD in the Global Science Forum.

We also greatly appreciate the excellent cooperation with Springer Verlag, Heidelberg in publishing this volume – the third one to collect and present results from the BMBF mathematics programme – in its internationally renowned series of scientific publications.

Heidelberg, Jülich *Willi Jäger*
March 2008 *Hans-Joachim Krebs*

Contents

Part I Microelectronics

Numerical Simulation of Multiscale Models for Radio Frequency Circuits in the Time Domain
U. Feldmann 3

Numerical Simulation of High-Frequency Circuits in Telecommunication
M. Bodestedt, C. Tischendorf 9

Wavelet-Collocation of Multirate PDAEs for the Simulation of Radio Frequency Circuits
S. Knorr, R. Pulch, M. Günther 19

Numerical Simulation of Thermal Effects in Coupled Optoelectronic Device-Circuit Systems
M. Brunk, A. Jüngel 29

Efficient Transient Noise Analysis in Circuit Simulation
G. Denk, W. Römisch, T. Sickenberger, R. Winkler 39

Part II Thin Films

Numerical Methods for the Simulation of Epitaxial Growth and Their Application in the Study of a Meander Instability
F. Haußer, F. Otto, P. Penzler, A. Voigt 53

Micro Structures in Thin Coating Layers: Micro Structure Evolution and Macroscopic Contact Angle
J. Dohmen, N. Grunewald, F. Otto, M. Rumpf 75

Part III Biochemical Reactions and Transport

Modeling and Simulation of Hairy Root Growth
P. Bastian, J. Bauer, A. Chavarría-Krauser, C. Engwer, W. Jäger,
S. Marnach, M. Ptashnyk, B. Wetterauer 101

Simulation and Optimization of Bio-Chemical Microreactors
R. Rannacher, M. Schmich 117

Part IV Computeraided Medicine

Modeling and Optimization of Correction Measures for Human Extremities
R. Brandenberg, T. Gerken, P. Gritzmann, L. Roth 131

Image Segmentation for the Investigation of Scattered-Light Images when Laser-Optically Diagnosing Rheumatoid Arthritis
H. Gajewski, J. A. Griepentrog, A. Mielke, J. Beuthan, U. Zabarylo,
O. Minet 149

Part V Transport, Traffic, Energy

Dynamic Routing of Automated Guided Vehicles in Real-Time
E. Gawrilow, E. Köhler, R. H. Möhring, B. Stenzel 165

Optimization of Signalized Traffic Networks
E. Köhler, R. H. Möhring, K. Nökel, G. Wünsch 179

Optimal Sorting of Rolling Stock at Hump Yards
R. S. Hansmann, U. T. Zimmermann 189

Stochastic Models and Algorithms for the Optimal Operation of a Dispersed Generation System Under Uncertainty
E. Handschin, F. Neise, H. Neumann, R. Schultz 205

Parallel Adaptive Simulation of PEM Fuel Cells
R. Klöfkorn, D. Kröner, M. Ohlberger 235

Part VI Risk Management in Finance and Insurance

Advanced Credit Portfolio Modeling and CDO Pricing
E. Eberlein, R. Frey, E. A. von Hammerstein 253

**Contributions to Multivariate Structural Approaches
in Credit Risk Modeling**
S. Becker, S. Kammer, L. Overbeck 281

**Economic Capital Modelling and Basel II Compliance
in the Banking Industry**
K. Böcker, C. Klüppelberg 295

**Numerical Simulation for Asset-Liability Management
in Life Insurance**
T. Gerstner, M. Griebel, M. Holtz, R. Goschnick, M. Haep 319

**On the Dynamics of the Forward Interest Rate Curve
and the Evaluation of Interest Rate Derivatives
and Their Sensitivities**
C. Croitoru, C. Fries, W. Jäger, J. Kampen, D.-J. Nonnenmacher 343

List of Contributors

Peter Bastian
Universität Stuttgart, IPVS
Universitätsstraße 38
D-70569 Stuttgart
peter.bastian@ipvs.uni-stuttgart.de

Jenny Bauer
Universität Heidelberg, IPMB
INF 364
D-69120 Heidelberg
jenny_bauerbiol@yahoo.de

Swantje Becker
Universität Gießen
Arndtstraße 2
D-35392 Gießen
swantje.becker@math.uni-giessen.de

Jürgen Beuthan
Charité Universitätsmedizin Berlin
Campus Benjamin Franklin
Institut für Medizinische Physik und
Lasermedizin
Fabeckstraße 60-62
D-14195 Berlin
juergen.beuthan@charite.de

Klaus Böcker
Risk Integration, Reporting &
Policies
Risk Analytics and Methods
UniCredit Group, Munich Branch
klaus.boecker@hvb.de

Martin Bodestedt
Universität zu Köln
Weyertal 86-90
D-50931 Köln
mbodeste@math.uni-koeln.de

René Brandenberg
Technische Universität München
Zentrum für Mathematik
Boltzmannstr. 3
D-85747 Garching bei München
brandenb@ma.tum.de

Markus Brunk
Universität Mainz
Staudingerweg 9
D-55099 Mainz
brunk@mathematik.uni-mainz.de

Andrés Chavarría-Krauser
Universität Heidelberg, IAM
INF 294
D-69120 Heidelberg
andres.chavarria.Krauser@iwr.uni-heidelberg.de

Cristian Croitoru
Universität Heidelberg, IWR
INF 368
D-69120 Heidelberg
*cristian.croitoru
@iwr.uni-heidelberg.de*

List of Contributors

Georg Denk
Qimonda AG
Am Campeon 1–12
D-85579 Neubiberg
georg.denk@qimoda.com

Julia Dohmen
Universität Bonn
Institut für Numerische Simulation
Nussallee 15
D-53115 Bonn
julia.dohmen@ins.uni-bonn.de

Ernst Eberlein
Universität Freiburg
Abteilung für Mathematische
Stochastik
Eckerstraße 1
D-79104 Freiburg
eberlein@stochastik.uni-freiburg.de

Christian Engwer
Universität Stuttgart, IPVS
Universitätsstraße 38
D-70569 Stuttgart
christian.engwer@ipvs.uni-stuttgart.de

Rüdiger Frey
Universität Leipzig
Abteilung für Mathematik
D-04081 Leipzig
ruediger.frey@math.uni-leipzig.de

Uwe Feldmann
Qimoda AG
Am Campeon 1–12
D-85579 Neubiberg
uwe.feldmann@online.de

Christian Fries
DZBank AG
Platz der Republik
D-60265 Frankfurt am Main
mail@christian-fries.de

Herbert Gajewski
Forschungsverbund Berlin e.V.
WIAS
Mohrenstraße 39
D-10117 Berlin
gajewski@wias-berlin.de

Ewgenij Gawrilow
Technische Universität Berlin
Institut für Mathematik, MA 6-1
Straße des 17. Juni 136
D-10623 Berlin
gawrilow@math.tu-berlin.de

Jens A. Griepentrog
Forschungsverbund Berlin e.V.
WIAS
Mohrenstraße 39
D-10117 Berlin
griepent@wias-berlin.de

Ralf Goschnik
Zürich Gruppe Deutschland
Poppelsdorfer Allee 25–33
D-53115 Bonn
r.goschnik@zurich.com

Michael Günther
Bergische Universität Wuppertal
Fachbereich C, Gaußstr. 20
D-42119 Wuppertal
guenther@math.uni-wuppertal.de

Michael Griebel
Universität Bonn
Institut für Numerische Simulation
Nussallee 15
D-53115 Bonn
griebel@ins.uni-bonn.de

Natalie Grunewald
Universität Bonn, IAM
Wegelerstr. 10
D-53115 Bonn
grunewald@ins.uni-bonn.de

Tobias Gerken
Technische Universität München
Zentrum für Mathematik
Boltzmannstr. 3
D-85747 Garching bei München
brandenb@ma.tum.de

Peter Gritzmann
Technische Universität München
Zentrum für Mathematik
Boltzmannstr. 3
D-85747 Garching bei München
brandenb@ma.tum.de

Thomas Gerstner
Universerität Bonn
Institut für Numerische Simulation
Nussallee 15
D-53115 Bonn
gerstner@ins.uni-bonn.de

Ronny Hansmann
Technische Universität Braunschweig
Institut für Mathematische
Optimierung
Pockelstraße 14
D-38106 Braunschweig
r.hansmann@tu-bs.de

Edmund Handschin
Universität Dortmund
Institut für Energiesysteme und
Energiewirtschaft
Emil-Figge-Straße 70
D-44227 Dortmund
edmund.handschin@udo.edu

Markus Holtz
Universität Bonn
Institut für Numerische Simulation
Nussallee 15
D-53115 Bonn
holtz@ins.uni-bonn.de

Frank Haußer
Technische Fachhochschule Berlin,
FB II
Luxemburger Str. 10
D-13353 Berlin
hausser@tfh-berlin.de

Marcus Haep
Zürich Gruppe Deutschland
Poppelsdorfer Allee 25–33
D-53115 Bonn
marcus.haep@zurich.com

Ernst August von Hammerstein
Universität Freiburg
Abteilung für Mathematische
Stochastik
Eckerstraße 1
D-79104 Freiburg
hammer@stochastik.uni-freiburg.de

Willi Jäger
IWR, Universität Heidelberg
Im Neuenheimer Feld 368
D-69120 Heidelberg
jaeger@iwr.uni-heidelberg.de

Ansgar Jüngel
Technische Universität Wien
Institut für Analysis und Scientific
Computing
Wiedner Hauptstr. 8–10
1040 Wien, Austria
juengel@anum.tuwien.ac.at

Jörg Kampen
Forschungsverbund Berlin e.V.
WIAS
Mohrenstraße 39
D-10117 Berlin
kampen@wias-berlin.de

Stefanie Kammer
Universität Gießen
Arndtstraße 2
D-35392 Gießen
*stefanie.kammer
@math.uni-giessen.de*

List of Contributors

Robert Klöfkorn
Mathematisches Institut
Abteilung für Angewandte
Mathematik
Universität Freiburg
Hermann-Herder-Str. 10
D-79104 Freiburg i. Br.
robertk@mathematik.uni-freiburg.de

Claudia Klüppelberg
Technische Universität München
Zentrum für Mathematische
Wissenschaften
D-85747 Garching bei München
cklu@ma-tum.de

Stephanie Knorr
Bergische Universität Wuppertal
Fachbereich C, Gaußstr. 20
D-42119 Wuppertal
knorr@math.uni-wuppertal.de

Ekkehard Köhler
Brandenburgische Technische
Universität Cottbus
Institut für Mathematik
Postfach 10 13 44
D-03013 Cottbus
ekoehler@math.tu-cottbus.de

Dietmar Kröner
Mathematisches Institut
Abteilung für Angewandte
Mathematik
Universität Freiburg
Hermann-Herder-Str. 10
D-79104 Freiburg i. Br.
dietmar@mathematik.uni-freiburg.de

Sven Marnach
Universität Stuttgart, IPVS
Universitätsstraße 38
D-70569 Stuttgart
sven.marnach@ipvs.uni-stuttgart.de

Alexander Mielke
Forschungsverbund Berlin e.V.
WIAS
Mohrenstraße 39
D-10117 Berlin
mielke@wias-berlin.de

Olaf Minet
Charité Universitätsmedizin Berlin
Campus Benjamin Franklin
Institut für Medizinische Physik und
Lasermedizin
Fabeckstraße 60-62
D-14195 Berlin
olaf.minet@charite.de

Rolf Möhring
Technische Universität Berlin
Institut für Mathematik, MA 6-1
Straße des 17. Juni 136
D-10623 Berlin
moehring@math.tu-berlin.de

Klaus Nökel
PTV AG
Stumpfstraße 1
D-76131 Karlsruhe
klaus.noekel@ptv.de

Frederike Neise
Universität Duisburg-Essen
Abteilung für Mathematik
Forsthausweg 2
D-47048 Duisburg
neise@math.uni-duisburg.de

Hendrik Neumann
Universität Dortmund
Institut für Energiesysteme und
Energiewirtschaft
Emil-Figge-Straße 70
D-44227 Dortmund
hendrik.neumann@udo.edu

List of Contributors

Dirk-Jens Nonnenmacher
HSH Nordbank AG
Gerhart-Hauptmann-Platz 50
D-20095 Hamburg
dirk.jens.nonnenmacher@hsh-nordbank.com

Mario Ohlberger
Institut für Numerische
und Angewandte Mathematik
Universität Münster
Einsteinstraße 62
D-48149 Münster
*mario.ohlberger
@math.uni-muenster.de*

Felix Otto
Universität Bonn, IAM
Wegelerstr. 10
D-53115 Bonn
otto@iam.uni-bonn.de

Ludger Overbeck
Universität Gießen
Arndtstraße 2
D-35392 Gießen
ludger.overbeck@math.uni-giessen.de

Mariya Ptashnyk
Universität Heidelberg, IAM
INF 294
D-69120 Heidelberg
*bernhard.wetterauer
@urz.uni-heidelberg.de*

Patrick Penzler
Universität Bonn, IAM
Wegelerstr. 10
D-53115 Bonn
penzler@iam.uni-bonn.de

Roland Pulch
Bergische Universität Wuppertal
Fachbereich C, Gaußstr. 20
D-42119 Wuppertal
pulch@math.uni-wuppertal.de

Werner Römisch
Humboldt-Universität zu Berlin
Institute of Mathematics
D-10099 Berlin
romisch@math.hu-berlin.de

Martin Rumpf
Universität Bonn
Institut für Numerische Simulation
Nussallee 15
D-53115 Bonn
martin.rumpf@ins.uni-bonn.de

Lucia Roth
Technische Universität München
Zentrum für Mathematik
Boltzmannstr. 3
D-85747 Garching bei München
brandenb@ma.tum.de

Rolf Rannacher
Universität Heidelberg, IAM
Im Neuenheimer Feld 293/294
D-69120 Heidelberg
rolf.rannacher@iwr.uni-heidelberg.de

Michael Schmich
Universität Heidelberg, IAM
Im Neuenheimer Feld 293/294
D-69120 Heidelberg
*michael.schmich
@iwr.uni-heidelberg.de*

Rüdiger Schulz
Universität Duisburg-Essen
Abteilung für Mathematik
Forsthausweg 2
D-47048 Duisburg
schultz@math.uni-duisburg.de

Thorsten Sickenberger
Humboldt-Universität zu Berlin
Institute of Mathematics
D-10099 Berlin
sickenberger@math.hu-berlin.de

Björn Stenzel
Technische Universität Berlin
Institut für Mathematik, MA 6-1
Straße des 17. Juni 136
D-10623 Berlin
stenzel@math.tu-berlin.de

Caren Tischendorf
Universität zu Köln
Weyertal 86-90
D-50931 Köln
ctischen@math.uni-koeln.de

Axel Voigt
Technische Universität Dresden
IWR
Zellescher Weg 12–14
D-01062 Dresden
axel.voigt@tu-dresden.de

Renate Winkler
Humboldt-Universität zu Berlin
Institute of Mathematics
D-10099 Berlin
winkler@math.hu-berlin.de

Bernhard Wetterauer
Universität Heidelberg, IPMB
INF 364
D-69120 Heidelberg
*bernhard.wetterauer
@urz.uni-heidelberg.de*

Gregor Wünsch
Technische Universität Berlin
Institut für Mathematik, MA 6-1
Straße des 17. Juni 136
D-10623 Berlin
wuensch@math.tu-berlin.de

Urszula Zabarylo
Charité Universitätsmedizin Berlin
Campus Benjamin Franklin
Institut für Medizinische Physik und
Lasermedizin
Fabeckstraße 60–62
D-14195 Berlin
urszula.zabarylo@charite.de

Uwe T. Zimmermann
Technische Universität Braunschweig
Institut für Mathematische
Optimierung
Pockelstraße 14
D-38106 Braunschweig
u.zimmermann@tu-bs.de

Part I

Microelectronics

Numerical Simulation of Multiscale Models for Radio Frequency Circuits in the Time Domain

Uwe Feldmann

Qimonda AG, Am Campeon 1–12, 85579 Neubiberg, Germany
uwe.feldmann@online.de

1 Introduction

Broadband data communication via high frequent (RF) carrier signals has become a prerequisite for successful introduction of new applications and services in the hightech domain, like cellular phones, broadband internet services, GPS, and radar sensors for automotive collision control. It is driven by the progress in microelectronics, i.e. by scaling down from micrometer dimensions into the nanometer range. Due to decreasing feature size and increasing operating frequency very powerful electronic systems can be realized on integrated circuits, which can be produced for mass applications at very moderate cost. However, technological progress also opens clearly a design gap, and in particular a simulation gap:

- Systems get much more complex, with stronger interaction between digital and analog parts.
- Parasitic effects become predominant, and neither mutual interactions nor the spatial extension of circuit elements can be further neglected.
- The signal-to-noise ratio decreases and statistical fluctuations in the fabrication lines increase, thus enhancing the risk of circuit failures and yield reduction.

Currently available industrial simulation tools can not cope with all of these challenges, since they are mostly decoupled, and adequate models are not yet completey established, or too expensive to evaluate. The purpose of this paper is to demonstrate that joint mathematical research can significantly contribute to improve simulation capabilities, by proper mathematical modelling and development of numerical methods which exploit the particular structure of the problems. The depth of mathematical research for achieving such progress is beyond industrial capabilities; so academic research groups are strongly involved. However, industry takes care that the problems being solved are of industrial relevance, and that the project results are driven into industrial use.

The challenging simulation problems mentioned above accumulate in RF transmitter-receiver pairs (transceivers), which constitute the core functionality of most RF communication systems. Although not being very large in terms of transistor count, transceivers are often extremely hard to simulate in practice. Mathematically, they exhibit widely separated timescales, they require coupling of semiconductor device equations with thermal equations and standard circuit equations, and they are sensitive with respect to device noise, which is stochastic by nature. For the purpose of this project, a simplified, but typical representative of a CMOS transceiver was chosen as a common benchmark. So this transceiver constitutes a common framework for the research activities within this project. The global objective was to extend standard simulation methods towards more accurate, comprehensive and efficient simulation of this transceiver in a large digital circuit environment.

Since in this setting frequency domain methods are not very helpful, research is focused on time domain models and methods.

For a further discussion of the mathematical problems involved, the transceiver is shortly described in the next Subsect.

1.1 RF Transceiver Blocks

Systems for RF data transmission usually have comprehensive parts for digital system processing which work on a sufficiently large number of parallel bits at conventional clock rates. For data transmission the signals are condensed by multiplexing and modulation onto a very high frequent analog carrier signal. This is illustrated in the upper part of Fig. 1[1].

Modulation is done by the multiplexer. The latter gets the high frequent carrier signal from an RF clock generator, which is usually built as a phase locked loop (PLL). The RF signal being modulated with the data is fed into the RF transmitter, which – for optical data transmission – may be a laser diode or – for wireless data transmission – an amplifier with a resonator and antenna.

A rough scheme for the clock generating PLL is given in Fig. 2. Its core is a voltage controlled oscillator (VCO), which generates the high frequent harmonic clock pulses. For stabilization, this RF clock is frequency divided down onto system frequency and fed into the phase detector. The latter compares it with the system clock and generates a controlling signal for the VCO: If the VCO is ahead then its frequency is reduced, and if the VCO lags behind then its frequency is increased. The number of cycles for the PLL to 'lock in' is usually rather large (up to $10^4 \ldots 10^5$), which gives rise to very challenging simulation tasks.

---

[1] In the Figs. bold lines denote RF signals (with frequencies in the range of $10 \ldots 50$ GHz), while the thinner lines denote signals at standard frequencies (currently about $1 \ldots 2$ GHz).

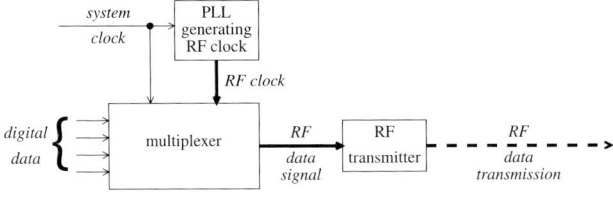

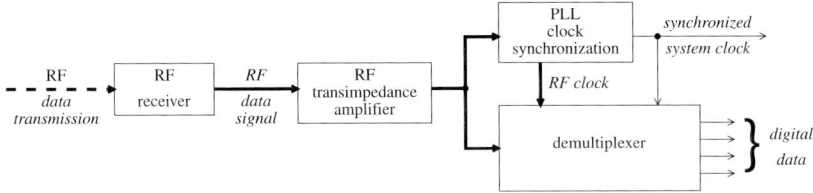

Fig. 1. RF data transmission with a pair of transmitter (*top*) and receiver (*bottom*)

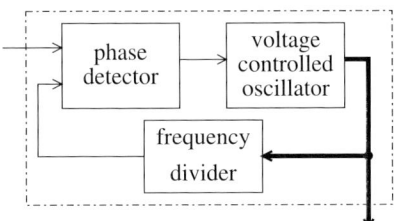

Fig. 2. PLL for generating RF clock

The receiver part of a transceiver system first has to amplify the signal and to synchronize itself with the incoming signal, before it can re-extract the digital information for further processing at standard clock rates, see the bottom part of Fig. 1.

The RF receiver will be a photo diode in case of optical transmission, and an antenna with resonator in case of electromagnetic transmission. The high frequent and noisy low power input signal is amplified in a transimpedance amplifier and then fed into the demultiplexer, which extracts the digital data from it and puts them on a low frequent parallel bus for the digital part. A second PLL takes care that the incoming signal and the receiver's system clock are well synchronized. Typically, in this PLL both VCO and phase detector operate at high frequency, see Fig. 3.

Finally, the VCO's output signal is down converted onto the base frequency for delivering a synchronized system clock into the digital part.

Usually, transmitter and receiver are realized on one single chip, in order to enable a handshaking mode between the server and the host system. Fur-

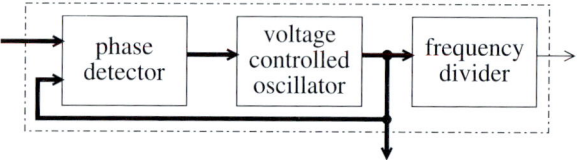

Fig. 3. PLL for clock synchronization

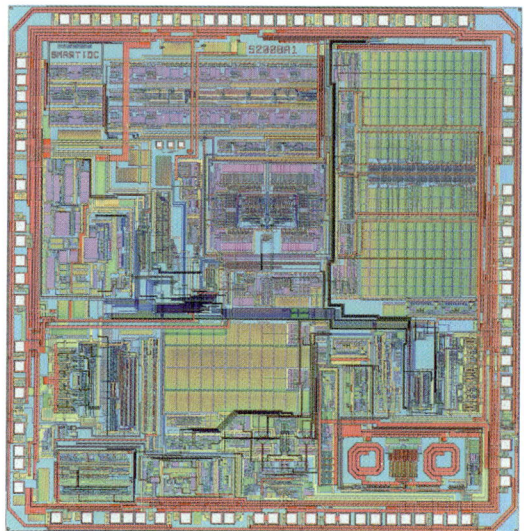

Fig. 4. A GSM transceiver circuit

thermore, both make use of the same building blocks, such that e. g. only one VCO is needed. Figure 4 shows a photo of a GSM transceiver circuit, inclusive digital signal processing. The inductor windings of the VCO can be clearly identified in the bottom right corner.

1.2 New Mathematics for Simulating Transceivers

Traditionally, transceivers have been simulated with a standard circuit simulator, which employs Kirchhoffs equations to set up a system of ordinary nonlinear differential algebraic equations of DAE index 2, and solves them either in the time or in the nonlinear frequency domain. However, technological progress with decreasing feature sizes and increasing frequency bandwith drives this approach beyond its limits.

Efficient solvers for multiscale problems in the time domain.
  Due to widely separated frequencies of the VCO and the digital system clock we have a multiscale system, which has to be analyzed over many

clock cycles for complete verification e. g. of the lock in of the PLL. With single rate integration, this may require weeks of simulation time. Recently developed schemes for separating time scales are more promising, but they cannot directly be used due to very nonharmonic shape of the digital clock signals.

Coupling of device and circuit simulation.

Spatial extension of some of the devices (transistors, diodes, ...) operating in the critical path at their technological limits can no longer be neglected. Hence it is advisable to substitute their formerly used compact models by sets of semiconductor equations, thus requiring coupled circuit device simulation. The latter is already available in some commercial packages, but here the focus is different: The new objective is to simulate some few transistors and diodes on the device level efficiently together with thousands of circuit level transistors. This requires extremely robust coupling schemes.

Interaction with thermal effects in device modelling.

In particular for the emitting laser diode and the receiving photo diode in opto-electronic data transmission there is interaction with thermal effects, whose impact on the transceiver chain has not yet been taken into account. So thermal feedback has to be incorporated into device simulation models and schemes.

Efficient solvers for stochastic differential algebraic equations.

Widely opened eye-diagrams of the noisy signals on the receiver side are essential for reliable signal detection and low bit error rates. Hence noise effects are to be considered very carefully in particular on the receiver part. Up to now, only frequency domain methods have been developed for this purpose. However, time domain based methods should be more generally applicable and hence more reliable in the setting to be considered here. Therefore, efficient numerical solvers for large systems of stochastic differential algebraic equations of index 2 are necessary. The focus is here on efficient calculation of some hundred or even thousand solution pathes, as are necessary for getting sufficient numerical confidence about opening of eye-diagrams, etc.

Although the items of mathematical activity look rather widespread here, all of them serve just to improve simulation capabilities for the RF transceiver circuitry in advanced CMOS technologies. Therefore all of their results will be combined in one single simulation environment which is part of or linked to industrial circuit simulation.

Numerical Simulation of High-Frequency Circuits in Telecommunication

Martin Bodestedt and Caren Tischendorf

Universität zu Köln, Mathematisches Institut, Weyertal 86–90, 50931 Köln, Germany, {mbodeste,ctischen}@math.uni-koeln.de

Summary. Driving circuits with very high frequencies may influence the standard switching behavior of the incorporated semiconductor devices. This motivates us to use refined semiconductor models within circuit simulation. Lumped circuit models are combined with distributed semiconductor models which leads to coupled systems of differential-algebraic equations and partial differential equations (PDAEs). Firstly, we present results of a detailed perturbation analysis for such PDAE systems. Secondly, we explain a robust simulation strategy for such coupled systems. Finally, a multiphysical electric circuit simulation package (MECS) is introduced including results for typical circuit structures used in chip design.

1 Introduction

In the development of integrated circuit so-called compact models are used to describe the transistors in the circuit simulations that are performed by the chip designer. Compact models are small circuits whose voltage-current characteristics are fitted to the ones of the real device by parameter tuning. Unfortunately, not even the computationally most expensive compact models are able to fully capture the switching behavior of field effect transistors when very high frequencies are used in RF-transceivers.

A remedy to this problem is to employ a more physical modeling approach and model the critical transistors with distributed device equations. Here we consider the drift-diffusion equations which is a system of mixed elliptic/parabolic partial differential equations (PDEs) describing the development of the electrostatic potential and the charge carrier densities in the transistor region.

For the non-critical devices a lumped modeling approach is followed with the modified nodal analysis (MNA) [13]. The electric network equations are in this case a differential algebraic equation (DAE) with node potentials and some of the branch currents as unknowns. In this article we discuss the analytical and numerical treatment of the partial differential algebraic equation

(PDAE) that is obtained when the MNA network equations are coupled with the drift-diffusion device equations.

This article is organize as follows. We start by discussing the the refined circuit PDAE in particular and the perturbation sensitivity of PDAEs in general. We state results categorizing the perturbations sensitivity of the PDAE in terms of the circuit topology. Then, we present the software package MECS (Multiphysical Electric Circuit Simulator) which makes it possible to perform transient simulations of circuits with a mixed lumped/distributed modeling approach for the semiconductor devices. The numerical approximation is obtained by the method of lines (MOL) combined with time-step controlled time-integration scheme especially suited for circuit simulations.

2 The Refined Circuit Model

For brevity the refined circuit model consisting of the MNA network equations coupled with the drift-diffusion device equations is summarized in Box 1 [2].

Electric Network Equations:
$$A_C q_C \left(A_C^T e\right)' + A_R g \left(A_R^T e\right) + A_L j_L + A_V j_V + A_{I_c} i_c \left(A_{I_c}^T e\right) + \bar{A}_S J_S = -A_{I_s} i_s, \quad (1a)$$
$$\phi(j_L)' - A_L^T e = 0, \quad (1b)$$
$$A_V^T e = v_s. \quad (1c)$$

Semiconductor Device Equations:
$$\varepsilon_m \Delta \psi = q(n - p - N), \quad (1d)$$
$$q\partial_t n - \operatorname{div} j_n = -qR, \quad \text{where} \quad j_n = qU_T \mu_n \operatorname{grad} n - q\mu_n n \operatorname{grad} \psi, \quad (1e)$$
$$q\partial_t p + \operatorname{div} j_p = -qR, \quad \text{where} \quad j_p = -qU_T \mu_p \operatorname{grad} p - q\mu_p p \operatorname{grad} \psi. \quad (1f)$$

Coupling interface:
$$J_S = \int_\Omega (j_n + j_p - \varepsilon_m \partial_t \operatorname{grad} \psi) \cdot \operatorname{grad} h_i \, dx \quad (1g)$$
$$\psi|_{\Gamma_D} = \left(h\bar{A}_S^T e + \psi_{\text{bi}}\right)|_{\Gamma_D}, \quad n|_{\Gamma_D} = n_D|_{\Gamma_D}, \quad p|_{\Gamma_D} = p_D|_{\Gamma_D}, \quad (1h)$$
$$\operatorname{grad} \psi \cdot \nu|_{\Gamma_N} = \operatorname{grad} n \cdot \nu|_{\Gamma_N} = \operatorname{grad} p \cdot \nu|_{\Gamma_N} = 0. \quad (1i)$$

Initial conditions:
$$(e(t_0), j_L(t_0), j_V(t_0)) = (e^0, j_L^0, j_V^0). \quad (1j)$$
$$n(x, t_0) = n^0(x), \quad p(x, t_0) = p^0(x), \quad \text{a.e. in } \Omega, \quad (1k)$$

Box 1. The refined circuit model

The topology of the circuit is described by a network graph consisting of network nodes and element branches. It contains the element types: resistors R, inductors L, capacitors C, independent voltage sources V, independent current sources I_s, controlled current sources I_c and semiconductor devices S. The positions of the different elements are described by the incidence matrices A accompanied by the corresponding subscript.

The unknowns in the network equations (1a–1c) are node potentials $e(t)$, inductor currents $j_L(t)$ and independent voltage source currents $j_V(t)$. The functions $q_C(A_C^T e)$, $g(A_R^T e)$ and $i_c(A_{I_c}^T e)$ model capacitor charges, resistor currents and currents through controlled current sources. They are dependent of their respective applied voltage. The function $\phi(j_L)$ describe the electromagnetic fluxes in the inductors. Data are the functions describing the independent voltage and current sources $v_s(t)$ and $i_s(t)$.

For the drift-diffusion equations (1d–1e), after the insertion of the expressions for the charge carrier densities $j_n(x,t)$ and $j_p(x,t)$ into the two balance equations we have the unknowns electrostatic potential $\psi(x,t)$, electron density $n(x,t)$ and hole density $p(x,t)$. The function $R(n,p)$ is a source term modeling the recombination of the charge carriers with the semiconductor substrate. The carrier mobilities $\mu_n(x)$ and $\mu_p(x)$ and the doping profile of the semiconductor substrate $N(x)$ are considered to be space dependent functions. The dielectric permittivity ε_m, the unit charge q, the intrinsic carrier density n_i and the thermal potential U_T are constants.

The semiconductor region $\Omega \in \mathbb{R}^d$, $d \in \{1,2,3\}$ has a boundary Γ which is the union of Dirichlet $\dot{\cup}_{i=1}^{n_D} \Gamma_{D_i}$ and Neumann segments $\dot{\cup}_{i=1}^{n_N} \Gamma_{N_i}$. In the simulations, we additionally to Neumann and Dirichlet conditions consider mixed boundary conditions at the so-called gate contacts.

The network and device equations are coupled in two ways at the Dirichlet boundaries. The currents flowing over the semiconductor contacts $J_S(t)$ are evaluated in (1g) and accounted for in Kirchoff's current law (1a). There, the functions h_i fulfill the Laplace equation and homogeneous Neumann respectively Dirichlet boundary conditions, except at Γ_{D_i} where they equal 1. We have a coupling in the other direction as well as the values of electrostatic potential ψ at the Dirichlet boundaries Γ_{D_i} depend on the node potentials e in the network (1h).

Beside the dynamic drift-diffusion model (1d–1f) we also consider the stationary version, which is obtained by putting $\partial_t n = \partial_t p = 0$ in (1e–1f):

$$\varepsilon_m \Delta \psi = q(n - p - N), \tag{1.1d'}$$
$$q \operatorname{div}(U_T \mu_n \operatorname{grad} n - \mu_n n \operatorname{grad} \psi) = qR, \tag{1.1e'}$$
$$q \operatorname{div}(U_T \mu_p \operatorname{grad} p + \mu_p p \operatorname{grad} \psi) = qR. \tag{1.1f'}$$

In this case the dynamical term in the current evaluation is neglected. Instead of (1g) we have

$$J_S = \int_\Omega (j_n + j_p) \cdot \operatorname{grad} h_i \, dx \tag{1.1g'}$$

The complete circuit model consisting of the MNA equations coupled with the stationary drift-diffusion model is (1.1a–c, 1.1d'–g', 1.1h–k).

3 Perturbation and Index Analysis

Differential algebraic equations (DAEs), but also PDAEs, which can be seen as abstract DAEs, may have solution operators that contain differential operators. Now, numerical differentiation is an ill-posed problem, wherefore numerical differentiation of small errors that arise in the approximation process can lead to large deviations of the approximative solution form the exact one. In order to successfully integrate PDAEs in time it is important to have knowledge of perturbation sensitivity of the solutions. A measure for this sensitivity is the perturbation index.

Definition 1. *Let X, Z and Y be real Hilbert spaces, $\mathcal{I} = [t_0, T]$, $F : Z \times X \times \mathcal{I} \to Y$, $\delta(t) \in Y$ for all $t \in \mathcal{I}$ and let w_δ and w solve the perturbed, respectively unperturbed problem:*

$$F(\partial_t w_\delta(t), w_\delta(t), t) = \delta(t), \qquad F(\partial_t w(t), w(t), t) = 0.$$

$F = 0$ is said to have the perturbation index ν if it is the lowest integer such that an estimate of the form

$$\max_{t \in \mathcal{I}} \|w_\delta(t) - w(t)\|_X \le c \left(\|w_\delta(t_0) - w(t_0)\|_X + \sum_{i=0}^{\nu-1} \max_{t \in \mathcal{I}} \|(\partial_t)^i (w_\delta(t) - w(t))\|_Y \right)$$

holds for some constant c.

Example 1. Consider the PDAE

$$\partial_t u_2(t) - \partial_{xx}^2 u_1(x,t) = 0, \qquad u_2(t) = 0, \qquad (x,t) \in \Omega \times \mathcal{I}, \qquad (2)$$

where $u_1(t) \in \mathbb{R}^n$ and $u_2(\cdot, t) \in H_0^1(\Omega)$. If the two equations are perturbed with $\delta_1 \in C(\mathcal{I}, H^{-1}(\Omega))$ and $\delta_2 \in C^1(\mathcal{I}, \mathbb{R}^n)$ we can derive the following bounds for the deviation from the exact solution

$$\max_{t \in \mathcal{I}} \|u_{1\delta}(t) - u_1(t)\|_{H_0^1} \le c \max_{t \in \mathcal{I}} \left(\|\delta_1(t)\|_{H^{-1}} + |\partial_t \delta_2(t)| \right),$$
$$\max_{t \in \mathcal{I}} |u_{2\delta}(t) - u_2(t)| = \max_{t \in \mathcal{I}} |\delta_2(t)|.$$

According to Definition 1 the perturbation index of (2) is two since a first order time-derivative of δ_2 appears in the first bound.

Before we turn to the results, we briefly summarize the assumptions needed for the mathematical analysis of the coupled system.

Assumption 1. *The electric network is consistent, its elements are passive and its data smooth. The controlled sources are shunted parallel with capacitors. The recombination is of Shockley-Read-Hall type and the doping fulfills $N \in H_0^1(\Omega \cap \Gamma_N) \cap W^{1,4}(\Omega) \cap H^2(\Omega)$.*

The domain $\Omega \in \mathbb{R}^d$ for $d \in \{1,2,3\}$ has a Lipschitz boundary. The Neumann boundary $\Gamma_N = \dot{\cup}_{i=1}^{n_N} \Gamma_{N_i}$ is the union of C^2-parameterizable segments. Further, the $(d-1)$-dimensional Lebesgue measure of the Dirichlet boundary $\Gamma_D = \dot{\cup}_{i=1}^{n_D} \Gamma_{D_i}$ is positive. The angles between Neumann and Dirichlet segments do not succeed $\pi/2$.

For a more exhaustive discussion of Assumption 1 we refer to [2] and the references therein. Next we will see that the perturbation sensitivity of the refined network model strongly depends on the topology of the electric network graph.

Theorem 1. *Let Assumption 1 hold and assume that the contacts of the device is connected by a capacitive paths.*
Then, the perturbation index of the PDAE (1.1a–c, 1.1d'–g', 1.1h–k) is 1 if and only if the network graph contains neither loops of capacitors and at least one voltage source nor cutsets of inductors and independent current sources. Otherwise, the perturbation index is 2.

The proof can be found in [2]. This result generalizes known index criteria for the MNA equations [13, 14, 11, 3] of that PDAE with stationary drift-diffusion equations.

Since we are interested in high-frequency applications it is important to account for the dynamical behavior of the devices, especially when a stationary description is used. This is done by the assumption that the contacts are connected by a capacitive path, which models the reactive or charge storing behavior of the device. In our next theorem we give a first perturbation result for a refined circuit model with dynamical device equations.

Theorem 2. *Let Assumption 1 hold, the network equations be linear and without loops of capacitors, semiconductors and at least one voltage source and also without cutsets of inductors and independent current sources. Assume that the domain Ω is one-dimensional. If the network equations are perturbed by continuous sufficiently small perturbations the perturbed solutions exist and the deviations fulfill the bounds*

$$\max_{t \in \mathcal{I}} \left(|\bar{y}(t)|^2 + \|\bar{n}(t)\|_{L^2}^2 + \|\bar{p}(t)\|_{L^2}^2 \right) + \int_{t_0}^{T} \left(\|\partial_x \bar{n}(\tau)\|_{L^2}^2 + \|\partial_x \bar{p}(\tau)\|_{L^2}^2 \right) d\tau$$

$$\leq C_{ynp} \left(|\delta_y^0|^2 + \|n_\delta^0\|_{L^2}^2 + \|p_\delta^0\|_{L^2}^2 + \int_{t_0}^{T} |\tilde{\delta}_P|^2 d\tau \right)$$

$$\max_{t \in \mathcal{I}} |\bar{z}(t)|^2 \leq C_z \left(|\delta_y^0|^2 + \|n_\delta^0\|_{L^2}^2 + \|p_\delta^0\|_{L^2}^2 + \int_{t_0}^{T} |\tilde{\delta}_p|^2 d\tau + \max_{t \in \mathcal{I}} |\delta_Q|^2 \right)$$

$$\max_{t \in \mathcal{I}} \max_{x \in \Omega} |\partial_x \bar{\psi}(x,t)| \leq C_\psi \left(|\delta_y^0|^2 + \|n_\delta^0\|_{L^2}^2 + \|p_\delta^0\|_{L^2}^2 + \int_{t_0}^{T} |\tilde{\delta}_p|^2 d\tau \right).$$

In order to split the variables into dynamical and algebraic parts we have put $y := (P_{CS}e, j_L)$ and $z := (Q_{CS}e, j_V)$ where Q_{CS} is a projector onto

$\ker(A_C A_S)^T$ and $P_{CS} = I - Q_{CS}$. Here, the bar denotes the deviation between the exact and perturbed solution.

The proof can be found in [2]. This result is in good correspondence with the index criteria in the previous theorem as well as with the ones in [13, 14, 11, 3].

If perturbations in the drift-diffusion equations are allowed one cannot (at least in the standard way [5, 1]) prove the non-negativity of the charge carriers and the charge preservation in the diode anymore. These properties are essential for the a priori estimates of the perturbed solutions, which in turn are necessary for the estimation of the nonlinear drift currents.

4 Numerical Simulation

Concerning the existence of well-established circuit simulation and device simulation packages alone, the first natural idea would be to couple these simulation packages in order to solve the circuit PDAE system described by (1). However, this approach turned out to involve persistent difficulties. The main problem consists of the adaption of the different time step controls within both simulations. This can be handled for low frequency circuits since time constants for circuits on the one hand and devices on the other hand differ by several magnitudes in such cases. Our main goal is to investigate high frequency circuits. Here, the pulsing of the circuit is driven near to the switching time of the device elements. Our coupling of circuit and device simulations often failed for such cases. Whereas the time step control of the circuit simulation works well, the device simulation does not find suitable stepsizes to provide sufficiently accurate solutions when higher frequencies are applied.

Therefore, we pursue a different strategy to solve the circuit PDAE system described by (1) numerically. In order to control the stepsize for the whole system at once we choose a method of lines approach. First, we discretize the system with respect to space. Then we use an integration method for DAEs for the resulting differential-algebraic equation system. Consequently, we take the same time discretization for the circuit as for the device part.

Space discretization is needed for the device equations (1d)–(1f) as well as for the coupling interface equations (1g)–(1i). The former ones represent a coupled system of one elliptic and two parabolic equations for each semiconductor. We use here finite element methods leading to

$$\varepsilon_m T_h \psi_h + q S_h (n_h - p_h - N_h) = 0, \tag{3a}$$

$$M_{n,h} \frac{\partial n}{\partial t} + g_{n,h}(j_{n.h}, n_h, p_h) = 0, \tag{3b}$$

$$M_{p,h} \frac{\partial n}{\partial t} + g_{p,h}(j_{p.h}, n_h, p_h) = 0, \tag{3c}$$

The coupling interface equations are handled as follows. The Neumann boundary conditions (1i) are already considered in (3a)–(3c) by choosing proper test functions leading to the discrete defined functions ψ_h, n_h and p_h. The Dirichlet boundary conditions (1h) are evaluated at the grid points of the Dirichlet boundary. For the approximation of the current equation (1g) we use Gauss quadrature.

Under the assumptions of Theorem 1, we obtain as DAE index for the resulting differential-algebraic equation (after space discretization of the coupled problem) exactly the same as the perturbation index for the system (1.1a–c, 1.1d'–g', 1.1h–k). Furthermore, under the assumptions of Theorem 2, we obtain DAE index 1 for the system (1.1a–k) as expected concering the perturbation result given in Theorem 2. For a proof we refer to [11].

4.1 The Software MECS

The multiphysical electric circuit simulator MECS [7] allows the time integration of electrical circuits using different models for semiconductor devices. Beside the standard use of lumped models the user can choose distributed models. So far, drift diffusion models are implemented. On the one hand, the standard model equations are used as described in (1d)–(1f). On the other hand one can also select the drift diffusion model where the Poisson equation (1d) is replaced by the current conservation equation

$$\mathrm{div}\left(j_n + j_p - \varepsilon_m \frac{\partial}{\partial t}\nabla\psi\right) = 0. \qquad (4)$$

For the space discretization, the standard finite element method as well as a mixed finite element [8] is implemented. For the time integration of the whole system, the user can choose between BDF methods [4], Runge Kutta methods [12] and a general linear method [15].

4.2 Flip Flop Circuitry

Flip flop circuits represent a basic circuit structure of digital circuits serving one bit of memory. Depending on the two input signals (set and reset), a flip flop switches between two stable states. The circuit in Fig. 1 shows a realization containing four MOSFETs (Metal Oxid Field Effect Transistors).

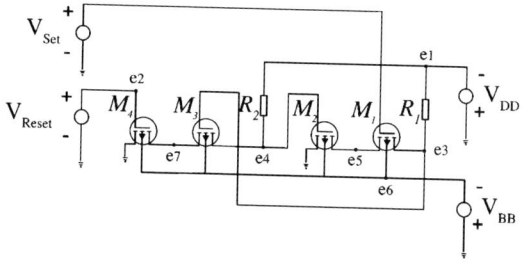

Fig. 1. Schematic diagram of a flip flop circuit

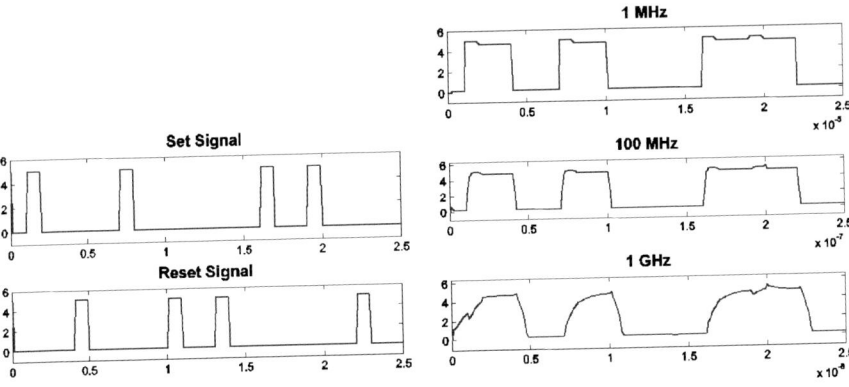

Fig. 2. Flip flop simulation results. On the *left*, the input signals Set and Reset. On the *right*, the output realising two more or less stable states when applying different frequencies

In Fig. 2, we see the output voltage for different applied frequencies depending on the two given input signals. It shows that the stable states 0V and 5V are not so stable when increasing higher frequencies, namely when applying 1GHz. The simulation results may provide a more stable behavior when decreasing the gate length or changing the doping profile. The advantage of the simulation here is the possibility to study the influence of the dimensions/doping/geometry of the semiconductors onto the switching behavior.

4.3 Voltage Controlled Oscillator (VCO)

As real benchmark we have tested a voltage controlled oscillator from the transceiver described in the preceding introduction by Uwe Feldmann. It generates a 1.8 GHz signal with the amplitude of 2.5V (see simulation result in Fig. 3 on the right). Continuing the simulation over longer time periods shows the expected behavior to hold the frequency and amplitude as the last periods on the figure indicate. The tested circuitry contained six MOSFETs that have been simulated using the drift diffusion model.

On the left of Fig. 3, the electrostatic potential of one of the transistors is shown at a randomly chosen time point. The simulation package MECS provides the data at each time discretization point for each semiconductor for the electrostatic potential and the charge carrier densities.

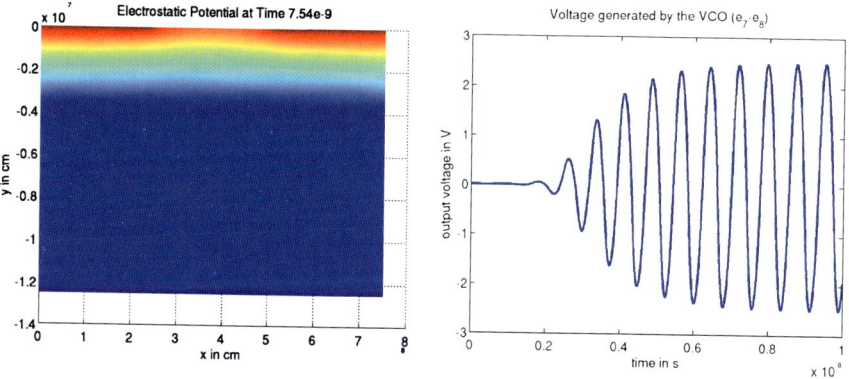

Fig. 3. VCO simulation results. On the *left*, the electrostatic potential of the third MOSFET at a certain time point. On the *right*, the generated oscillator voltage

References

1. Alì G., Bartel A. and Günther M.: Parabolic differential-algebraic models in electrical network design, Multiscale Model. Simul. 4, No. 3, 813–838 (2005).
2. Bodestedt M.: Perturbation Analysis of Refined Models in Circuit Simulation. Ph.D. thesis, Technical University Berlin, Berlin (2007).
3. Bodestedt M. and Tischendorf C.: PDAE models of integrated circuits and index analysis. Math. Comput. Model. Dyn. Syst. 13, No. 1, 1–17 (2007).
4. Hanke M.: A New Implementation of a BDF method Within the Method of Lines. http://www.nada.kth.se/~hanke/ps/fldae.1.ps.
5. Gajewski H.: On existence, uniqueness and asymptotic behavior of solutions of the basic equations for carrier transport in semiconductors. Z. Angew. Math. Mech. 65, 101–108 (1985).
6. Griepentrog E. and März R.: Differential-algebraic equations and their numerical treatment. Teubner, Leipzig (1986).
7. Guhlke C., Selva M.: Multiphysical electric circuit simulation (MECS) manual. http://www.mi.uni-koeln.de/~mselva/MECS.html.
8. Guhlke C.: Ein gemischter Finite Elemente-Ansatz zur gekoppelten Schaltungs- und Bauelementsimulation. Diploma thesis, Humboldt University of Berlin (2006).
9. März. R.: Solvability of linear differential algebraic equations with properly stated leading terms. Results Math. 45, pp. 88–105 (2004).
10. R. März.: Nonlinear differential-algebraic equations with properly formulated leading term. Technical Report 01-3, Institute of Mathematics, Humboldt-University of Berlin (2001).
11. Selva M., Tischendorf C.: Numerical Analysis of DAEs from Coupled Circuit and Semiconductor Simulation. Appl. Numer. Math. 53, No. 2–4, 471–488 (2005).
12. Teigtmeier S.: Numerische Lösung von Algebro-Differentialgleichungen mit proper formuliertem Hauptterm durch Runge-Kutta-Verfahren. Diploma thesis. Humboldt University of Berlin (2002).

13. Tischendorf C.: Topological index calculation of DAEs in circuit simulation, Surv. Math. Ind. 8, No. 3–4, 187–199 (1999).
14. Tischendorf C.: Coupled Systems of Differential Algebraic and Partial Differential Equations in Circuit and Device Simulation. Modeling and Numerical Analysis, Habilitation thesis, Humboldt-University of Berlin (2004).
15. Voigtmann S.: GLIMDA solver. General LInear Methods for Differential Algebraic equations. http://www.math.hu-berlin.de/~steffen/software.html.

Wavelet-Collocation of Multirate PDAEs for the Simulation of Radio Frequency Circuits

Stephanie Knorr, Roland Pulch, and Michael Günther

Bergische Universität Wuppertal, Fachbereich C, Gaußstr. 20, 42119 Wuppertal, Germany, {knorr,pulch,guenther}@math.uni-wuppertal.de

1 Motivation

As already explained in the preceding introduction, radio frequency (RF) circuits introduce several difficulties for their numerical simulation, e.g. widely separated time scales and a nonharmonic shape of digital signals. Multiscale signals require huge computational effort in numerical integration schemes, since the fast time scale restricts the step sizes, whereas the slow time scale determines a relatively long integration interval. The occurrence of steep gradients in digital signal structures demands an additional refinement of grids in time domain methods. Moreover, the low smoothness of pulsed signals possibly causes further difficulties in the numerical simulation.

We present a wavelet-collocation scheme based on a multivariate modeling of different time scales. Simulation results for a switched-capacitor circuit show the efficient adaptive grid generation.

2 Multivariate Signal Model

In this section, a multivariate model is introduced for amplitude modulated (AM) signals. It is applied to the differential-algebraic network equations and the special structure of the resulting system is investigated.

2.1 Multivariate Model for AM Signals

We consider circuits with quasiperiodic steady state responses including widely separated time scales. The core idea to efficiently describe multitone signals is to decouple the time scales by associating a corresponding variable to each of them. An *m-tone quasiperiodic* signal $\mathbf{x} : \mathbb{R} \to \mathbb{C}^d$ reads

$$\mathbf{x}(t) = \sum_{j_1=-\infty}^{+\infty} \cdots \sum_{j_m=-\infty}^{+\infty} \mathbf{X}_{j_1,\ldots,j_m} \exp\bigl(i(j_1\omega_1 + \cdots + j_m\omega_m)\,t\bigr) \qquad (1)$$

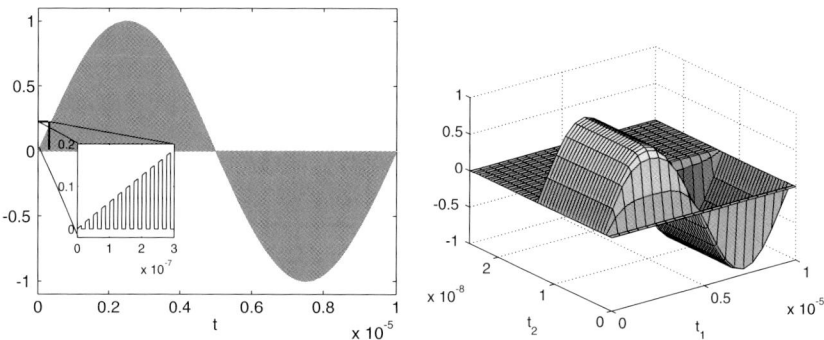

Fig. 1. AM signal x (*left*) and its multivariate function $\hat{x}$ (*right*)

with Fourier coefficients $\mathbf{X}_{j_1,\ldots,j_m} \in \mathbb{C}^d$, the imaginary unit $i = \sqrt{-1}$ and the fundamental frequencies $\omega_l = 2\pi/T_l$ with according time scales T_l, $l = 1, \ldots, m$. The multitone structure of the quasiperiodic signal (1) naturally leads to the corresponding m-periodic *multivariate function (MVF)* $\hat{\mathbf{x}} : \mathbb{R}^m \to \mathbb{C}^d$ with

$$\hat{\mathbf{x}}(t_1,\ldots,t_m) = \sum_{j_1=-\infty}^{\infty} \cdots \sum_{j_m=-\infty}^{\infty} \mathbf{X}_{j_1,\ldots,j_m} \exp\bigl(i(j_1\omega_1 t_1 + \cdots + j_m\omega_m t_m)\bigr). \quad (2)$$

The MVF (2) is periodic in each time variable and the original signal can be reconstructed following the 'diagonal' direction ($t_1 = t_2 = \cdots = t_m$):

$$\mathbf{x}(t) = \hat{\mathbf{x}}(t,\ldots,t). \quad (3)$$

To illustrate the efficiency of this multivariate modeling, Fig. 1 (left) shows a high frequency pulse function $x : \mathbb{R} \to \mathbb{R}$, whose amplitude is modulated by a low frequency oscillation. Its MVF $\hat{x}$ (right) exhibits a simple structure and is determined in the rectangle $[0, T_1] \times [0, T_2]$ with $T_1 \gg T_2$.

2.2 MPDAE Model

The above multidimensional signal model is now applied to represent solutions of the differential-algebraic network equations [GF99]. Below, we write the system in the compact form

$$\tfrac{\mathrm{d}}{\mathrm{d}t}\mathbf{q}(\mathbf{x}(t)) = \mathbf{f}(\mathbf{b}(t), \mathbf{x}(t)), \quad (4)$$

where $\mathbf{q} : \mathbb{R}^d \to \mathbb{R}^d$ denotes charges and fluxes, $\mathbf{b} : \mathbb{R} \to \mathbb{R}^s$ comprises independent input signals (voltage and current sources) and unknown node potentials and branch currents are collected in $\mathbf{x} : \mathbb{R} \to \mathbb{R}^d$. We are interested in systems (4) with quasiperiodic solutions (1).

Assuming m different time scales, we introduce MVFs $\hat{\mathbf{x}} : \mathbb{R}^m \to \mathbb{R}^d$ of the state variables and $\hat{\mathbf{b}} : \mathbb{R}^m \to \mathbb{R}^s$ of the input signals. Regarding the DAE (4), Brachtendorf et al. [BWLB96] introduce the corresponding *multirate partial differential-algebraic equation* (MPDAE)

$$\frac{\partial \mathbf{q}(\hat{\mathbf{x}})}{\partial t_1} + \cdots + \frac{\partial \mathbf{q}(\hat{\mathbf{x}})}{\partial t_m} = \mathbf{f}\big(\hat{\mathbf{b}}(t_1, \ldots, t_m), \hat{\mathbf{x}}(t_1, \ldots, t_m)\big). \tag{5}$$

Given a solution $\hat{\mathbf{x}}$ of the MPDAE (5), a solution of the original DAE (4) can be obtained via the reconstruction formula (3). For a detailed description of the relation between DAE and MPDAE solutions, we refer to [Roy01].

To solve the MPDAE (5), we have to impose boundary conditions (BCs), which determine the structure of the corresponding solution. Looking for an m-tone quasiperiodic solution (1) of the DAE with periods $T_1, \ldots, T_m$ known from input signals, we solve the MPDAE for an m-periodic MVF with BCs

$$\hat{\mathbf{x}}(t_1, \ldots, t_m) = \hat{\mathbf{x}}(t_1 + k_1 T_1, \ldots, t_m + k_m T_m) \text{ for all } t_1, \ldots, t_m \in \mathbb{R} \\ \text{and all } k_1, \ldots, k_m \in \mathbb{Z}. \tag{6}$$

2.3 Characteristic System

Starting from network equations that only comprise ordinary differential equations (ODEs), the corresponding multirate partial differential equations (MPDEs) are of hyperbolic type, where each component of the system exhibits a derivative in the direction of the diagonal, see [PG02]. Thus, the information transport takes place along *characteristic projections*, which are straight lines in diagonal direction. The algebraic constraints in case of DAEs do not affect this information transport and we are able to formulate the characteristic system of the MPDAE:

$$\begin{aligned} \tfrac{\mathrm{d}}{\mathrm{d}\tau} t_l(\tau) &= 1, \quad l = 1, \ldots, m, \\ \tfrac{\mathrm{d}}{\mathrm{d}\tau} \mathbf{q}\big(\bar{\mathbf{x}}(\tau)\big) &= \mathbf{f}\big(\hat{\mathbf{b}}(t_1(\tau), \ldots, t_m(\tau)), \bar{\mathbf{x}}(\tau)\big). \end{aligned} \tag{7}$$

Thereby, the restrictions $\bar{\mathbf{x}}(\tau) = \hat{\mathbf{x}}(t_1(\tau), \ldots, t_m(\tau))$ as well as the time variables depend on a parameter $\tau \in \mathbb{R}$. The part corresponding to the time variables can be solved explicitly, leading to the characteristic projections $(t_1(\tau), \ldots, t_m(\tau)) = (\tau + c_1, \ldots, \tau + c_m)$ for arbitrary $c_1, \ldots, c_m \in \mathbb{R}$. These characteristic projections represent a continuum of parallel straight lines in the domain of dependence. Inserting this result in the last equation of the characteristic system (7) yields

$$\tfrac{\mathrm{d}}{\mathrm{d}\tau} \mathbf{q}\big(\bar{\mathbf{x}}(\tau)\big) = \mathbf{f}\big(\hat{\mathbf{b}}(\tau + c_1, \ldots, \tau + c_m), \bar{\mathbf{x}}(\tau)\big). \tag{8}$$

This family of DAE systems completely describes the transport of information in the MPDAE (5).

The special hyperbolic structure of the MPDAE can be exploited to set up a method of characteristics, see [PG02, Pul03]. In the following section, this scheme will be demonstrated in detail. The well-posedness of the MPDAE system is investigated in [KG06].

Furthermore, the idea of the multivariate modeling can also be carried forward to frequency modulated signals. For a detailed survey on the MPDAE-modeling, see [PGK07].

3 Wavelet-Collocation

In this section, we present the method of characteristics to obtain m-periodic solutions $\hat{\mathbf{x}}$ of (5,6). Corresponding boundary value problems along the characteristic projections are solved by a collocation scheme, where the time-frequency localization of a wavelet basis is used to generate adaptive grids.

3.1 Method of Characteristics

To shorten notations, we restrict ourselves to $m = 2$ different time scales and consider the biperiodic boundary value problem (BVP) of the MPDAE

$$\frac{\partial \mathbf{q}(\hat{\mathbf{x}})}{\partial t_1} + \frac{\partial \mathbf{q}(\hat{\mathbf{x}})}{\partial t_2} = \mathbf{f}\bigl(\hat{\mathbf{b}}(t_1, t_2), \hat{\mathbf{x}}(t_1, t_2)\bigr)$$

with boundary conditions

$$\hat{\mathbf{x}}(t_1, t_2) = \hat{\mathbf{x}}(t_1 + T_1, t_2) = \hat{\mathbf{x}}(t_1, t_2 + T_2) \quad \text{for all } t_1, t_2 \in \mathbb{R}.$$

We exploit that the biperiodic solution is uniquely defined by its initial values on the manifold $\{(t_1, t_2) \in \mathbb{R}^2 : t_2 = 0\}$ and use the initial points

$$(t_1, t_2) = ((j_1 - 1)h_1, 0) \quad \text{for } h_1 := \frac{T_1}{n_1} \text{ and } j_1 = 1, \ldots, n_1.$$

As we have seen in the previous section, the information transport takes place along parallel straight lines in direction of the diagonal. Figure 2 shows the respective domain $[0, T_1] \times [0, T_2]$ with $T_1 \gg T_2$, where these characteristic projections are indicated by dotted lines. Thus, we consider the unknown functions

$$\bar{\mathbf{x}}_{j_1}(\tau) := \hat{\mathbf{x}}((j_1 - 1)h_1 + \tau, \tau) \quad \text{for } j_1 = 1, \ldots, n_1, \qquad (9)$$

which extend along the diagonal direction. The corresponding systems (8) now read

$$\frac{\mathrm{d}\mathbf{q}(\bar{\mathbf{x}}_{j_1})}{\mathrm{d}\tau}(\tau) = \mathbf{f}\bigl(\hat{\mathbf{b}}((j_1 - 1)h_1 + \tau, \tau), \bar{\mathbf{x}}_{j_1}(\tau)\bigr) \quad \text{for } j_1 = 1, \ldots, n_1 \qquad (10)$$

and have to be solved for $\tau \in [0, T_2]$.

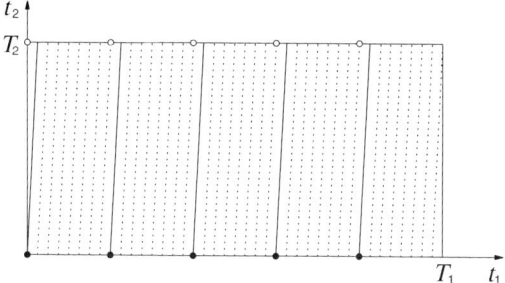

Fig. 2. Characteristic projections of the MPDAE in the domain of dependence

The periodicity in the second time scale leads to linear boundary conditions

$$\left(\bar{\mathbf{x}}_1(T_2)^\top, \ldots, \bar{\mathbf{x}}_{n_1}(T_2)^\top\right)^\top = B \left(\bar{\mathbf{x}}_1(0)^\top, \ldots, \bar{\mathbf{x}}_{n_1}(0)^\top\right)^\top, \qquad (11)$$

where the constant matrix $B \in \mathbb{R}^{n_1 d \times n_1 d}$ describes an interpolation scheme to determine the values $\bar{\mathbf{x}}_{j_1}(T_2)$ using the values $\hat{\mathbf{x}}((j_1 - 1)h_1, T_2) = \bar{\mathbf{x}}_{j_1}(0)$.

As the separate characteristic systems (10) are only coupled by the boundary conditions (11), this approach is much more efficient than a discretization on a uniform grid, which performs an unnecessary strong coupling in both coordinate directions, see [Pul03].

As we assume a strongly nonlinear behavior in the fast time scale t_2 (recall the MVF in Fig. 1 (right) with a pulsed oscillation in t_2-direction), an adaptive discretization along the characteristic projections is essential to obtain an efficient simulation technique.

3.2 Wavelet-Based Grid Generation

We aim at the detection of steep gradients in the solution for the generation of an adaptive grid. The use of wavelets allows us both time and frequency localization (whereas Fourier transforms are designed for frequency extraction, only): the discrete wavelet transform of a signal $x : \mathbb{R} \to \mathbb{R}$ reads

$$w_{j,k}(x) := \int_\mathbb{R} x(t)\, \psi_{j,k}(t)\, \mathrm{d}t, \quad \text{with} \quad \psi_{j,k}(t) := 2^{j/2}\, \psi(2^j t - k). \qquad (12)$$

Thereby, the dilation and translation of the 'mother-wavelet' ψ is controlled by the parameters $j, k \in \mathbb{Z}$, which specify the frequency range and the time localization, respectively.

Since we consider pulsed waveforms, wavelets of low order are an adequate choice. We use

$$\psi(t) = \begin{cases} 1.5 - 4\,|t|, & |t| \leq 0.5 \\ 0.5\,|t| - 0.75, & 0.5 < |t| \leq 1.5 \\ 0, & \text{elsewhere} \end{cases},$$

which we call hat-wavelet in the following. It is obtained from a so-called scaling function, which is the linear B-spline in our case.

Note that in contrast to frequently used orthonormal bases, our wavelet basis gives rise to a biorthogonal system. The main difference is that for the synthesis of a wavelet-transformed signal, the dual basis applies[1]. So in our case, the dual wavelet basis in fact undertakes the time-frequency localization of the coefficients in the wavelet-representation of the respective signal x:

$$x(t) = \sum_{j,k \in \mathbb{Z}} \widetilde{w}_{j,k} \psi_{j,k},$$

where $\widetilde{w}_{j,k}$ is the discrete wavelet transform (12) of x with respect to the dual wavelet $\widetilde{\psi}$. For more details on biorthogonal wavelets, see e.g. [Coh92].

As already mentioned in the previous section, we have to solve BVPs (10,11) and thus, we restrict our basis functions to a compact interval $[0, L]$ by 'folding' the hat-functions at the interval boundaries, see [CDV93]. In this way, we obtain a so-called multiresolution analysis of $L^2([0, L])$. Without loss of generality we choose $L \in \mathbb{R}$ such that $\operatorname{supp} \psi \subseteq [0, L]$. Then, L defines the number of dilated basis functions on the bounded domain, which of course can be transformed to the desired interval $[0, T_2]$.

For our numerical approximation, we regard a subspace $V_J^{[0,L]} \subset L^2([0, L])$ of finite dimension, which is composed by a direct sum of a central space $V_0^{[0,L]}$ (spanned by integer translates of the scaling function[2]) and hat-wavelet spaces $W_j^{[0,L]}$, $j = 0, \ldots, J - 1$ with according frequency localizations:

$$V_J^{[0,L]} = V_0^{[0,L]} \oplus \bigoplus_{j=0}^{J-1} W_j^{[0,L]}.$$

Due to the bounded domain, the number of translated wavelets $\psi_{j,k}$ (12) for $k \in \mathcal{I}_j \subset \mathbb{N}$, spanning the spaces $W_j^{[0,L]}$ is finite:

$$W_j^{[0,L]} = \operatorname{span}\{\psi_{j,k}(t) \mid k \in \mathcal{I}_j\} \quad \text{for } j = 0, \ldots, J - 1.$$

To illustrate the capabilities of the hat-wavelets for our digital-like signals, we approximate a pulse function $p : [0, 1] \to \mathbb{R}$ in the subspace $V_2^{[0,20]}$, i.e., using the refinement of $L + 1 = 21$ basis functions in the central space. The pulse function p is given in Fig. 3 as dashed line. There we show in three plots the coefficients for the corresponding spaces $V_0^{[0,20]}$, $W_0^{[0,20]}$ and $W_1^{[0,20]}$, respectively. The different markers indicate the size of these coefficients versus the corresponding time localization centers, which are the locations of the basis functions' maximal values.

---

[1] We choose the lowest order wavelet $\widetilde{\psi}$ out of the set of possible dual wavelets, which supplement the hat-wavelets to a biorthogonal basis.

[2] The B-spline scaling functions give us the coarsest approximation.

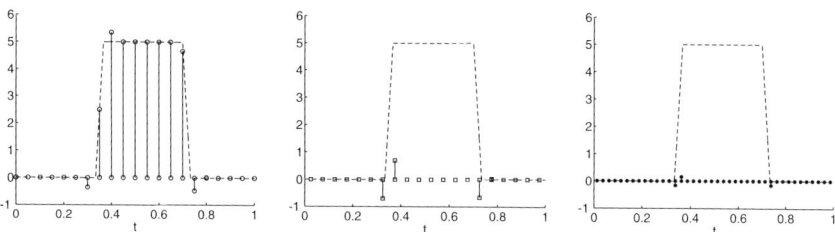

Fig. 3. Coefficients in $V_0^{[0,20]}$ (*left*), $W_0^{[0,20]}$ (*middle*) and $W_1^{[0,20]}$ (*right*)

Notice that the shape of the pulse is well represented by $V_0^{[0,20]}$ (scaling functions are normalized to 1), only at the location of the steep gradients the refinement levels $W_0^{[0,20]}$, $W_1^{[0,20]}$ have contributions, but attenuate fast. Thus, we can employ this localization of the steep gradients to define an adaptive grid: given a guess of the solution, we 'simply' have to inspect the size of the respective wavelet transforms.

Grid points associated with 'small' coefficients (smaller than a given threshold) are left out and grid points are added, where the coefficients exceed a given upper threshold. Of course, the relative level of these coefficients (compared within the same subspace) is most instructive. A more detailed description of the grid generation algorithm can be found in [BKP07].

3.3 Wavelet-Based Collocation

Now, we combine the method of characteristics introduced in Sect. 3.1 and the wavelet-based grid generation described in Sect. 3.2 by solving the DAE BVP (10, 11) with a wavelet-collocation scheme.

The hat-wavelets serve as ansatz functions to approximate the solutions (9) along each characteristic projection for $\tau \in [0, T_2]$:

$$\bar{\mathbf{x}}_{j_1}(\tau) \doteq \tilde{\mathbf{x}}_{j_1}(\tau) := \sum_{i=1}^{N} \mathbf{c}_{j_1,i} \Psi_i(\tau). \tag{13}$$

This approximation holds for each component of $\bar{\mathbf{x}}_{j_1} : [0, T_2] \to \mathbb{R}^d$, $\mathbf{c}_{j_1,i} \in \mathbb{R}^d$ for all $j_1 = 1, \ldots, n_1$ and $i = 1, \ldots, N$. Thereby, all $N = 2^J L + 1$ basis functions of $V_J^{[0,L]}$ (see Sect. 3.2) are denoted by $\Psi_i : [0, T_2] \to \mathbb{R}$, $i = 1, \ldots, N$.

We demand that for each characteristic projection, the DAE (10) has to be fulfilled in $n_2 = N - 1$ collocation points $\tau_m \in [0, T_2]$, $m = 1, \ldots, n_2$:

$$\left.\frac{\mathrm{d}}{\mathrm{d}\tau} \mathbf{q}(\tilde{\mathbf{x}}_{j_1}(\tau))\right|_{\tau=\tau_m} = \mathbf{f}\big(\hat{\mathbf{b}}((j_1 - 1)h_1 + \tau_m, \tau_m), \tilde{\mathbf{x}}_{j_1}(\tau_m)\big) \tag{14}$$

for $j_1 = 1, \ldots, n_1$. Together with the n_1 BCs (11) for the approximations (13),

$$\big(\tilde{\mathbf{x}}_1(0)^\top, \ldots, \tilde{\mathbf{x}}_{n_1}(0)^\top\big)^\top = B \big(\tilde{\mathbf{x}}_1(T_2)^\top, \ldots, \tilde{\mathbf{x}}_{n_1}(T_2)^\top\big)^\top, \tag{15}$$

we end up with $n_1 \cdot N$ nonlinear equations for the unknown coefficients $c_{j_1,i}$.

Of course, we want to use the time-frequency localization property of our wavelets to determine an adaptive set of collocation points for the numerical simulation. We establish an equidistant start grid in $[0, T_2]$:

$$\tau_m = (m-1)h_2, \quad h_2 = \frac{T_2}{N-1}, \quad m = 1, \ldots, N = 2^J L + 1,$$

which corresponds to the time localization centers of the basis functions, cf. Sect. 3.2. Then, we solve the nonlinear system (14, 15) using a Newton-type method. Starting values have to be determined for all grid points in the domain $[0, T_1] \times [0, T_2]$. After a few iterations (in our test examples two iterations were sufficient), the wavelet coefficients already contain enough information about the structure of the solution to determine an adaptive grid as outlined in Sect. 3.2. Then, the Newton iteration is continued on the new mesh to solve for the respective wavelet coefficients.

4 Simulation Results

We investigate a switched-capacitor circuit, the Miller integrator, which is depicted in Fig. 4 (left). It contains two MOS-transistors[3] M_1 and M_2 driven by two complementary pulse functions p_a and p_b. The output at node 3 approximates the negative integral[4] of the input v_{in}. Applying a sinusoidal input signal, the respective output voltage u_3^{ref} (reference solution) can be seen in Fig. 4 (right). The discrete sampling via the pulse functions causes the signal to be rough, which is revealed by the zoom in this figure.

The index-1 differential-algebraic model describing the network behavior can be found in [KF06].

The sinusoidal input signal v_{in} determines the slow time scale $T_1 = 10^{-5}$ s, whereas the pulses exhibit a period of $T_2 = 2.5 \cdot 10^{-8}$ s.

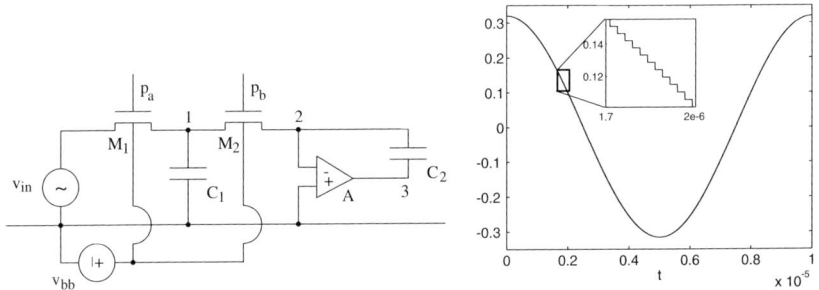

Fig. 4. Miller integrator circuit (*left*) and output voltage u_3^{ref} (*right*)

---

[3] MOS stands for *metal oxide semiconductor*.
[4] The scaling of the output voltage depends on the design parameters.

We apply the multidimensional signal model from Sect. 2 and solve the resulting biperiodic BVP (10, 11) of the MPDAE by the wavelet-collocation introduced in Sect. 3. We discretize on $n_1 = 30$ characteristic projections and use an equidistant start grid of $N = 121$ points (with $J = 2$ and $L = 30$). A different adaptive grid is determined for each characteristic projection, which results in an average of $n_2 = 60$ mesh points (with a finest resolution as for an equidistant grid with 241 points).

In Fig. 5 (left), we display the first component $\hat{u}_1$ of the MPDAE solution, which shows a strong influence of the pulse functions. The steep gradients are sharply detected by the wavelet basis, which results in the adaptive grid depicted in Fig. 5 (right). The MVF $\hat{u}_1$ is plotted on a common grid for all characteristic projections (using respective evaluations of the basis functions).

In comparison to the reference solution u_3^{ref} in Fig. 4 (right), the reconstructed DAE solution u_3 is depicted in Fig. 6 (right). The reference solution to the reconstructed DAE solution u_1 (Fig. 6, left) is shown in Fig. 1 (left) in Sect. 2.1. Both components of the approximative solution show a good agreement with the reference solution, which is confirmed by a discrete L^2-error of only about 2 %. The pulsed structure of the signals is sharply resolved and no undesired peaks occur.

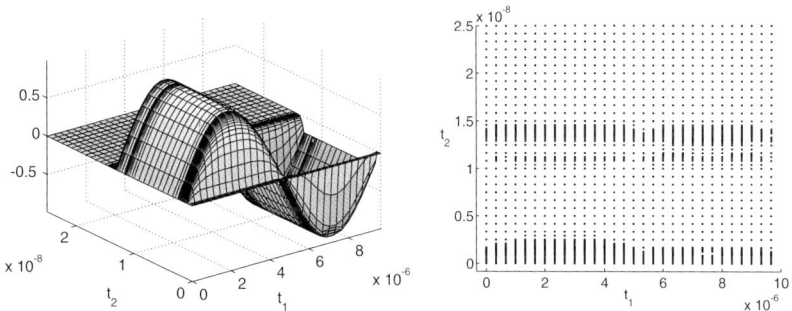

Fig. 5. Miller Integrator: MPDAE solution $\hat{u}_1$ (*left*) and adaptive grid (*right*)

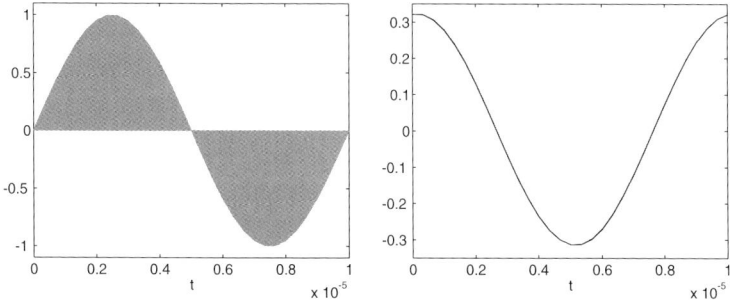

Fig. 6. Miller Integrator: Reconstructed DAE solutions u_1 (*left*) and u_3 (*right*)

5 Conclusions

A multivariate modeling is applied to decouple widely separated time scales. We presented a novel algorithm to solve the resulting MPDAE system. An adaptive wavelet-based collocation scheme yields an efficient simulation of RF circuits including analog and digital signal structures. The capability of the wavelet basis to detect steep gradients and the efficiency of the numerical scheme are verified by simulation results for a switched-capacitor circuit.

References

[BKP07] Bartel, A., Knorr, S., Pulch, R.: Wavelet-based adaptive grids for the simulation of multirate partial differential-algebraic equations. To appear in: Appl. Numer. Math.

[BWLB96] Brachtendorf, H.G., Welsch, G., Laur, R., Bunse-Gerstner, A.: Numerical steady state analysis of electronic circuits driven by multi-tone signals. Electrical Engineering, **79**, 103–112 (1996)

[Coh92] Cohen, A.: Biorthogonal wavelets. In: Chui, C. (ed.) Wavelet Analysis and its Applications II: Wavelets – A Tutorial in Theory and Applications, Academic Press, New York, 123–152 (1992)

[CDV93] Cohen, A., Daubechies, I., Vial, P.: Wavelets on the interval and fast wavelet transforms. Appl. Comput. Harm. Anal., **1**, 54–81 (1993)

[GF99] Günther, M., Feldmann, U.: CAD based electric circuit modeling I: mathematical structure and index of network equations. Surv. Math. Ind. **8**:1, 97–129 (1999)

[KF06] Knorr, S., Feldmann, U.: Simulation of pulsed signals in MPDAE-modelled SC-circuits. In: Di Bucchianico, A., Mattheij, R.M.M., Peletier, M.A. (eds.) Progress in Industrial Mathematics at ECMI 2004, Mathematics in Industry 8, Springer, Berlin, 159–163 (2006)

[KG06] Knorr, S., Günther, M.: Index analysis of multirate partial differential-algebraic systems in RF-circuits. In: Anile, A.M., Alì, G., Mascali, G. (eds.) Scientific Computing in Electrical Engineering, Mathematics in Industry 9, Springer, Berlin, 93–100 (2006)

[Pul03] Pulch, R.: Finite difference methods for multi time scale differential algebraic equations. Z. Angew. Math. Mech., **83**:9, 571–583 (2003)

[PG02] Pulch, R., Günther, M.: A method of characteristics for solving multirate partial differential equations in radio frequency application. Appl. Numer. Math., **42**, 397–409 (2002)

[PGK07] Pulch, R., Günther, M., Knorr, S.: Multirate partial differential algebraic equations for simulating radio frequency signals. To appear in: Euro. Jour. Appl. Math.

[Roy01] Roychowdhury, J.: Analyzing circuits with widely-separated time scales using numerical PDE methods. IEEE Trans. CAS I, **48**, 578–594 (2001)

Numerical Simulation of Thermal Effects in Coupled Optoelectronic Device-Circuit Systems

Markus Brunk[1] and Ansgar Jüngel[2]

[1] Institut für Mathematik, Universität Mainz, Staudingerweg 9, 55099 Mainz, Germany, brunk@mathematik.uni-mainz.de
[2] Institut für Analysis und Scientific Computing, Technische Universität Wien, Wiedner Hauptstr. 8–10, 1040 Wien, Austria
juengel@anum.tuwien.ac.at

1 Introduction

The control of thermal effects becomes more and more important in modern semiconductor circuits like in the simplified CMOS transceiver representation described by U. Feldmann in the above article *Numerical simulation of multiscale models for radio frequency circuits in the time domain*. The standard approach for modeling integrated circuits is to replace the semiconductor devices by equivalent circuits consisting of basic elements and resulting in so-called compact models. Parasitic thermal effects, however, require a very large number of basic elements and a careful adjustment of the resulting large number of parameters in order to achieve the needed accuracy.

Therefore, it is preferable to model those semiconductor devices which are critical for the parasitic effects by semiconductor transport equations. The transport of electrons in the devices is modeled here by the one-dimensional energy-transport model allowing for the simulation of the electron temperature. The electric circuits are described by modified nodal analysis. Thus, the devices are modeled by (nonlinear) partial differential equations, whereas the circuit is described by differential-algebraic equations. The coupled model, which becomes a system of (nonlinear) partial differential-algebraic equations, is numerically discretized in time by the 2-stage backward difference formula (BDF2), since this scheme allows to maintain the M-matrix property, and the semi-discrete equations are approximated by a mixed finite-element method.

The objective is the simulation of a benchmark high-frequency transceiver circuit, using a laser diode as transmitter and a photo diode as receiver. The optical field in the laser diode is modeled by recombination terms and a rate equation for the number of photons in the device. The optical effects in the photo diode are described by generation terms. The numerical results show that the thermal effects can modify significantly the behavior of the transmitter circuit.

2 Modeling

Circuit Modeling

A well-established mathematical description of electric circuits, consisting of resistors, capacitors, and inductors (RCL circuit) is the modified nodal analysis (MNA) which can be easily extended to circuits containing semiconductor devices. In the following, the circuit model is described.

The circuit is replaced by a directed graph. The RLC branches are characterized by the incidence matrix A, and the semiconductor branches are characterized by the semiconductor incidence matrix A_S. The basic tools for the MNA are the Kirchhoff laws and the current-voltage characteristics for the basic elements,

$$Ai + A_S j_S = 0, \quad v = A^\top e, \quad i_R = g(v_R), \quad i_C = \frac{dq}{dt}(v_C), \quad v_L = \frac{d\Phi}{dt}(i_L),$$

where i, v, and e are the vectors of branch currents, branch voltages, and node potentials, respectively, and j_S denotes the semiconductor current (see below). The variable g denotes the conductivity of the resistor, q is the charge of the capacitor, and Φ the flux of the inductor. The incidence matrix A is assumed to consist of the block matrices A_R, A_C, A_L, A_i, and A_v, where the indices i and v indicate the current source and voltage source branches, respectively.

Denoting by $i_s = i_s(t)$ and $v_s = v_s(t)$ the given input functions for the sources, we obtain the system for the charge-oriented MNA [13],

$$A_C \frac{dq}{dt}(A_C^\top e) + A_R g(A_R^\top e) + A_L i_L + A_v i_v + A_S j_S = -A_i i_s, \quad (1)$$

$$\frac{d\Phi}{dt}(i_L) - A_L^\top e = 0, \quad A_v^\top e = v_s, \quad (2)$$

for the unknowns $e(t)$, $i_L(t)$, and $i_v(t)$. Equation (1) expresses the Kirchhoff current law, the first equation in (2) is the voltage-current characteristic for inductors, and the last equation allows to compute the node potentials.

Semiconductor Device Modeling

The flow of minority charge carriers (holes) in the device is modeled by the drift-diffusion model for the hole density p. The electron flow is described by the energy-transport equations [8]. The first model consists of the conservation law for the hole mass, together with a constitutive relation for the hole current density. The latter model also includes the conservation law for the electron energy and a constitutive relation for the energy flux. Both models can be derived from the semiconductor Boltzmann equation (see [8] and references therein). They are coupled through recombination-generation terms and the

Poisson equation for the electric potential. More precisely, the electron density n, the hole density p, and the electron temperature T are obtained from the parabolic equations

$$\mu_n^{-1}\partial_t g_1 - \operatorname{div} J_n = -R(\mu_n^{-1}g_1, p), \quad \partial_t p + \operatorname{div} J_p = -R(\mu_n^{-1}g_1, p) \quad (3)$$

$$\mu_n^{-1}\partial_t g_2 - \operatorname{div} J_w = -J_n \cdot \nabla V + W(\mu_n^{-1}g_1, T) - \frac{3}{2}TR(\mu_n^{-1}g_1, p), \quad (4)$$

where $g_1 = \mu_n n$ and $g_2 = \mu_n w$ are auxiliary variables allowing for a drift-diffusion-type formulation of the fluxes [8], $w = \frac{3}{2}nT$ is the thermal energy, and μ_n and μ_p are the electron and hole mobilities, respectively. The electron current density J_n, the energy flux J_w, and the hole current density J_p are given by

$$J_n = \nabla g_1 - \frac{g_1}{T}\nabla V, \quad J_w = \nabla g_2 - \frac{g_2}{T}\nabla V, \quad J_p = -\mu_p(\nabla p + p\nabla V). \quad (5)$$

The equations are coupled self-consistently to the Poisson equation for the electric potential V,

$$\lambda^2 \Delta V = \mu_n^{-1}g_1 - p - C(x), \quad (6)$$

where λ is the scaled Debye length and the given function $C(x)$ models the doping profile. The functions

$$W(n,T) = -\frac{3}{2}\frac{n(T-T_L)}{\tau_0} \quad \text{and} \quad R(n,p) = \frac{np - n_i^2}{\tau_p(n + n_i) + \tau_n(p + n_i)} \quad (7)$$

with the (scaled) energy relaxation time τ_0 and lattice temperature $T_L = 1$ describe the relaxation to the equilibrium energy and Shockley-Read-Hall recombination-generation processes with intrinsic density n_i and electron and hole lifetimes τ_n and τ_p, respectively.

Equations (3)–(6) are solved in the bounded semiconductor domain Ω, where some initial values n_I, p_I, and T_I are imposed. The boundary of Ω is assumed to split into two parts. On the insulating parts of the boundary Γ_N, it is assumed that the normal components of the current densities and of the electric field vanish. For the temperature, homogenous Neumann boundary conditions are assumed as in [1]. We have shown in [5] that boundary layers for the particle densities can be avoided if Robin-type boundary conditions similar as in [14] are employed on the remaining boundary parts,

$$n - \theta_n J_n \cdot \nu = n_a \quad \text{and} \quad p + \theta_p J_p \cdot \nu = p_a \quad \text{on } \partial\Omega\backslash\Gamma_N, \quad (8)$$

where θ_n and θ_p are some parameters and n_a and p_a are ambient particle densities. Notice that in the one-dimensional simulations presented below, $\Gamma_N = \emptyset$.

Coupling to the Circuit

The boundary conditions for the electric potential at the contacts are determined by the circuit and are given as

$$V = e_i + V_{\text{bi}} \quad \text{on } \Gamma_k, \; t > 0, \quad \text{where} \quad V_{\text{bi}} = \operatorname{arsinh}\left(\frac{C}{2n_i}\right), \qquad (9)$$

if the terminal k of the semiconductor is connected to the circuit node i.

The semiconductor current entering the circuit consists of the electron current J_n, the hole current J_p, and the displacement current $J_d = -\lambda^2 \partial_t \nabla V$, guaranteeing charge conservation. The current leaving the semiconductor device at terminal k, corresponding to the boundary part Γ_k, is defined by

$$j_k = \int_{\Gamma_k} (J_n + J_p + J_d) \cdot \nu \, ds,$$

where ν is the exterior unit normal vector to $\partial \Omega$. We denote by j_S the vector of all terminal currents except the reference terminal. In the one-dimensional case, there remains only one terminal, and the current through the terminal at $x = 0$ is given by

$$j_S(t) - (J_n(0,t) + J_p(0,t) - \partial_t j_{d,S}(0,t)) = 0, \quad j_{d,S} - \lambda^2 V_x = 0, \qquad (10)$$

where the circuit equations (1)–(2) have to be appropriately scaled [6].

The complete coupled system consists of equations (1)–(10) forming an initial boundary-value problem of partial differential-algebraic equations. The system resulting from the coupled circuit drift-diffusion equations has at most index 2 and it has index 1 under some topological assumptions [3, 13]. No analytical results are available for the coupled circuit energy-transport system.

Optoelectronic Device Modeling

The interaction between optical and electrical effects is modeled by recombination-generation terms appearing in (7). In the following, we present the model used in the numerical simulations and we refer to [6] for a discussion about the model simplifications.

For a vertical photo diode, the supplied photons generate free charge carriers generating the photo current. We model this effect by adding to the Shockley-Read-Hall term (7) the generation rate $G_{\text{opt}}(x)$ of free carriers at depth x, caused by the (scaled) optical irradiation power P_{in} with angular frequency ω [10],

$$G_{\text{opt}}(x) = \eta(1-r)\frac{P_{\text{in}}}{\hbar \omega A}\alpha_{\text{ab}} e^{-\alpha_{\text{ab}} x}, \qquad (11)$$

where the physical parameters are the quantum efficiency η, the reflectivity r of the irradiated surface with area A, the reduced Planck constant $\hbar = h/2\pi$, and the optical absorption α_{ab}.

The laser diode is modeled by a *pin* heterostructure diode in which the intrinsic (active) region consists of a low-band gap material causing carrier confinement. The active region works as a Fabry-Perot laser cavity and can be modeled as a single mode laser. The band discontinuities are simply described by adding a constant band potential to the electric potential V in the active region [9]. Additionally to (7), spontaneous and stimulated recombination is introduced,

$$R_{\text{spon}} = Bnp \quad \text{and} \quad R_{\text{stim}} = \frac{c}{\mu_{\text{opt}}} g(n) |\Xi|^2 S, \qquad (12)$$

respectively, where B is the spontaneous recombination parameter, c the speed of light μ_{opt} the refractive index of the material, $g(n)$ the optical gain depending on the electron density, $|\Xi|^2$ the intensity distribution of the optical field, which is a solution of the waveguide equation [6], and $S = S(x,t)$ is the number of photons in the device.

The optical gain is approximated by $g(n) = g_0(n - n_{\text{th}})$ [7], with differential gain g_0 and threshold density n_{th}. In the lasing mode we can employ the quasi-neutral assumption $n \approx p$ in the active region such that the gain becomes approximately $g(p) = g_0(p - n_{\text{th}})$ in the recombination term occurring in the hole equation (3). This allows for a discretization that guarantees positivity of the discretized hole density [6]. The number of photons S is balanced by the rate equation

$$\partial_t S = v_g (\beta - \alpha) S + R_{\text{spon}}, \quad \text{where} \quad \beta = \int_{\Omega_a} (g(n) - \alpha_{\text{bg}}) |\Xi|^2 \, ds,$$

α is the total loss by external output and scattering, α_{bg} denotes the background loss, and Ω_a is the transverse cross section of the active region. We prescribe the initial condition $S(\cdot, 0) = S_I$ in Ω. Finally, the output power is computed from of the number of photons by

$$P_{\text{out}} = \hbar \omega \frac{c}{\mu_{\text{opt}}} \alpha_f |\Xi|^2 S \qquad (13)$$

(see [2]), where α_f denotes the facet loss of the laser cavity.

3 Numerical Simulations

The system of coupled partial differential-algebraic equations is first discretized in time by the BDF2 method since this scheme allows to maintain the M-matrix property of the final discrete system. The Poisson equation is discretized in space by the linear finite-element method. Then the discrete electric potential is piecewise linear and the approximation of the electric field $-V_x$ is piecewise constant.

The semi-discrete continuity equations at one time step are of the form

$$-J_{j,x} + \sigma_j g_j = f_j, \quad J_j = g_{j,x} - \frac{g_j}{T} V_x, \quad j = 1, 2, \qquad (14)$$

with the current densities $J_1 = J_n$ and $J_2 = J_w$ and some expressions σ_j and f_j. These equations are discretized in space by a hybridized exponentially fitted mixed finite-element method [8]. We employ the finite elements of [12] since they guarantee the positivity of the discrete variables if positive initial and Dirichlet boundary data are prescribed and if $\sigma_j \geq 0$, $f_j \geq 0$ for $j = 1, 2$. This property also holds for the Robin conditions (8) [6]. Finally, the nonlinear discrete system is solved by Newton's method.

Rectifying Circuit

As a test example we consider a rectifying circuit containing four silicon pn diodes as in [5] (Fig. 1). Each of the diodes has the length $L = 0.1\,\mu$m (and $L_y = 0.1\,\mu$m, $L_z = 2\,\mu$m) or $L = 1\,\mu$m (and $L_y = 1\,\mu$m, $L_z = 20\,\mu$m) and a maximum doping of $10^{22}\,\mathrm{m}^{-3}$. We have chosen the resistance $R = 100\,\Omega$ and the voltage source $v(t) = U_0 \sin(2\pi\omega t)$ with $U_0 = 5\,\mathrm{V}$ and $\omega = 1\,\mathrm{GHz}$ or $\omega = 10\,\mathrm{GHz}$. The remaining physical parameters are listed in Table 1. As initial conditions we take thermal equilibrium densities in the device and vanishing node potentials and branch currents in the circuit. The initial value for the displacement current is determined by (10). A computation according to [11] shows that these values are consistent for the coupled DAE system.

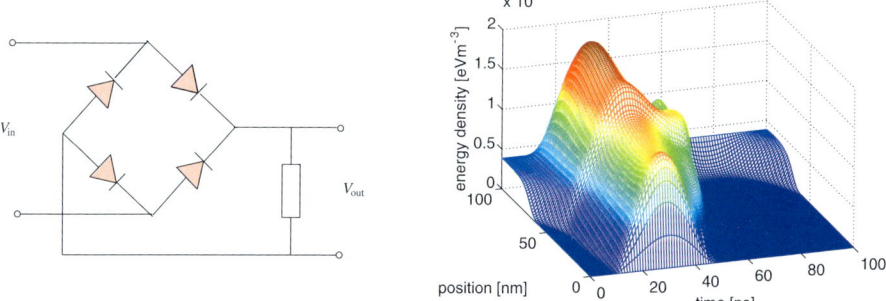

Fig. 1. *Left:* Graetz circuit. *Right:* Thermal energy in a pn diode during one oscillation of V_{in}

Table 1. Physical parameters for a silicon pn-junction diode

Parameter	Physical meaning	Numerical value
q	elementary charge	$1.6 \cdot 10^{-19}$ As
ϵ_s	permittivity constant	$1.05 \cdot 10^{-10}$ As/Vm
U_T	thermal voltage at 300 K	0.026 V
μ_n/μ_p	low-field carrier mobilities	1500/450 cm^2/Vs
τ_n/τ_p	carrier lifetimes	$10^{-6}/10^{-5}$ s
n_i	intrinsic density	10^{16} m^{-3}
τ_0	energy relaxation time	$4 \cdot 10^{-13}$ s

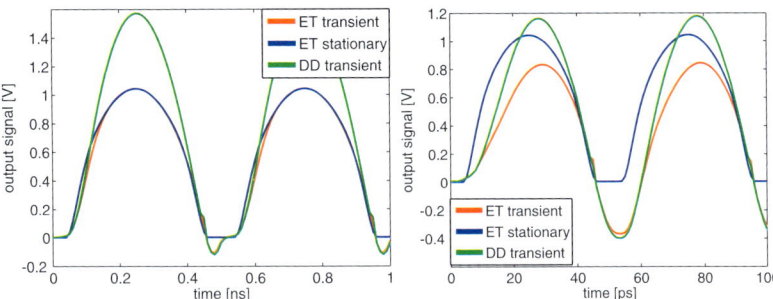

Fig. 2. Output signal of the Graetz circuit for two frequencies of the voltage source. *Left:* $\omega = 1\,\text{GHz}$ and $L = 0.1\,\mu\text{m}$. *Right:* $\omega = 10\,\text{GHz}$ and $L = 1\,\mu\text{m}$

In Fig. 1 the energy density in one of the diodes during one oscillation is presented. As expected, we observe a high thermal energy in forward bias ($t \in [0, 50\,\text{ps}]$), whereas it is negligible in backward bias ($t \in [50\,\text{ps}, 100\,\text{ps}]$) although the electron temperature (not shown) may be very large around the junction [4].

The impact of the thermal effects on the electrical behavior of the circuit is shown in Fig. 2. The figure clearly shows the rectifying behaviour of the cuircuit. The largest current is obtained from the drift-diffusion model since we have assumed a constant electron mobility such that the drift is unbounded with respect to the modulus of the electric field. The stationary energy-transport model is not able to catch the capacitive effect at the junction which is particularly remarkable at higher frequencies.

Optoelectronic Circuit

Next we consider a AlGaAs/GaAs laser diode with a digital input signal. The transmitted signal is received by a silicon photo diode coupled to a high-pass filter (see Fig. 3). We have taken a capacitance of 10 pF, the resistances

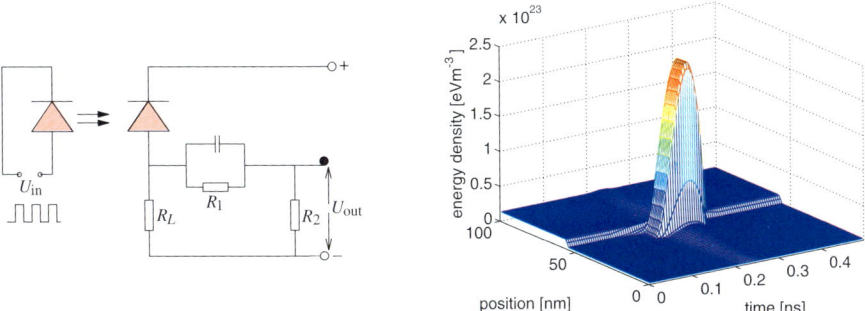

Fig. 3. *Left:* Laser and photo diode with a high-pass filter. *Right:* energy density in the laser diode for signal $v(t) = 2\,\sin(2\pi t 10^9)\,\text{V}$

Table 2. Physical parameters for a laser diode of $Al_{0.7}Ga_{0.3}As$ (superscript A) and GaAs (superscript G). Parameters without superscript are taken for both materials

Parameter	Physical meaning	Numerical value
L_y/L_z	extension of device in y/z-direction	$10^{-6}/10^{-5}$ m
U_n/U_p	band potentials in active region	$0.1/-0.1$ V
B	spontaneous recombination parameter	10^{-16} m³/s
n_{th}	threshold density	10^{24} m⁻³
α_f/α_{bg}	mirror/optical background loss	$5000/4000$ m⁻¹
$\epsilon_s^A/\epsilon_s^G$	material permittivity	$1.08 \cdot 10^{-10}/1.14 \cdot 10^{-10}$ As/Vm
μ_n^A/μ_n^G	electron mobilities	$2300/8300$ cm²/Vs
μ_p^A/μ_p^G	hole mobilities	$145/400$ cm²/Vs
μ_{opt}^A/μ_{opt}^G	refractive index	$3.3/3.15$
n_i^A/n_i^G	intrinsic density	$2.1 \cdot 10^9/2.1 \cdot 10^{12}$ m⁻³
g_0^G	differential gain in GaAs	10^{-20} m²

$R_1 = 1$ MΩ, $R_2 = 100\,\Omega$, and $R_3 = 1$ kΩ, and a backward bias of 0.2 V. The laser diode has the length of 1 μm with an intrinsic region of 0.1 μm length in the center of the device. The doping concentration is -10^{24} m⁻³ in the p-doped region, 10^{24} m⁻³ in the n-doped region, and 10^{18} m⁻³ in the intrinsic region. The photo diode has a size of $L = 6$ μm, $L_y = 10^{-5}$ m and $L_z = 10^{-4}$ m. For the quantum efficiency we assume $\eta = 0.5$, for the surface reflectivity $r = 0.3$ and $\alpha = 5000$ m⁻¹ for the absorption. The remaining parameters are taken from Tables 1 and 2.

In Fig. 3 the energy density in the laser diode during one half oscillation is shown. After having passed the threshold, the energy density increases tremendously in the active region. This is due to carrier confinement in the heterostructure, as in the lasing mode the carrier density is very high in the active region.

Finally, we operate the transmitter with a 1 GHz digital signal of 2 V. In Fig. 4 the light output signal and the received signal by the high-pass

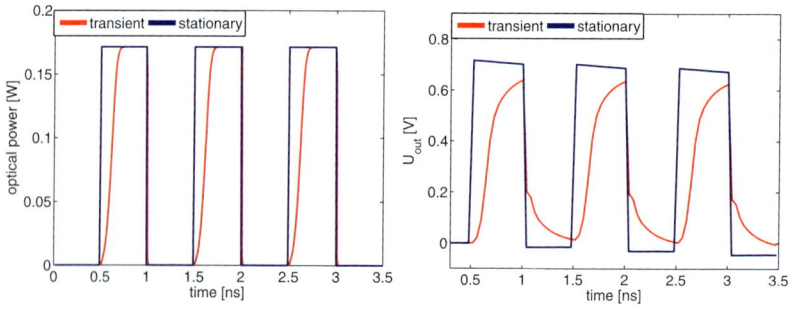

Fig. 4. Output of the laser diode and the high-pass filter for the stationary and transient energy-transport model with a digital input signal of 1 GHz

filter is presented. Again we observe that the transient energy-transport model responds better to the capacitive effects in high-frequency circuits than the stationary model.

4 Conclusion

We have presented a coupled model consisting of the circuit equations from modified nodal analysis and the energy-transport model for semiconductor devices, resulting in a system of nonlinear partial differential-algebraic equations. This system allows for a direct simulation of thermal effects and can help to improve compact models of integrated circuits. The coupled model is tested on a Graetz circuit and a high-frequency transmitter with laser and photo diodes. The results show the impact of the thermal energy on the circuit. Compared to the constant-temperature drift-diffusion model, the output signal is smaller due to thermal effects.

With decreasing size of the basic components in integrated circuits and special power devices, the thermal interaction between circuit elements will increase in importance in the near future. Therefore, we need to model not only the carrier temperature but also the device temperature and the interaction between the circuit elements. Thus, a heat equation for the temperature of the semiconductor lattice needs to be included in the presented model. This extension is currently under investigation. We expect that the resulting model will improve significantly the prediction of hot-electron effects and hot spots in integrated circuits.

References

1. A. M. Anile, V. Romano, and G. Russo. Extendend hydrodynamic model of carrier transport in semiconductors. *SIAM J. Appl. Math.* 61 (2000), 74–101.
2. U. Bandelow, H. Gajewski, and R. Hünlich. Fabry-Perot lasers: thermodynamic-based modeling. In: J. Piprek (ed.), *Optoelectronic Devices. Advanced Simulation and Analysis.* Springer, Berlin (2005), 63–85.
3. M. Bodestedt. *Index Analysis of Coupled Systems in Circuit Simulatiom*. Licentiate Thesis, Lund University, Sweden, 2004.
4. F. Brezzi, L. Marini, S. Micheletti, P. Pietra, R. Sacco, and S. Wang. Discretization of semiconductor device problems. In: W. Schilders and E. ter Maten (eds.), *Handbook of Numerical Analysis. Numerical Methods in Electromagnetics.* Elsevier, Amsterdam, Vol. 13 (2005), 317–441.
5. M. Brunk and A. Jüngel. Numerical coupling of electric circuit equations and energy-transport models for semiconductors. To appear in *SIAM J. Sci. Comput.*, 2007.
6. M. Brunk and A. Jüngel. Simulation of thermal effects in optoelectronic devices using energy-transport equations. In preparation, 2007.
7. S. L. Chuang. *Physics of Optoelectronic Devices.* Wiley, New York, 1995.

8. P. Degond, A. Jüngel, and P. Pietra. Numerical discretization of energy-transport models for semiconductors with non-parabolic band structure. *SIAM J. Sci. Comp.* 22 (2000), 986–1007.
9. M. Friedrich. *Ein analytisches Modell zur Simulation von Silizium-Germanium Heterojunction-Bipolartransistoren in integrierten Schaltungen.* PhD Thesis, Ruhr-Universität Bochum, Germany, 2002.
10. G. R. Jones, R. J. Jones, and W. French. Infrared HgCdTe Optical Detectors. In: J. Piprek (ed.), *Optoelectronic Devices. Advanced Simulation and Analysis.* Springer, Berlin (2005), 381–403.
11. R. Lamour. Index determination and calculation of consistent initial values for DAEs. *Computers Math. Appl.* 50 (2005), 1125–1140.
12. L. D. Marini and P. Pietra. New mixed finite element schemes for current continuity equations. *COMPEL* 9 (1990), 257–268.
13. C. Tischendorf. Modeling circuit systems coupled with distributed semiconductor equations. In: K. Antreich, R. Bulirsch, A. Gilg, and P. Rentrop (eds.), *Modeling, Simulation, and Optimization of Integrated Circuits*, Internat. Series Numer. Math. 146 (2003), 229–247.
14. A. Yamnahakki. Second-order boundary conditions for the drift-diffusion equations for semiconductors. *Math. Models Meth. Appl. Sci.* 5 (1995), 429–455.

Efficient Transient Noise Analysis in Circuit Simulation

Georg Denk[1], Werner Römisch[2], Thorsten Sickenberger[2], and Renate Winkler[2]

[1] Qimonda AG, Products, München, Germany, `georg.denk@qimonda.com`
[2] Institute of Mathematics, Humboldt-Universität zu Berlin, Germany
`{romisch,sickenberger,winkler}@math.hu-berlin.de`

Transient noise analysis means time domain simulation of noisy electronic circuits. We consider mathematical models where the noise is taken into account by means of sources of Gaussian white noise that are added to the deterministic network equations, leading to systems of stochastic differential algebraic equations (SDAEs). A crucial property of the arising SDAEs is the large number of small noise sources that are included. As efficient means of their integration we discuss adaptive linear multi-step methods, in particular stochastic analogues of the trapezoidal rule and the two-step backward differentiation formula, together with a new step-size control strategy. Test results including real-life problems illustrate the performance of the presented methods.

1 Transient Noise Analysis in Circuit Simulation

The increasing scale of integration, high clock frequencies and low supply voltages cause smaller signal-to-noise ratios. Reduced signal-to-noise ratio means that the difference between the wanted signal and noise is getting smaller. A consequence of this is that the circuit simulation has to take noise into account. In several applications the noise influences the system behaviour in an essentially nonlinear way such that linear noise analysis is no longer satisfactory and transient noise analysis, i.e., the simulation of noisy systems in the time domain, becomes necessary (see [4, 16]). For an implementation of an efficient transient noise analysis in an analog simulator, both an appropriate modelling and integration scheme is necessary (see [3]).

Here we deal with the thermal noise of resistors as well as the shot noise of semiconductors that are modelled by additional sources of additive or multiplicative Gaussian white noise currents that are shunt in parallel to the noise-free elements. Thermal noise i_{th} of resistors is caused by the thermal motion of electrons and is described by Nyquist's theorem. Shot noise i_{shot} of

pn-junctions, caused by the discrete nature of currents due to the elementary charge, is modelled by Schottky's formula and inherits noise intensities that depend on the deterministic currents:

$$i_{th} = \sqrt{\frac{2kT}{R}}\xi(t), \qquad i_{shot} = \sqrt{q_e|i_{det}(u)|}\xi(t). \tag{1}$$

Here $\xi(t)$ is a standard Gaussian white noise process, R denotes the resistance, T is the temperature, $k = 1.38 \cdot 10^{-23}$ is Boltzmann's constant, $i_{det}(u)$ is the characteristic of the noise-free current through the pn-junction and $q_e = 1.60 \cdot 10^{-19}$ is the elementary charge.

Combining Kirchhoff's current law with the element characteristics and using the charge-oriented formulation yields a stochastic differential-algebraic equation (SDAE) of the form (see e.g. [15], or for the deterministic case [6])

$$A\frac{d}{dt}q(x(t)) + f(x(t),t) + \sum_{r=1}^{m} g_r(x(t),t)\xi_r(t) = 0, \tag{2}$$

where A is a constant singular incidence matrix determined by the topology of the dynamic circuit parts, the vector $q(x)$ consists of the charges and the fluxes, and x is the vector of unknowns consisting of the nodal potentials and the branch currents through voltage-defining elements. The term $f(x,t)$ describes the impact of the static elements, $g_r(x,t)$ denotes the vector of noise intensities for the r-th noise source, and ξ is an m-dimensional vector of independent Gaussian white noise sources (see e.g. [4, 16]). One has to deal with a large number of equations as well as of noise sources, where one can and has to exploit the fact that compared to the other quantities the noise intensities $g_r(x,t)$ are small.

Though the system (2) formally differs only by the additional noise term from the deterministic system, a completely different mathematical framework has to be applied. A serious mathematical description begins by introducing the Brownian motion or the Wiener process that is caused by integrating the white noise "$W(t) = \int_0^t \xi(s)ds = \int_0^t dW(s)$" (see e.g. [1]). Problem (2) is then understood as a stochastic integral equation

$$A\,q(X(s))\Big|_{t_0}^{t} + \int_{t_0}^{t} f(X(s),s)ds + \sum_{r=1}^{m}\int_{t_0}^{t} g_r(X(s),s)dW_r(s) = 0, \quad t \in [t_0,T], \tag{3}$$

where the second integral is an Itô-integral, and W denotes an m-dimensional Wiener process (or Brownian motion) given on the probability space $(\Omega, \mathcal{F}, P)$ with a filtration $(\mathcal{F}_t)_{t \geq t_0}$. The solution is a stochastic process depending on the time t and on the random sample ω. The value at fixed time t is a random variable $X(t,\cdot) = X(t)$ whose argument ω is usually dropped. For a fixed sample ω representing a fixed realization of the driving Wiener process, the function $X(\cdot,\omega)$ is called a realization or a path of the solution. Due to the influence of the Gaussian white noise, typical paths of the solution are nowhere differentiable.

The theory of stochastic differential equations distinguishes between the concepts of strong, i.e., pathwise solutions and weak, i.e., the distribution law of solutions. We decided to aim at the simulation of solution paths, i.e., strong solutions that reveal the phase noise that is of particular interest in case of oscillating solutions. From the solution paths statistical data of the phase as well as moments of the solution can be computed in a post-processing step. We therefore use the concept of strong solutions and strong (mean-square) convergence of approximations.

By the implicitness of the systems (2) or (3) and the singularity of the matrix A the model is not an SDE, but an SDAE. We refer to [15] for analytical results as well as convergence results for certain drift-implicit methods.

In this paper we discuss adaptive linear multi-step methods, in particular stochastic analogues of the trapezoidal rule and the two-step backward differentiation formula, see Sect. 2. The applied step-size control strategy is described in Sect. 3. Here we extensively use the smallness of the noise. In Sect. 4 new ideas for the control both of time and chance-discretization are discussed. Test results including real-life problems that illustrate the performance of the presented methods are given in Sect. 5.

2 Adaptive Numerical Methods

The key idea to design methods for SDAEs is to force the iterates to fulfill the constraints of the SDAE at the current time-point. Here we consider stochastic analogues of methods that have proven very useful in the deterministic circuit simulation. Paying attention to the DAE structure, the discretization of the deterministic part (drift) is implicit, whereas the discretization of the stochastic part (diffusion) is explicit.

We consider stochastic analogues of the variable coefficient two-step backward differentiation formula (BDF$_2$) and the trapezoidal rule, where only the increments of the driving Wiener process are used to discretize the diffusion part. Analogously to the Euler-Maruyama scheme we call such methods multi-step Maruyama methods. The variable step-size BDF$_2$ Maruyama method for the SDAE (3) has the form (see [11] and, for constant step-sizes, e.g. [2])

$$A\frac{q(X_\ell) + \alpha_{1,\ell} q(X_{\ell-1}) + \alpha_{2,\ell} q(X_{\ell-2})}{h_\ell} + \beta_{0,\ell} f(X_\ell, t_\ell)$$
$$+ \sum_{r=1}^{m} g_r(X_{\ell-1}, t_{\ell-1}) \frac{\Delta W_r^\ell}{h_\ell} - \alpha_{2,\ell} \sum_{r=1}^{m} g_r(X_{\ell-2}, t_{\ell-2}) \frac{\Delta W_r^{\ell-1}}{h_\ell} = 0, \quad (4)$$

$\ell = 2, \ldots, N$. Here, X_ℓ denotes the approximation to $X(t_\ell)$, $h_\ell = t_\ell - t_{\ell-1}$, and $\Delta W_r^\ell = W_r(t_\ell) - W_r(t_{\ell-1}) \sim N(0, h_\ell)$ on the grid $0 = t_0 < t_1 < \cdots < t_N = T$. The coefficients $\alpha_{1,\ell}, \alpha_{2,\ell}, \beta_{0,\ell}$ depend on the step-size ratio $\kappa_\ell =$

$h_\ell/h_{\ell-1}$ and satisfy the conditions for consistency of order one and two in the deterministic case. By construction the scheme has order $1/2$ in the stochastic case (see [11]). A correct formulation of the stochastic trapezoidal rule for SDAEs requires more structural information (see [12]). It should implicitly realize the stochastic trapezoidal rule for the so called inherent regular SDE of (3) that governs the dynamical components. Both the BDF$_2$ Maruyama method and the stochastic trapezoidal rule of Maruyama type have only an asymptotic order of strong convergence of $1/2$, i.e.,

$$\|X(t_\ell) - X_\ell\|_{L_2(\Omega)} := \max_{\ell=1,\ldots,N}(E|X(t_\ell) - X_\ell|^2)^{1/2} \leq c \cdot h^{1/2}, \qquad (5)$$

where $h := \max_{\ell=1,\ldots,N} h_\ell$ is the maximal step-size of the grid. For additive noise the order may be 1. This holds true for all numerical schemes that include only information on the increments of the Wiener process.

However, the noise densities given in Section 1 contain small parameters and the error behaviour is much better. In fact, the errors are dominated by the deterministic terms as long as the step-size is large enough [2, 11]. In more detail, the error of the given methods behaves like $\mathcal{O}(h^2 + \varepsilon h + \varepsilon^2 h^{1/2})$, when ε is used to measure the smallness of the noise, i.e., $g_r(x,t) = \varepsilon \hat{g}_r(x,t)$, $r=1,\ldots,m$ where $\varepsilon \ll 1$. Thus we can expect order 2 behaviour if $h \gg \varepsilon$. Higher numerical effort for higher deterministic order pays off only if the noise is *very* small.

3 Local Error Estimates

The smallness of the noise allows special estimates of the local error terms, which can be used to control the step-size. We aim at an efficient estimate of the mean-square of local errors by means of a number of simultaneously computed solution paths. This leads to an adaptive step-size sequence that is identical for all paths. For the drift-implicit Euler-Maruyama scheme this step-size control has been presented in [10], see also [4, 16].

In [13, 14] the authors extended this strategy to stochastic linear multi-step methods with deterministic order 2 and provided a reliable error estimate. Let $\widetilde{L}_\ell$ approximates the dominating local error in AX_ℓ by

$$\widetilde{L}_\ell = c_\ell h_\ell \cdot \left[\frac{2\kappa_\ell}{\kappa_\ell + 1} f(X_\ell, t_\ell) - 2\kappa_\ell f(X_{\ell-1}, t_{\ell-1}) + \frac{2\kappa_\ell^2}{\kappa_\ell + 1} f(X_{\ell-2}, t_{\ell-2}) \right], \qquad (6)$$

where c_ℓ is the error constant of the related deterministic scheme. This estimate is based on already computed values of the drift term. Recall that $\widetilde{L}_\ell$ is a vector valued random variable as is the solution X_ℓ. For the measurement of errors we use the mean-square norm in $L_2(\Omega)$. In dependence on the small parameter ε and the step-size h_ℓ the L_2-norm of the local error behaves like $\mathcal{O}(h_\ell^3 + \varepsilon h_\ell^{3/2} + \varepsilon^2 h_\ell)$. The term of order $\mathcal{O}(h_\ell^3)$ dominates the local error behaviour as long as h_ℓ^3 is much larger than $\varepsilon h_\ell^{3/2}$, i.e., $\varepsilon^{2/3} \ll h_\ell$. Under this condition also the expression $\|\widetilde{L}_\ell\|_{L_2}$ approximates the local error.

Depending on the available information we will monitor different quantities to satisfy accuracy requirements,

i) control $\|(A + h_\ell \beta_{0,\ell} J_\ell)^{-1} \widetilde{L}_\ell\|_{L_2}$ to match a given tolerance for X_ℓ,
ii) control $\|\widetilde{L}_\ell\|_{L_2}$ to match a given tolerance for AX_ℓ, or
iii) control $\|A^{-}\widetilde{L}_\ell\|_{L_2}$ to match a given tolerance for PX_ℓ.

Here J_ℓ is the Jacobian of the drift function f w.r.t. the first variable, and A^- denotes the pseudo inverse of A, and P is an appropriate projector. Since $(A/h_\ell + \beta_{0,\ell} J_\ell) = 1/h_\ell \cdot (A + h_\ell \beta_{0,\ell} J_\ell)$ is the Jacobian of the discrete scheme (4) this matrix (or a good approximation to it) and its factorization are usually available. In case of M sampled paths, the L_2-norm in i)–iii) is estimated by using the M values $\widetilde{L}_\ell^i$, $i = 1, \ldots, M$. For example, in case i) we use

$$\left\|(A + h_\ell \beta_{0,\ell} J_\ell)^{-1} \widetilde{L}_\ell\right\|_{L_2} \approx \left(\frac{1}{M} \sum_{i=1}^{M} \left|(A + h_\ell \beta_{0,\ell} J_\ell^i)^{-1} \widetilde{L}_\ell^i\right|^2\right)^{1/2}. \quad (7)$$

4 A Solution Path Tree Algorithm

In the analysis so far, the number M of sample paths has not been specified yet. It influences the sampling error in the approximation of the L_2-norm in the error estimate (7). We have $\|\widetilde{L}_\ell\|_{L_2} = \hat{\eta}_\ell + \vartheta_\ell$, where $\hat{\eta}_\ell$ is the approximation of the dominating local error term based on M sample paths and ϑ_ℓ is the sampling error.

Our aim in tuning the number of paths is to balance the local error and the sampling error. Let d_ℓ be a given upper bound for the sampling error ϑ_ℓ at time t_ℓ, e.g. calculated as an approximation of the higher deterministic error term of order $O(h_\ell^4)$. We then derive the best number M_ℓ of paths by

$$M_\ell = \left\lfloor \frac{1}{d_\ell^2} \frac{\hat{\mu}_\ell^2 \cdot \hat{\sigma}_\ell^2}{\hat{\mu}_\ell^2 + \hat{\sigma}_\ell^2} \right\rfloor, \quad (8)$$

where $\hat{\mu}_\ell$ and $\hat{\sigma}_\ell^2$ are estimates of the mean and the standard deviation of the error estimate at time-point t_ℓ, respectively. Here $\lfloor x \rfloor$ denotes the smallest integer greater or equal to x.

The best number of paths M_ℓ depends on the time-point t_ℓ and is realized by approximate solutions generated on a tree of paths that is extended, reduced or kept fixed adaptively. In [9] the authors describe the construction of a solution path tree in detail. The method uses probabilities π_ℓ^i ($\ell = 1, \ldots, N$; $i = 1, \ldots, M_\ell$) to weight the solution paths. Figure 1 gives an impression, how a solution path tree looks like. At each time-step the optimal expansion or reduction problem is formulated by means of combinatorial optimization models. The path selection is modelled as a mass transportation problem in terms of the L_2-Wasserstein metric (see [5] in context of scenario reduction in stochastic programming). The algorithm has been implemented in practice. The results presented in the next section show its performance.

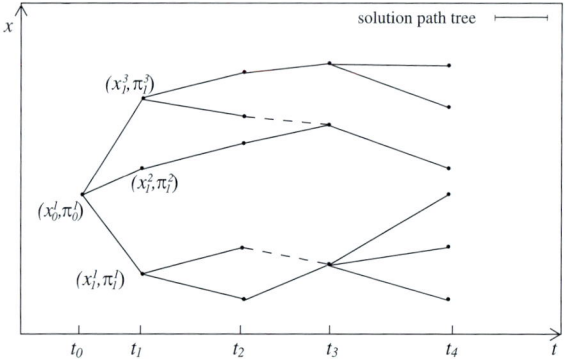

Fig. 1. A solution path tree: Variable time-points t_ℓ, solution states x_ℓ^i and path weights π_ℓ^i

5 Numerical Results

Here we present numerical experiments for the stochastic BDF$_2$ applied to two circuit examples. The first one is a small test problem, for which we have used an implementation of the adaptive methods discussed in the previous sections in Fortran code. To be able to handle real-life problems, a slightly modified version of the schemes has been implemented in Qimonda's in-house analog circuit simulator TITAN. The second example shows the performance of this industrial implementation.

A MOSFET Inverter Circuit

We consider a model of an inverter circuit with a MOSFET transistor, under the influence of thermal noise. The related circuit diagram is given in Fig. 2. The MOSFET is modelled as a current source from source to drain that is controlled by the nodal potentials at gate, source and drain.

The thermal noise of the resistor and of the MOSFET is modelled by additional white noise current sources that are shunt in parallel to the original,

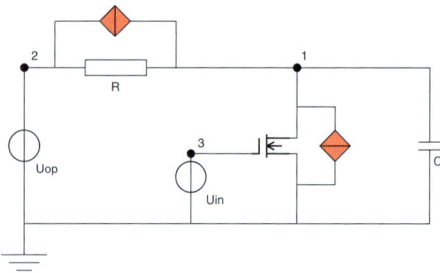

Fig. 2. Thermal noise sources in a MOSFET inverter circuit

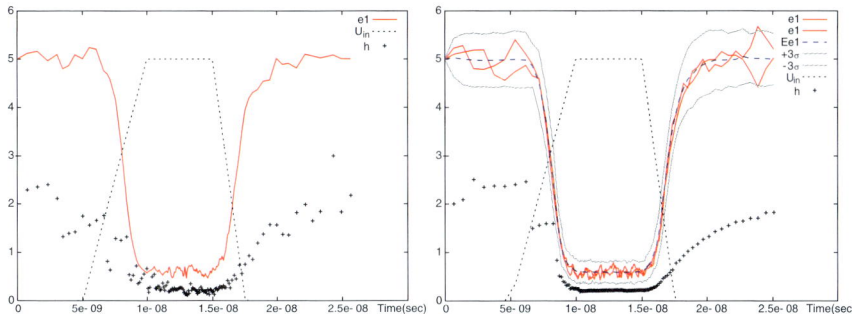

Fig. 3. Simulation results for the noisy inverter circuit: (*left*) 1 path 127(+29 rejected) steps; (*right*) 100 paths 134(+11 rejected) steps

noise-free elements. To highlight the effect of the noise, we scaled the diffusion coefficient by a factor of 1000.

In Fig. 3 we present simulation results, where we plotted the input voltage U_{in} and values of the output voltage e_1 versus time. Moreover, the applied step-sizes, suitably scaled, are shown by means of single crosses. We compare the results for the computation of a single path (left picture) with those for the computation of 100 simultaneously computed solution paths (right picture). The additional solid lines show two different solution paths, the dashed line gives the mean of 100 paths and the outer thin lines the 3σ-confidence interval for the output voltage e_1. We observe that using the information of an ensemble of simultaneously computed solution paths smoothes the step-size sequence and considerably reduces the number of rejected steps, when compared to the simulation of a single path. The computational cost that is mainly determined by the number of computed (accepted + rejected) steps is reduced.

We have applied the solution path tree algorithm to this example. The upper graph in Fig. 4 shows the computed solution path tree together with the applied step-sizes. The lower graph shows the simulation error (solid line), its error bound (dashed line) and the used number of paths (marked by ×), vs. time. The maximal number of paths was set to 250.

The results indicate that there exists a region from nearly $t = 1 \cdot 10^{-8}$ up to $t = 1.5 \cdot 10^{-8}$ where we have to use much more than 100 paths. This is exactly the area in which the MOSFET is active and the input signal is inverted. Outside this region the algorithm proposes approximately 70 simultaneously computed solution paths.

A Voltage Controlled Oscillator

As an industrial test application we us a voltage controlled oscillator that is a simplified version of a fully integrated 1.3 GHz VCO for GSM in 0.25 μm standard CMOS (see [8]). For simulation, the oscillator is embedded in a test environment. The VCO is tunable from about 1.2 GHz up to 1.4 GHz. The

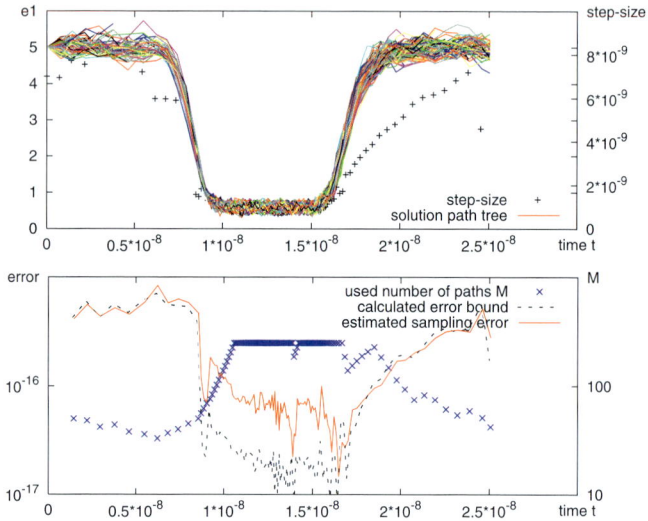

Fig. 4. Simulation results for the noisy inverter circuit: Solution path tree and step-sizes (*top*), sampling error, its error bound and the number of paths (*bottom*)

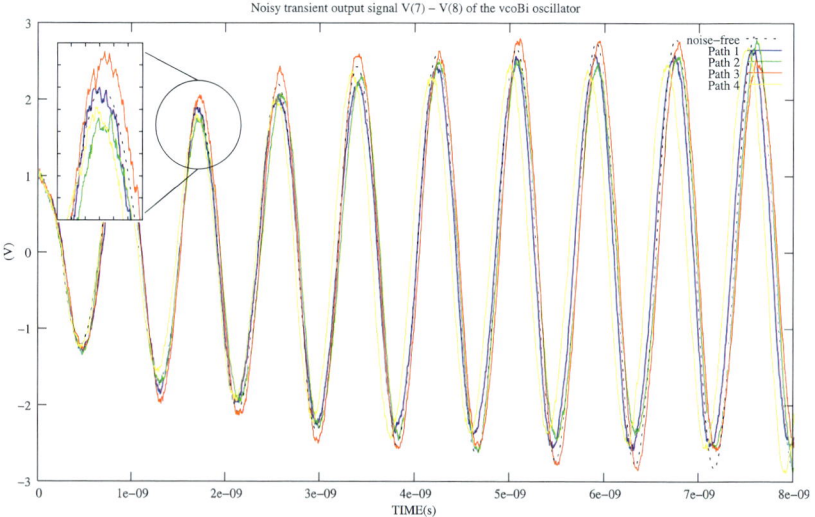

Fig. 5. Noisy transient output signal of a VCO

unknowns of the VCO in the MNA system are the charges of the six capacities, the fluxes of the four inductors, the 15 nodal potentials and the currents through the voltage sources. This circuit contains 5 resistors and 6 MOSFETs, which induce 53 sources of thermal or shot noise. To make the differences

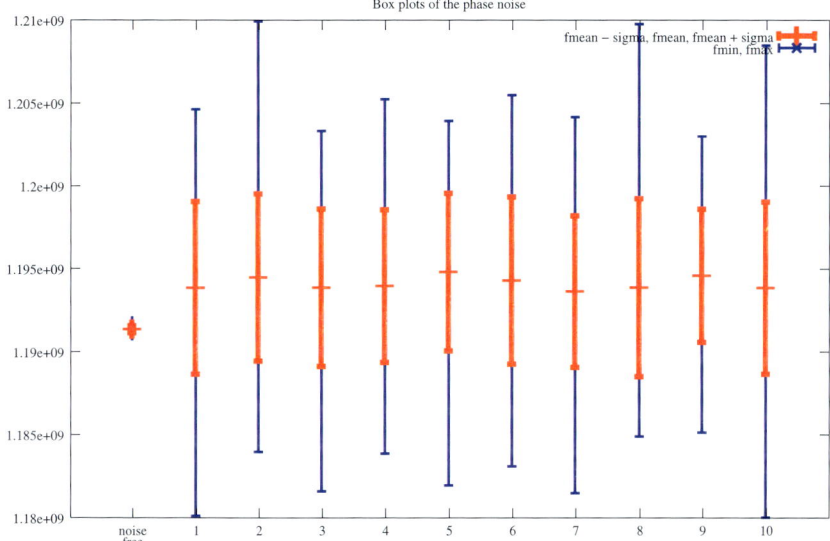

Fig. 6. Boxplots of the phase noise, scaled by a factor of 500

between the solutions of the noisy and the noise-free model more visible, the noise intensities had been scaled by a factor of 500.

Numerical results obtained with a combination of the BDF_2 and the trapezoidal rule are shown in Fig. 5, where we plotted the difference of the nodal potential $V(7) - V(8)$ of node 7 and 8 versus time. The solution of the noise-free system is given by a dashed line. Four sample paths (dark solid lines) are shown. They cannot be considered as small perturbations of the deterministic solution, phase noise is highly visible.

To analyze the phase noise we performed 10 simultaneous simulations with different initializations of the pseudo-random numbers. In a postprocessing step we computed the length of the first 50 periods for each solution path and then from these the corresponding frequencies. In Fig. 6 the mean μ of the frequencies (horizontal lines), the smallest and the largest frequencies (boundaries of the vertical thin lines) and the boundaries of the confidence interval $\mu \pm \sigma$ (the plump lines) are presented, where σ was computed as the empirical estimate of the standard deviation. The mean appears increased and differs by about $+0.25\%$ from the noiseless, deterministic solution.

Further on, the frequencies vacillate from 1.18 GHz (-0.95%) up to 1.21 GHz ($+1.55\%$). So the transient noise analysis shows that the voltage controlled oscillator runs in a noisy environment with increased frequencies and smaller phases, respectively.

6 Conclusions

Quite similar to deterministic circuit simulation, it is essential to have the application in mind while developing new efficient algorithms. We have presented variable step-size two-step schemes for SDAEs which require only the increments of the driving Wiener process. Though these schemes possess only convergence order $1/2$ from a theoretical point of view, they show order 2 in circuit simulation, as the deterministic terms dominate the errors. This can be considered in the step-size control. Taking the stochastic properties of the circuit into account leads to an increased efficiency of the methods.

An important application of transient noise analysis is to get insight into the statistical properties of the solution paths. We showed that the number of paths necessary for this purpose varies with the time-points. By implementing a solution path tree algorithm, it is possible to save computing time or to get more accurate outputs compared to a naive approach which would require to calculate all paths over the complete integration interval.

These results allow an efficient transient noise simulation which helps the designer to cope with challenges due to technology progress. Further improvements may include parallelisation and the handling of flicker noise.

References

1. Arnold, L.: Stochastic differential equations: Theory and Applications. Wiley, New York, 1974.
2. Buckwar, E., Winkler, R.: Multi-step methods for SDEs and their application to problems with small noise. SIAM J. Num. Anal., **44**(2), 779–803 (2006)
3. Denk, G.: Circuit simulation for nanoelectronics. In: M. Anile, G. Alì, G. Mascali (Eds.): Scientific Computing in Electrical Engineering (Mathematics in Industry 9), 13–20. Springer, Berlin, 2006.
4. Denk, G., Winkler, R.: Modeling and simulation of transient noise in circuit simulation. Math. and Comp. Modelling of Dyn. Systems, **13**(4), 383–394 (2007)
5. Dupačová J., Gröwe-Kuska, N., Römisch, W.: Scenario reduction in stochastic programming. Math. Program., Ser. A **95**, 493–511 (2003)
6. Günther, F., Feldmann, U.: CAD-based electric-circuit modeling in industry I. Mathematical structure and index of network equations. Surv. Math. Ind., **8**, 97–129 (1999)
7. Higham, D.J.: An algorithmic introduction to numerical simulation of stochastic differential equations. SIAM Review, **43**, 525–546 (2001)
8. Tiebout, M.: A fully integrated 1.3 GHz VCO for GSM in 0.25 µm standard CMOS with a phasenoise of -142 dBc/Hz at 3MHz offset. In: Proceedings 30th European Microwave Conference, Paris (2000)
9. Römisch, W., Sickenberger, Th.: On generating a solution path tree for efficient step-size control. In preparation.
10. Römisch, W., Winkler, R.: Stepsize control for mean-square numerical methods for SDEs with small noise. SIAM J. Sci. Comp., **28**(2), 604–625 (2006)

11. Sickenberger, T.: Mean-square convergence of stochastic multi-step methods with variable step-size. J. Comput. Appl. Math., to appear
12. Sickenberger, T., Winkler, R.: Efficient transient noise analysis in circuit simulation. In: Proceedings of the GAMM Annual Meeting 2006, Berlin, Proc. Appl. Math. Mech. **6**(1), 55–58 (2006)
13. Sickenberger, T., Weinmüller, E., Winkler, R.: Local error estimates for moderately smooth problems: Part I – ODEs and DAEs. BIT Numerical Mathematics, **47**(2), 157–187 (2007).
14. Sickenberger, T., Weinmüller, E., Winkler, R.: Local error estimates for moderately smooth problems: Part II – SDEs and SDAEs. Preprint 07-07, Institut für Mathematik, Humboldt-Universität zu Berlin (2007). Submitted for publication.
15. Winkler, R.: Stochastic differential algebraic equations of index 1 and applications in circuit simulation. J. Comput. Appl. Math., **157**(2), 477–505 (2003)
16. Winkler, R.: Stochastic differential algebraic equations in transient noise analysis. In: M. Anile, G. Alì, G. Mascali (Eds.): Scientific Computing in Electrical Engineering (Mathematics in Industry 9), 151–158. Springer, Berlin, 2006.

Part II

Thin Films

Numerical Methods for the Simulation of Epitaxial Growth and Their Application in the Study of a Meander Instability

Frank Haußer[1], Felix Otto[2], Patrick Penzler[2], and Axel Voigt[3]

[1] FB II, Technische Fachhochschule Berlin, Luxemburger Str. 10, 13353 Berlin, Germany, hausser@tfh-berlin.de
[2] IAM, Universität Bonn, Wegelerstr. 10, 53115 Bonn, Germany {otto,penzler}@iam.uni-bonn.de
[3] IWR, Technische Universität Dresden, Zellescher Weg 12–14, 01062 Dresden, Germany, axel.voigt@tu-dresden.de

Summary. The surface morphology of thin crystalline films grown by molecular beam epitaxy (MBE), a technique to produce high-quality, almost defect-free crystals, is strongly influenced by kinetic processes on an atomistic scale. To incorporate these effects in a continuum model requires some care. Here we use a step flow model, which is a free boundary problem for the position of atomic height steps on the crystalline surface. We present two complementary approaches to derive a numerical method for solving this problem: a front tracking ansatz and a diffuse interface approximation. The numerical methods are used to study the nonlinear regime of a step meandering instability

1 Epitaxial Growth

1.1 Introduction

Epitaxial growth is a modern technology of growing crystalline films that inherit atomic structures from substrates. It produces almost defect-free, high quality materials that have a wide range of device applications. Microscopic processes in epitaxial growth include the deposition of atoms or molecules, atom adsorption and desorption, adatom (adsorbed atom) diffusion, adatom island nucleation, the attachment and detachment of adatoms to and from island boundaries or terrace steps, and island coalescence.

There are various models for epitaxial growth of thin films that are distinguished by different scales in time and space, see Voigt [2005] for an overview. These models range from full atomistic descriptions using molecular dynam-

Fig. 1. Epitaxial growth of Si(001). The figure shows atomistically flat terraces, which are separated by steps of atomic height. Courtesy of Polop, Bleikamp, and Michely

ics (MD) and kinetic Monte Carlo (KMC) methods over discrete-continuous models in which only the growth direction is resolved on an atomistic scale and the lateral direction is coarse-grained, to fully continuous models which describe the thin film as a smooth hypersurface.

An initially atomistically flat surface typically does not remain flat during growth, but is subject to various instabilities. There are essentially three types of instabilities which influence the film morphology during growth: step bunching, step meandering and mound formation, see e.g. Politi et al. [2000], Krug [2005]. They all have their origin on the atomistic scale and result from asymmetries in the energy barriers for individual hops of atoms on the surface. However, a fully atomistic description of the film is limited to sample sizes of several nm and thus far off from any feature size in semiconductor devices. In order to predict the surface morphology on larger length scales, continuum models are required which incorporate the instabilities generated on the atomistic scale. Fully continuous models with these properties still have to be derived. On a mesoscopic scale, discrete-continuum models – so called step flow models – are promising candidates. These models are discrete in the growth direction but continuous in the lateral directions. Figure 1 shows a scanning tunneling microscopy (STM) image of a Si(001) surface, consisting of large terraces separated by atomistic height steps, which motivates this modeling approach. Atomistic hops on terraces are modeled by a continuum diffusion equation. The atomistic processes of attachment and detachment at the atomic height steps are incorporated by appropriate boundary conditions. Moreover, the atomistically rough steps are treated as smooth curves and the local geometry enters via the curvature.

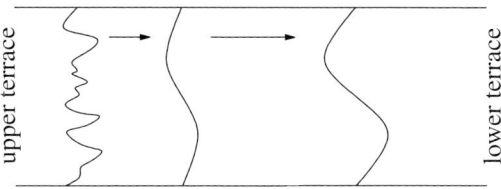

Fig. 2. Linear instability: Some wave numbers of an initially small perturbation of a straight step grow and the step starts to meander

1.2 Step Meandering Instability

In this work we are concerned with a meandering instability of a step. The seminal work of Bales and Zangwill [1990] showed the influence of a (terrace) Ehrlich–Schwoebel barrier on the morphology of a step. The Ehrlich–Schwoebel barrier is an atomistic energy barrier atoms have to overcome if they attach to a step from the upper terrace [Ehrlich and Hudda, 1966, Schwoebel and Shipsey, 1966]. This barrier leads to an effective uphill current resulting in a meander instability. This means that given an initially straight step with a perturbation of small magnitude, there are some wave numbers of the perturbation which will grow, see Fig. 2 for a sketch. Based on a linear stability analysis, an explicit formula for the growth rate $\omega(k)$ depending on the wave number k has been given in Bales and Zangwill [1990].

Experimentally observed step meanders do not fall into the linear regime, as they show large amplitudes. Thus, they cannot be described within the analytic treatment. Our goal is therefore to use numerical tools to study step meandering in the nonlinear regime, where the step flow might even break down.

1.3 Step Flow Model

The layer-by-layer growth of the crystalline surface is described by the classical Burton–Cabrera–Frank (BCF) model [Burton et al., 1951], which is a semicontinous model – discrete in the height, but continuous in the lateral directions. Here, the surface is assumed to consist of flat terraces separated by atom-high steps; the steps are modeled as continuous curves, see Fig. 3.

Vapor atoms arriving at the surface become adatoms (ad-sorbed atoms) and diffuse on the flat terraces. The diffusive fluxes at the steps lead to growth via propagating steps. If nucleation of new islands or steps on the terraces can be neglected, the growth dynamics are essentially described by the attachment kinetics at the steps, i.e., the boundary conditions of the adatom density at the terrace boundaries. This leads to a free boundary problem for the adatom densities (concentrations) on the terraces with free boundaries given

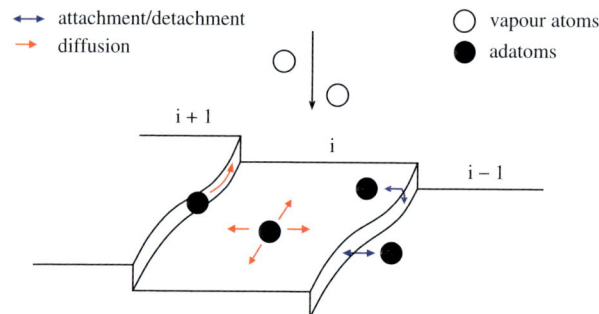

Fig. 3. Step flow model: Vapour atoms are deposited on a surface, where they become ad(sorbed)atoms and diffuse on atomistically flat terraces; eventually adatoms attach/detach at steps

by the step position. On each terrace, the adatom density c obeys the following diffusion equation:

$$\partial_t c + \nabla \cdot \mathbf{j} = F, \quad \mathbf{j} = -D\nabla c, \tag{1a}$$

where D is the diffusion constant and F is the deposition flux rate. Desorption of adatoms has been neglected, which is valid in typical MBE experiments [Maroutian et al., 1999]. The fluxes of adatoms to a step are given by

$$j_\pm := \pm(c_\pm v - \mathbf{j}_\pm \cdot \mathbf{n}), \tag{1b}$$

where subscripts "+" and "−" denote quantities at a step up (i.e. on the lower terrace) and a step down, respectively, $\mathbf{n}$ denotes the normal pointing from upper to lower terrace and v is the normal velocity of the step. Assuming first order kinetics for the attachment/detachment of adatoms at the steps, the diffusive fluxes at the step (terrace boundary) are proportional to the deviation of the adatom density from equilibrium c_{eq}, i.e., the adatom density satisfies the following *kinetic boundary conditions* at a step:

$$j_\pm = k_\pm(c_\pm - c_{eq}). \tag{1c}$$

With this notation, asymmetric attachment rates $0 < k_- < k_+$ model the (terrace) Ehrlich–Schwoebel (ES) effect. The equilibrium density c_{eq} at a step is given by the linearized Gibbs–Thomson relation

$$c_{eq} = c^*(1 + \Gamma\kappa), \quad \Gamma = a^2(\gamma + \gamma'')/(k_B T),$$

where c^* is the constant equilibrium density for a straight step, a^2 is the atomic area with a being the lattice spacing and κ denotes the curvature of the step (we define the curvature of a circular island as positive); k_B is Boltzmann's constant, T the temperature and $\gamma = \gamma(\theta)$ denotes the step free energy per unit length, which may depend on the local orientation $\theta = \theta(\mathbf{n})$.

Finally, the normal velocity of a step is given by

$$v = a^2(j_+ + j_-) - a\partial_s j_s, \qquad j_s = -D_{st}\partial_s(\Gamma\kappa), \tag{1d}$$

where ∂_s denotes the tangential derivative along the step and D_{st} is the diffusion constant of atoms along the step. The second term in the velocity law (1d) represents edge diffusion of edge atoms along the step, whereas the first term ensures mass conservation. For a more detailed description of the step flow model see e.g. Krug [2005].

2 Modeling and Numerical Methods

In this section we present two complementary powerful numerical methods to solve the free boundary problem as described in Sect. 1.3.

In the first part we describe a front-tracking-type method, where the free boundaries (the steps) and the adatom concentrations are discretized on two independent meshes. Using an operator splitting ansatz and a careful discretization based on linear finite elements leads to an accurate and efficient semi-implicit scheme. However, topological changes are not naturally included in this approach and would require awkward remeshing.

To overcome this limitation, in the second part, we present a diffuse-interface approximation. Via this approximation, we transform the (predominantly) third-order free boundary problem into a fourth-order problem – without boundaries – for an auxiliary quantity ϕ mimicking the shape of the crystal. The step positions are then given by level sets of ϕ. Both ϕ and the adatom concentration use the same mesh.

2.1 Front Tracking Method

We will review the front tracking type finite element method for discretizing the moving boundary problem (1a)–(1d) that has been developed in Bänsch et al. [2004]. First note that the moving boundary problem may be naturally divided into two subproblems: the adatom diffusion and the step evolution. Thus we proceed as follows:

(a) We derive a weak formulation for the time-dependent diffusion equation. To avoid the complexity in the spatial discretization near boundaries, in each time step, we extend the diffusion equation from terraces of same height to the whole computational domain. The extended equation is discretized using the linear finite element method. This leads to two diffusion equations to be solved in each time step: one for the adatom densities on terraces of even height and one for the adatom densities on terraces of odd height. Note that this properly reflects the fact, that the kinetic boundary conditions (1c) in general lead to discontinuities of the adatom densities at the steps, which are resolved by having two degrees of freedom at the terrace boundaries.

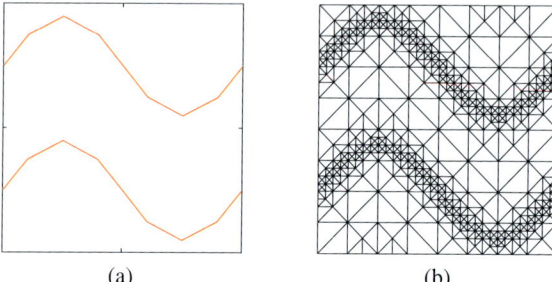

Fig. 4. (a) Finite element mesh (1d) for terrace boundaries (b) Locally refined Finite element mesh (2d) for adatom density

(b) The geometric motion of the island boundaries includes both the mean curvature flow originating from the Gibbs-Thomson relation and the line diffusion. It is treated in a variational formulation utilizing the curvature vector, and discretized by a semi-implicit front-tracking method using parametric finite elements. This method is adapted with modification from Bänsch et al. [2005b] and is generalized to allow for anisotropic line energies.

This operator splitting approach results in the following numerical scheme: in each time

(i) solve the geometric evolution equation for the terrace boundaries (steps) using the step position and the adatom densities of the last time step
(ii) solve the diffusion equation for the adatom density using the updated step position.

We remark that the two-dimensional (2d) and the one-dimensional (1d) finite element meshes are essentially independent from each other. To obtain satisfactory computational results, meshes with sufficiently fine resolutions are needed for both the adatom diffusion equation and the boundary evolution equation. Thus, it is indispensable to use adaptivity in order for the method to be efficient. We use simple error indicators within an h-adaptive method to locally increase the spatial resolution. In Fig. 4 we give a prototype example of a 1d mesh representing two steps and the corresponding locally refined 2d mesh for the adatom densities.

Adatom Diffusion

Let $c_i(x,t)$ denote the adatom density on the domain (terraces) $\Omega_i(t)$ of atomic height i, with boundaries $\Gamma_-(t)$ and $\Gamma_+(t)$ representing the downward and upward steps, respectively. Multiplying Equation (1a) with a smooth time-independent test function ϕ and integrating over the domain $\Omega_i(t)$ yields

$$\int_{\Omega_i} \partial_t c_i \phi + \int_{\Omega_i} D\nabla c_i \cdot \nabla \phi - \int_{\Gamma_-} D\nabla c_i \cdot \mathbf{n}\phi + \int_{\Gamma_+} D\nabla c_i \cdot \mathbf{n}\phi = \int_{\Omega_i} F\phi$$

Using the identity $\frac{d}{dt}\int_{\Omega_i(t)} c_i\phi = \int_{\Omega_i(t)} \partial_t c_i \phi - \int_{\Gamma_+(t)} c_i v\phi + \int_{\Gamma_-(t)} c_i v\phi$ and plugging in the boundary conditions (1c) leads to

$$\frac{d}{dt}\int_{\Omega_i(t)} c_i\phi + \int_{\Omega_i(t)} D\nabla c_i \cdot \nabla \phi + \int_{\Gamma_+(t)} k_+ c_i \phi + \int_{\Gamma_-(t)} k_- c_i \phi$$
$$= \int_{\Omega_i(t)} F\phi + \int_{\Gamma_+(t)} k_+ c^*(1+\Gamma\kappa)\phi + \int_{\Gamma_-(t)} k_- c^*(1+\Gamma\kappa)\phi \quad (2)$$

Now we use the first-order implicit scheme to discretize the time derivative: Consider discrete time instants $t_0 < t_1 < \cdots$ with time steps $\Delta t_m = t_{m+1} - t_m$ and denote $\Omega_i(t_m) = \Omega_i^m$, $c_i(t_m) = c_i^m$. Substituting

$$\frac{d}{dt}\int_{\Omega_i(t)} c_i \phi \longrightarrow \frac{1}{t_{m+1}-t_m}\left[\int_{\Omega_i^{m+1}} c_i^{m+1}\phi - \int_{\Omega_i^m} c_i^m \phi\right]$$

and using extended variables

$$\left(c_i^m(x), D_i^m(x), F_i^m(x)\right) = \begin{cases} (c_i^m(x), D, F) & : x \in \bar{\Omega}_i^m \\ (0,0,0) & : x \in \Omega \setminus \bar{\Omega}_i^m \end{cases}$$

leads to a weak formulation of the time discretized eq. (2) on the whole (time independent) domain Ω:

$$\int_\Omega \frac{c_i^{m+1} - c_i^m}{\Delta t_m}\phi + \int_\Omega D_i^{m+1}\nabla c_i^{m+1} \cdot \nabla \phi + \int_{\Gamma_+} k_+ c_i^{m+1}\phi + \int_{\Gamma_-} k_- c_i^{m+1}\phi$$
$$= \int_{\Gamma_+} k_+ c^*(1+\Gamma\kappa)\phi + \int_{\Gamma_-} k_- c^*(1+\Gamma\kappa))\phi + \int_\Omega F_i^{m+1}\phi \quad (3)$$

Equation (3) is discretized in space using linear finite elements leading to a symmetric positive definite system to be solved in each time step. However, on elements where coefficients are discontinuous, a careful integration is necessary when assembling the matrices. Also note, that the boundaries Γ_+, Γ_- are the solutions $\Gamma_+^{m+1}, \Gamma_-^{m+1}$ of the boundary evolution. For a more detailed description see Bänsch et al. [2004].

Boundary Evolution

To solve the evolution equation (1d) the boundary conditions (1c) are used to arrive at a geometric evolution equation (for each boundary $\Gamma_i := \bar{\Omega}_{i+1} \cap \bar{\Omega}_i$) of the form

$$v = f - \beta\Gamma\kappa + \alpha\partial_{ss}\Gamma\kappa, \quad (4)$$

where $\beta = a^2(k_+ + k_-)c^*$ and $\alpha = aD_{st}$ are positive constants, $\Gamma = \Gamma(\theta)$ is a possibly orientation dependent positive function and the adatom densities of the upper and lower terrace, c_+, c_- enter via $f = a^2 k_+(c_+ - c^*) + a^2 k_-(c_- - c^*)$.

Equation (4) may be interpreted as an equation for (1d) anisotropic surface diffusion with lower order terms and is solved following the ideas of Bänsch et al. [2005b], for an adaption to the anisotropic case see Bänsch et al. [2005a]. First a second order splitting of the fourth order equation is performed by introducing the curvature vector $\boldsymbol{\kappa} := \kappa \mathbf{n}$ and using the relation $-\partial_{ss}\mathbf{x} = \boldsymbol{\kappa}$ with the position vector $\mathbf{x}$. Then a semi-implicit time discretization is introduced by representing the next boundary $\Gamma(t_{m+1})$ in terms of the current boundary $\Gamma(t_m)$ by updating the position vector $\mathbf{x}^{m+1} = \mathbf{x}^m + \Delta t_m \mathbf{v}^{m+1}$. The resulting system of second order equations is discretized in space using linear parametric finite elements.

2.2 Diffuse Interface Approximation

We now discuss the second approach: a diffuse-interface approximation (DIA). Let us first show how to imagine such an approximation.

The BCF model is two-dimensional, but every domain Ω_i is labeled with a discrete height, so one can imagine it as a three-dimensional landscape with sharp jumps, see Fig. 5a. The DIA can now be thought of as a smeared-out version of this landscape, where the sharp jumps are replaced by a smooth transition region of width ε, see Fig. 5b.

Diffuse-interface models have been used for various applications, e.g. spinodal decomposition. The connection between diffuse-interface models and the nonlocal sharp-interface models of the type considered here were first shown by Pego [1989] and Caginalp [1989] using formal asymptotic analysis. Alikakos et al. [1994] calculated the error between the true solution of the Cahn–Hilliard equation and the formal expansion, thereby gaining a rigorous proof of the convergence $\varepsilon \to 0$, as long as the underlying solution stays smooth. For a general phase-field equation, the rigorous proof for a convergence to a sharp interface model was given by Caginalp and Chen [1998].

Diffuse-interface approximations for step flow growth have already been introduced in Liu and Metiu [1997] and Karma and Plapp [1998], but none of these models allowed the incorporation of the Ehrlich–Schwoebel-barrier.

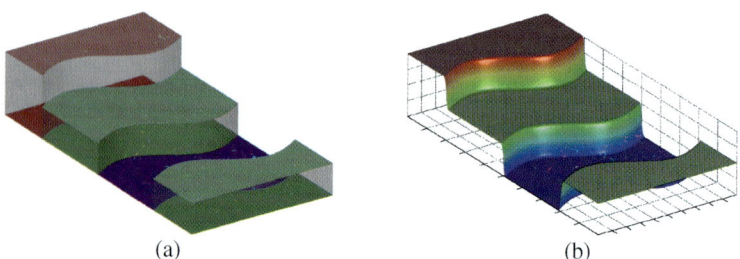

Fig. 5. (a) BCF model: each domain is associated with a discrete height, thus forming a three-dimensional landscape with sharp interfaces. (b) Diffuse-interface approximation: the sharp interfaces are "smeared out", resulting in a smooth function

Nondimensionalization

Before we proceed to the diffuse-interface approximation, we rewrite equations (1). We will focus on a regime where we have no edge diffusion and the adatom density has enough time to relax to its quasi-stationary equilibrium on the terraces. We state the equations in terms of the excess adatom density $w := c - c^*$. Rescaling time and space as in Otto et al. [2004], we get a *Mullins–Sekerka-type* free boundary problem:

$$-\Delta w = f \quad \text{in } \Omega_i \tag{5a}$$
$$w^\pm = \kappa \pm \zeta^\pm \nabla w^\pm \cdot \mathbf{n} \quad \text{on } \Gamma \tag{5b}$$
$$v = \nabla(w^+ - w^-) \cdot \mathbf{n} \quad \text{on } \Gamma \tag{5c}$$

The dimensionless parameters $\zeta^\pm$ are antiproportional to the attachment rate $k_\pm$. In the following, we will only consider the case of unlimited attachment to a step up, i.e. $\zeta^+ = 0$.

A Cahn–Hilliard-type Equation

Consider the equation

$$\partial_t \phi + \nabla \cdot \left[-M(\phi) \nabla \frac{\delta E_\varepsilon}{\delta \phi}(\phi) \right] = f. \tag{6}$$

The energy functional $E_\varepsilon(\phi)$ is the Ginzburg–Landau free energy with a double-well potential G

$$E_\varepsilon(\phi) = \int_\Omega \frac{\varepsilon}{2} |\nabla \phi|^2 + \varepsilon^{-1} G(\phi), \qquad G(\phi) = 18\phi^2(1-\phi)^2 \tag{7}$$

and $M(\phi)$ is a mobility function modeling the Ehrlich–Schwoebel barrier:

$$M(\phi) = (1 + \varepsilon^{-1} \zeta^- \sigma(\phi))^{-1},$$

see Fig. 6. Here $\sigma(\phi)$ is an asymmetric function in ϕ. The potential G is restricted to the interval $[0, 1]$ and then periodically continued, so that we get a multiwell potential with equally deep wells at the integers.

In Otto, Penzler, Rätz, Rump, and Voigt [2004] it was shown by formal asymptotic expansion that the above equation yields for $\varepsilon \to 0$ the BCF-model (5) including the Ehrlich–Schwoebel barrier.

The position of the boundary is given by the level sets

$$\{\phi = \mathbb{Z} + \tfrac{1}{2}\}$$

and the L^2–differential $\frac{\delta E_\varepsilon}{\delta \phi}(\phi) =: w$ is the approximation of the excess density from equations (5), see Fig. 7.

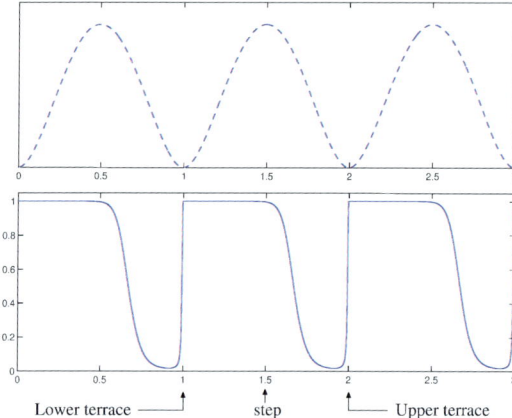

Fig. 6. The potential G (*dashed line*) and the mobility function M (*solid line*): coming from an upper terrace ("$\phi = 2$"), the atoms experience reduced mobility while attaching to the step ("$\phi = 1.5$"). On the other hand, coming from a lower terrace to the step, there is no reduction in mobility. This models the Ehrlich–Schwoebel barrier

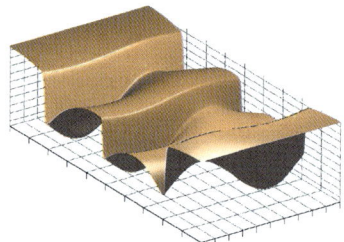

Fig. 7. Excess density w for data as in Fig. 5b. Clearly visible are the boundary values $w = \kappa$ at a step up, the jump due to the ES-barrier and the smooth solution of $-\Delta w = 1$ on the terraces

Discretization

In order to formulate a finite element approximation, we rewrite equation (6) by introducing the flux J:

$$\partial_t \phi + \nabla \cdot J = f \tag{8a}$$

$$\frac{1}{M(\phi)} J = -\nabla \frac{\delta E_\varepsilon}{\delta \phi}(\phi). \tag{8b}$$

As we shall see, this splitting yields a conformal spatial discretization and a time discretization which leads to a symmetric positive definite system. As boundary conditions we use

- equilibrium and no-flux boundary conditions: $\nabla \phi \cdot \nu = 0$, $J \cdot \nu = 0$ or
- periodic boundary conditions on a square domain $\Omega = [0, L_x] \times [0, L_y]$.

The weak form in either case is

$$\frac{d}{dt}\int_\Omega \phi\zeta + \int_\Omega (\nabla \cdot J)\zeta = \int_\Omega f\zeta \quad \forall \zeta \in L^2(\Omega) \tag{9a}$$

$$\int_\Omega \frac{1}{M(\phi)} J \cdot \tilde{J} - \int_\Omega \frac{\delta E_\varepsilon}{\delta \phi}(\phi)(\nabla \cdot \tilde{J}) = 0 \quad \forall \tilde{J} \in H(\nabla \cdot, \Omega). \tag{9b}$$

Here, the space $H(\nabla \cdot, \Omega)$ is defined as

$$H(\nabla \cdot, \Omega) := \left\{ F \in L^2(\Omega)^2 : \|\nabla \cdot F\|_{L^2(\Omega)} < \infty \right\}.$$

Time Discretization

The guideline for our discretization is a gradient flow formulation: consider the manifold

$$\mathcal{M} = \left\{ \phi \,\Big|\, \int_\Omega \phi = \mathrm{const} \right\}, \quad T_\phi \mathcal{M} = \left\{ v \,\Big|\, \int_\Omega v = 0 \right\}$$

with the metric

$$g_\phi(u, v) = \langle u, v \rangle_{H_M^{-1}(\Omega)} := \int_\Omega \nabla w \cdot M(\phi) \nabla \tilde{w}$$

where w and $\tilde{w}$ are the solutions of

$$\nabla \cdot (M(\phi)\nabla w) = u \quad \text{and} \quad \nabla \cdot (M(\phi)\nabla \tilde{w}) = v,$$

respectively, using Neumann boundary conditions. The gradient flow is then given by $\partial_t \phi = -\nabla E$. By definition of the gradient, this is equivalent to

$$g_\phi(\partial_t \phi, v) + \mathrm{diff}\, E_\varepsilon(\phi) v = 0 \quad \forall v \in T_\phi(M),$$

where we use the notation

$$\mathrm{diff}\, E_\varepsilon(\phi) v := \frac{d}{d\delta} E_\varepsilon(\phi + \delta v)\Big|_{\delta=0}.$$

Inserting the definition of the metric yields

$$\int_\Omega \nabla w \cdot M(\phi) \nabla \tilde{w} + \int_\Omega \frac{\delta E_\varepsilon}{\delta \phi}(\phi) v = 0 \quad \forall v \in T_\phi \mathcal{M} \tag{10}$$

with

$$\partial_t \phi + \nabla \cdot (-M(\phi)\nabla w) = 0 \quad \text{and} \quad v = \nabla \cdot (M(\phi)\nabla \tilde{w}).$$

Defining $J := -M(\phi)\nabla w$ and $\tilde{J} := -M(\phi)\nabla \tilde{w}$ finally gives

$$\partial_t \phi + \nabla \cdot J = 0$$

$$\int_\Omega \frac{1}{M(\phi)} J \cdot \tilde{J} = \int_\Omega \frac{\delta E_\varepsilon}{\delta \phi}(\phi)(\nabla \cdot \tilde{J}) \quad \forall \tilde{J} \in H(\nabla \cdot, \Omega)$$

which is equivalent to equation (9) with $f = 0$. Partial integration in (10) reveals $w = \frac{\delta E_\varepsilon}{\delta \phi}(\phi)$.

The idea is now to use a semi-implicit time discretization, where the only explicit quantity is the base point of the metric.

To allow for bigger time steps, we use a second-order timestep method. Our first choice was the Crank–Nicolson method, but since perturbations of high frequency are only damped very weakly (amplification factor $\to 1$), this method is not suitable for Cahn–Hilliard-type equations. Following Weikard [2002], we chose a method by Bristeau et al. [1987] (BGP-scheme). To solve an equation $\partial_t \phi = F(\phi)$, it uses two intermediate steps per time step τ:

$$\frac{1}{\theta \tau}\left(\phi^* - \phi^k\right) = \alpha F(\phi^*) + \beta F\left(\phi^k\right) \tag{11a}$$

$$\frac{1}{(1-2\theta)\tau}(\phi^{**} - \phi^*) = \beta F(\phi^{**}) + \alpha F(\phi^*) \tag{11b}$$

$$\frac{1}{\theta \tau}\left(\phi^{k+1} - \phi^{**}\right) = \alpha F\left(\phi^{k+1}\right) + \beta F\left(\phi^{**}\right) \tag{11c}$$

with

$$\theta = 1 - \frac{1}{\sqrt{2}}, \quad \alpha = 2 - \sqrt{2}, \quad \beta = \sqrt{2} - 1.$$

To keep notation simple, we present the main ideas for time discretization using the backward Euler scheme.

To handle the nonlinearity, we use Newton's method. Again to simplify notation, we show only one Newton-step, i.e. a linearization of the nonlinearity, and understand the following equations to be valid for all $\tilde{J} \in H(\nabla \cdot, \Omega)$. This yields

$$\frac{1}{\tau}\int_\Omega \left(\phi^{k+1} - \phi^k\right)\zeta + \int_\Omega (\nabla \cdot J)\zeta = \int_\Omega f\zeta$$

$$\int_\Omega \frac{1}{M(\phi^k)} J^{k+1} \cdot \tilde{J} = \int_\Omega \left(\frac{\delta E_\varepsilon}{\delta \phi}(\phi^k) + \frac{\delta^2 E_\varepsilon}{\delta \phi^2}(\phi^k)\left(\phi^{k+1} - \phi^k\right)\right)\left(\nabla \cdot \tilde{J}\right)$$

for equations (9). Using the first equation, we can eliminate ϕ^{k+1} completely from the second equation:

$$\int_\Omega \frac{1}{\tau M(\phi^k)} J^{k+1} \cdot \tilde{J} + \int_\Omega \frac{\delta^2 E_\varepsilon}{\delta \phi^2}(\phi^k)\left(\nabla \cdot J^{k+1}\right)\left(\nabla \cdot \tilde{J}\right)$$
$$= \int_\Omega \left(\frac{1}{\tau}\frac{\delta E_\varepsilon}{\delta \phi}(\phi^k) + \frac{\delta^2 E_\varepsilon}{\delta \phi^2}(\phi^k) f\right)\left(\nabla \cdot \tilde{J}\right). \tag{12}$$

So starting with ϕ^0, we solve equtaion (12) to get J^1 and then use

$$\int_\Omega \phi^{k+1}\zeta = \int_\Omega \phi^k \zeta + \tau\left(\int_\Omega f\zeta - \int_\Omega (\nabla \cdot J)\zeta\right) \quad \forall \zeta \in L^2(\Omega)$$

to get ϕ^1.

Spatial Discretization

Inserting the energy (7) into equation (12) yields

$$\int_\Omega \frac{1}{\tau M(\phi^k)} J^{k+1} \cdot \tilde{J} + \varepsilon \int_\Omega \left(\nabla\nabla \cdot J^{k+1}\right) \cdot \left(\nabla\nabla \cdot \tilde{J}\right)$$
$$+ \varepsilon^{-1} \int_\Omega G''\left(\phi^k\right) \left(\nabla \cdot J^{k+1}\right) \left(\nabla \cdot \tilde{J}\right) = \text{rhs}. \quad (13)$$

This is the equation we will now discretize using finite elements.

Given a triangulation of Ω, we denote by $\mathcal{E}_h$ the set of edges and by $\mathcal{T}_h$ the set of triangles. As finite-dimensional subspaces, we choose

- {piecewise constant functions} $=: \mathcal{L}_0(\mathcal{T}_h) \subset L^2(\Omega)$ and
- {lowest order Raviart–Thomas elements} $=: \mathcal{RT}_0(\mathcal{E}_h) \subset H(\nabla\cdot, \Omega)$.

Raviart–Thomas Elements

The Raviart–Thomas elements are linear on each triangle – more specifically they are of the form $V(\mathbf{x}) = a\mathbf{x} + \mathbf{b}$ (note that a is scalar, so this is *not* the full set of linear vector fields) – and their normal component is continuous across the edges. This makes the space $\mathcal{RT}_0(\mathcal{E}_h)$ a subspace of $H(\nabla\cdot, \Omega)$.

A general property of such vector fields is that their normal component is constant along straight lines, therefore prescribing the normal component on the edges yields a basis for $\mathcal{RT}_0(\mathcal{E}_h)$.

The basis elements are

$$\Psi_i(x) := \pm \frac{|E_i|}{2|T|} \left(\mathbf{x} - \mathbf{P}_i^T\right), \quad i = 1, \ldots, \#\text{edges},$$

with $|E_i|$ denoting the length of the edge, $|T|$ denoting the area of the triangle and $\mathbf{P}_i^T$ is the point in triangle T which does not belong to E_i.

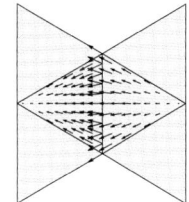

Weak Gradient

Since the divergence of a vector field in $\mathcal{RT}_0$ is piecewise constant and the operator $\nabla\nabla\cdot$ appears in equation (13), we have to define a discrete gradient $\nabla_h : \mathcal{L}_0 \to \mathcal{RT}_0$. It is natural to define ∇_h of a function $f \in \mathcal{L}_0(\mathcal{T}_h)$ by duality:

$$\int_\Omega \nabla_h f \cdot \tilde{J}_h = -\int_\Omega f \left(\nabla \cdot \tilde{J}_h\right) \quad \forall \tilde{J}_h \in \mathcal{RT}_0(\mathcal{E}_h).$$

This leads to a splitting of equation (13) into

$$\int_\Omega K_h \cdot \tilde{J}_h = \int_\Omega \left(\nabla \cdot J_h^{k+1}\right) \left(\nabla \cdot \tilde{J}_h\right) \quad \forall \tilde{J}_h \in \mathcal{RT}_0(\mathcal{E}_h)$$

and

$$\int_\Omega \frac{1}{\tau M\left(\phi_h^k\right)} J_h^{k+1} \cdot \tilde{J}_h + \varepsilon \int_\Omega (\nabla \cdot K_h) \cdot (\nabla \cdot \tilde{J}_h)$$
$$+ \varepsilon^{-1} \int_\Omega G''\left(\phi_h^k\right) \left(\nabla \cdot J_h^{k+1}\right) \left(\nabla \cdot \tilde{J}_h\right) = \text{rhs}_h \quad \forall \tilde{J}_h \in \mathcal{RT}_0(\mathcal{E}_h).$$

Matrix Representation

Taking the above basis for $\mathcal{RT}_0(\mathcal{E}_h)$, we get

$$\mathbf{B}_0 \underline{K} = \mathbf{A}_0 \underline{J}^{k+1} \tag{14a}$$

$$\frac{1}{\tau}\mathbf{B}_1^k \underline{J}^{k+1} + \varepsilon \mathbf{A}_0 \underline{K} + \varepsilon^{-1} \mathbf{A}_1^k = \underline{r}, \tag{14b}$$

where $\mathbf{B}_0$ is the mass matrix, $\mathbf{B}_1^k$ is the mass matrix weighted with the mobility and $\mathbf{A}_0$, $\mathbf{A}_1^k$ are the constant fourth order and non-constant second order stiffness matrices. The underlined quantities are coefficient vectors.

One might be tempted to lump masses and insert the first equation into the second. Unfortunately, this is not possible for Raviart–Thomas elements. Therefore, we have to find other ways to solve the system (14). After investigating in a number of possible approaches, we decided to follow Arnold and Brezzi [1985]: the idea is to search for vector fields not in $\mathcal{RT}_0(\mathcal{E}_h)$, but in a bigger space "$\mathcal{RT}_{-1}(\mathcal{T}_h)$" and enforce the solution to be in $\mathcal{RT}_0(\mathcal{E}_h)$ via a constraint. Then the mass matrices are block diagonal (with 3×3 blocks) and can be easily inverted. On the minus side, we get a Lagrange multiplier making the system to be solved bigger.

The resulting linear system is positive definite and is solved using the conjugate gradient method.

3 Results

In this section we give some examples of numerical simulations using the described numerical methods. In particular we will study step meandering in a nonlinear regime. Here, being the computationally more efficient method, the front tracking approach is used to explore a wide range of parameters in order to find interesting nonlinear behavior. If topological changes are encountered, the front-tracking simulations have to stop. Using the same parameters and initial conditions, the diffuse-interface approximation is used to go beyond the topological change.

3.1 Front Tracking

The numerical scheme resulting from the front tracking method as described in Sect. 2.1 has been implemented in the FEM-Package Alberta [Schmidt and

Siebert, 2005]. It has been shown to be quite accurate and efficient when simulating the growth of islands, see Bänsch et al. [2004]. In particular, the influence of capillary forces (strength of the line tension) and of the presence of edge diffusion as well the importance of anisotropy can be explored. Since neither topological changes nor a nucleation model are included so far, simulations of island dynamics are essentially restricted to the monolayer regime. In this context, an important application is Ostwald ripening of monolayer islands on a crystalline surface, where – as long as the coverage is not to large – only trivial topological changes (disappearing of islands) occur. The front tracking method has been used to simulate Ostwald ripening with a couple of hundreds of islands in Haußer and Voigt [2005].

Here we will present some results for the growth of vicinal surfaces. Using "skew periodic" boundary conditions for index of the step height an endless step train can be modeled. In this case, the growth of hundreds of atomic layers can be simulated.

We start with presenting a simulation of the linear instability caused by the ES-Effect as introduced in Sect. 1. This will allow to check the overall accuracy of the full numerical scheme by comparing theoretically obtained growth-rates with the numerical results. As a second example we investigate the nonlinear regime of the meander instability. Finally we present an example where anisotropic edge energy does lead to coarsening of the meander wave length in the nonlinear regime.

Linear Instability

We will use the linear instability and in particular the dispersion relation to validate the numerical scheme. To this end we consider a periodic step train modeled as two down steps with terrace width $l = 10$ on a periodic domain of size 100×20. Using the parameters $D = 10^2$, $c^* = 10^{-3}$, $k_- = 1$, $k_+ = 10$, $\Gamma = 10$, $D_{st} = 0$ and $F = 2 \cdot 10^{-3}$, the predicted most unstable wavelength is $\lambda_{max} \approx 102.7$. In the numerical simulations, the randomly perturbed steps synchronize very fast and then develop the predicted meander with a growth rate coinciding very well with the theoretical dispersion relation, see Fig. 8.

We also note, that in all numerical tests with a larger number of equally spaced steps, the step meander synchronized at an early stage of the evolution. Thus it is sufficient to simulate the evolution of two steps on a periodic domain to investigate the meandering instability in the nonlinear regime.

Nonlinear Regime

For practical purposes, the nonlinear regime of the instability is of much more importance, because meandering patterns observed during growth show large amplitudes. We used numerical simulations to explore the nonlinear behavior in various parameter regimes, for a detailed discussion we refer to Haußer and Voigt [2007].

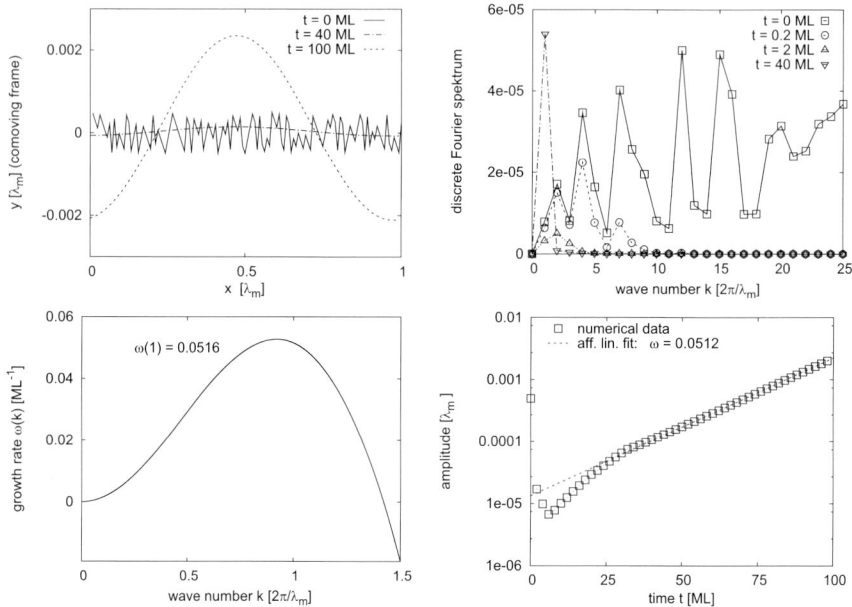

Fig. 8. Time evolution of two equidistant, initially straight steps with small-amplitude random perturbation on a periodic domain. (*Top left*) Profile of one of the two steps at different times (given in units of ML, i.e. monolayer) shows the emergence of a meander instability with a wavelength $\lambda_m = 100a$ corresponding to the most unstable wavelength $\lambda_{max} = 102.7$ of the linear instability. (*Top right*) The Fourier spectrum of the step profile clearly shows, that only the most unstable mode survives. (*Bottom left*) The growth rate as given in Bales and Zangwill [1990] is depicted for the parameters used in the simulation. As can be seen, there is only one unstable mode in the chosen domain size of length $\lambda_m = 100a$. (*Bottom right*) The predicted growth rate $\omega(\lambda_m) = 0.0516(ML)^{-1}$ compares very well with the numerically obtained value $\omega = 0.0512(ML)^{-1}$

As has been predicted by Pierre-Louis et al. [1998], Danker et al. [2003], Pierre-Louis et al. [2005] using a local amplitude equation, we observe endless growth of the meander amplitude in parameter regimes, where the meander wavelengths is large compared to the terrace width. In this regime the steps are strongly coupled. Passing to shorter meander wavelength being of similar size as the terrace width, the step profile starts to develop overhangs, which eventually lead to a self-crossing of the steps and thus to the formation of a closed loop, i.e., a vacancy island – a void of the depths of one atomic height. If the steps become even more isolated, i.e., if the meander wavelength is considerably smaller than the terrace width, we observe stationary step profiles with a fixed amplitude, see Fig. 9.

As the formation of the vacancy island and the subsequent evolution can not be simulated within this numerical approach, the parameters are passed

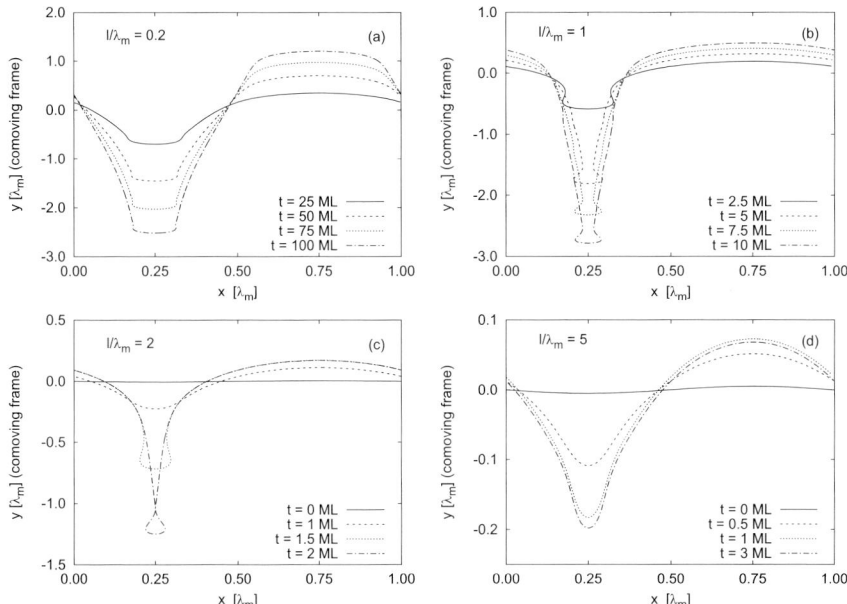

Fig. 9a–d. Evolution of a step meander; crossover to large l/λ_m. For intermediate $l/\lambda_m \approx 1$ the step profile starts to develop overhangs, but we still observe endless growth of the amplitude, see (**a**), (**b**). Further increase of l/λ_m leads to a pinch-off, i.e. the formation of a vacancy island as shown in (**c**). For even larger l/λ_m the step profile evolves to a steady state with a finite amplitude, see (**d**). The following parameters have been used in the simulations: $k_- = 1$, $k_+ = 100$, $F = 10^{-3}$, $D = 10^2$, $c^* = 10^{-3}$, $D_{st} = 0$, $l = 10$. The most unstable wavelength λ_m (and therefore the ratio l/λ_m) is varied by changing the stiffness Γ from $\Gamma = 1.4$ to $\Gamma = 10^{-3}$

to the diffuse interface approximation and used there to study the pinch-off, see Fig. 13.

Anisotropic Edge Energy

We finally give an example, where anisotropy does play role. Performing simulation on a larger domain, it appears, that no coarsening appears for isotropic edge energies, whereas for anisotropic edge energies interrupted coarsening can be observed, see Fig. 10, in agreement with Danker et al. [2003].

3.2 Diffuse-Interface-Approximation

We have developed a Finite Element software implementing the ideas of Sect. 2.2. We use adaptivity in space and time, see Fig. 11a. Another very

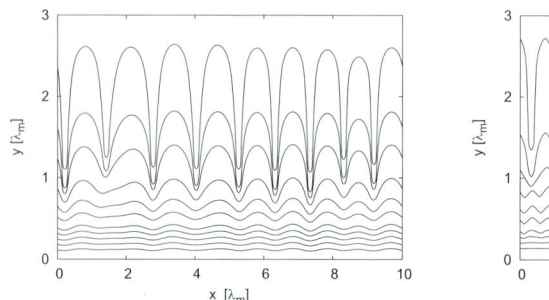 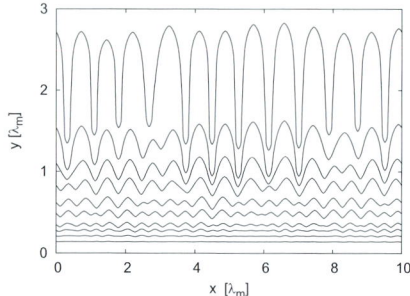

Fig. 10. Meandering on a larger domain of size $10\lambda_m$, λ_m being the most unstable wavelength in the isotropic case. (*Left*) Isotropic edge energy: The wavelength is fixed in the initial stage followed by endless growth of the amplitude. (*Right*) Anisotropic edge energy: after selection of the most unstable wavelength (being smaller as in the isotropic case, since $\Gamma(\theta = 0) < \Gamma_0$), one observes coarsening in the intermediate stage. In the late stage, the coarsening stops

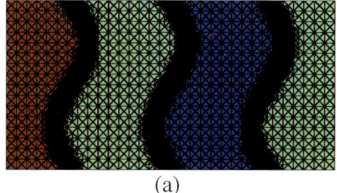 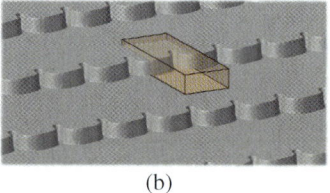

(a)   (b)

Fig. 11. (a) Grid belonging to Fig. 5b. Around the steps, the mesh is fine enough to resolve the diffuse interface and gets geometrically coarser with distance from the steps. Away from the steps, it is still fine enough to resolve the solution of the Poisson problem. (b) A step train. The computational domain is marked with the orange box

useful feature is the ability to use "skew–periodic" boundary conditions, i.e.

$$\phi(L_x, y) = \phi(0, y) + n, n \in \mathbb{Z}, \quad \phi(x, L_y) = \phi(x, 0).$$

This enables us to simulate step trains, see Fig. 11b, which are a common setup for studying the morphology of steps, e.g. the meander instability.

Coalescence

In this example, the Ehrlich–Schwoebel barrier is turned off ($\zeta^- = 0 \Rightarrow M(\phi) \equiv 1$) and there is no deposition ($f = 0$), so equation (6) becomes the Cahn–Hilliard equation, which is an approximation of the classical Mullis–Sekerka problem.

In that case, the Γ–limit of the energy E_ε for $\varepsilon \to 0$ is the total length of the interface. Therefore, the solutions will try to minimize the interface length.

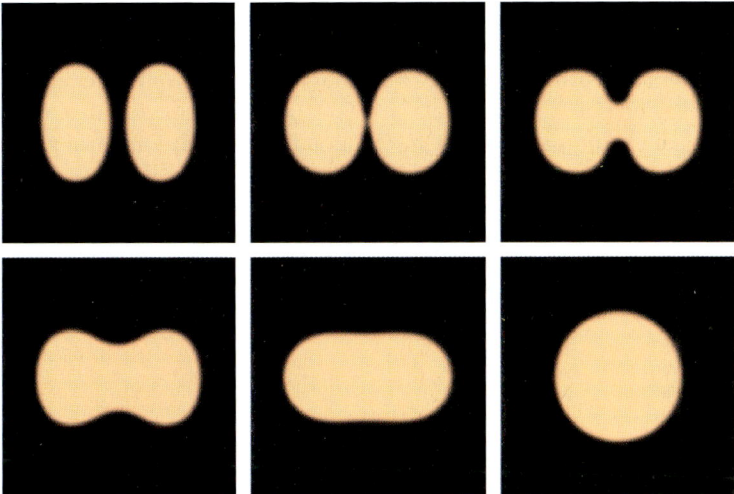

Fig. 12. Coalescence of two islands

We take two ellipses as initial datum. Since both ellipses change their shape to become a circle, they will touch after some time and then coalesce, see Fig. 12. The final shape will be a single circle.

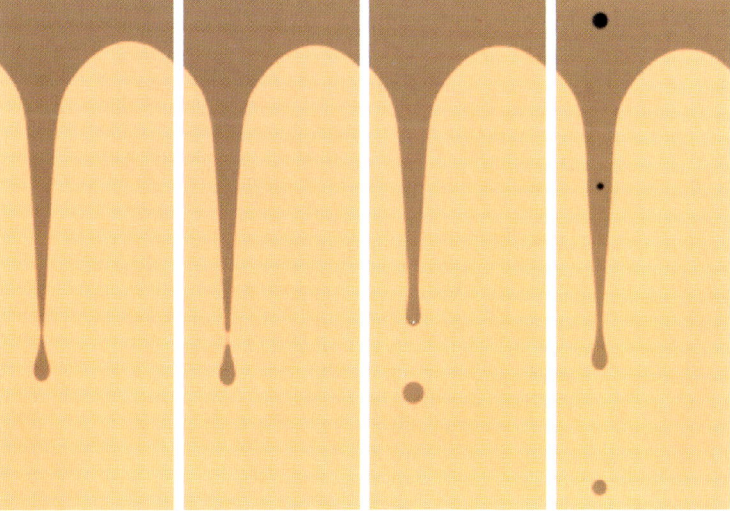

Fig. 13. Pinch-off of a vacancy. The leftmost picture corresponds to the last position in Fig. 9c. As the simulation continues, vacancies keep pinching off and then disappear due to deposition, see the rightmost picture

Pinch-off

In case (c) of the parameter study carried out with the front tracking method, see Fig. 9, a pinch-off became apparent. To study the behavior after the pinch-off, we use the same parameters. In the nondimensinoal form, they translate into

$$\zeta^- \approx 94.1, \quad \hat{l} \approx 9.41, \quad \text{and} \quad \hat{\lambda} \approx 3.64.$$

As expected, we see the same instability and can continue the simulation after the pinch-off, see Fig. 13.

References

N. Alikakos, P. Bates, and X. Chen. Convergence of the Cahn–Hilliard equation to the Hele–Shaw model. *Arch. Ration. Mech. An.*, 128:165–205, 1994.

D. N. Arnold and F. Brezzi. Mixed and nonconforming finite element methods: implementation, postprocessing and error estimates. *RAIRO Modél. Math. Anal. Numér.*, 19(1):7–32, 1985.

G. Bales and A. Zangwill. Morphological instability of a terrace edge during step-flow growth. *Phys. Rev. B*, 41:5500, 1990.

E. Bänsch, F. Haußer, O. Lakkis, B. Li, and A. Voigt. Finite element method for epitaxial growth with attachment-detachment kinetics. *J. Comp. Phys.*, 194: 409–434, 2004.

E. Bänsch, F. Haußer, and A. Voigt. Finite element method for epitaxial growth with thermodynamic boundary conditions. *SIAM J. Sci. Comp.*, 26:2029–2046, 2005a.

E. Bänsch, P. Morin, and R. H. Nochetto. A finite element method for surface diffusion: the parametric case. *J. Comput. Phys.*, 203:321–343, 2005b.

M. O. Bristeau, R. Glowinski, and J. Périaux. Numerical methods for the Navier-Stokes equations. Applications to the simulation of compressible and incompressible viscous flows. In *Finite elements in physics (Lausanne, 1986)*, pages 73–187. North-Holand, Amsterdam, 1987.

W. K. Burton, N. Cabrera, and F. C. Frank. The growth of crystals and the equilibrium of their surfaces. *Phil. Trans. Roy. Soc. London Ser. A*, 243(866):299–358, 1951.

G. Caginalp. Stefan and Hele-Shaw type models as asymptotic limits of the phase-field equations. *Phys. Rev. A*, 39(11):5887–5896, Jun 1989.

G. Caginalp and X. Chen. Convergence of the phase field model to its sharp interface limits. *Eur. J. Appl. Math.*, 9:417–445, Aug 1998.

G. Danker, O. Pierre-Louis, K. Kassner, and C. Misbah. Interrupted coarsening of anisotropic step meander. *Phys. Rev. E*, 68:020601, 2003.

G. Ehrlich and F. G. Hudda. Atomic view of surface diffusion: tungsten on tungsten. *J. Chem. Phys.*, 44:1036–1099, 1966.

F. Haußer and A. Voigt. Ostwald ripening of two-dimensional homoepitaxial islands. *Phys. Rev. B*, 72:035437, 2005.

F. Haußer and A. Voigt. Step meandering in epitaxial growth. *J. Cryst. Growth*, 303(1):80–84, 2007.

A. Karma and M. Plapp. Spiral surface growth without desorption. *Phys. Rev. Lett.*, 81:4444–4447, 1998.

J. Krug. Introduction to step dynamics and step instabilities. In A. Voigt, editor, *Multiscale modeling of epitaxial growth*, volume 149 of *ISNM*. Birkhäuser, 2005.

F. Liu and H. Metiu. Stability and kinetics of step motion on crystal surfaces. *Phys. Rev. E*, 49:2601–2616, 1997.

T. Maroutian, L. Douillard, and H. Ernst. Wavelength selection in unstable homoepitaxial step flow growth. *Phys. Rev. Lett.*, 83:4353, 1999.

F. Otto, P. Penzler, A. Rätz, T. Rump, and A. Voigt. A diffusive-interface aproximation for step flow in epitaxial growth. *Nonlinearity*, 17(2):477–491, 2004.

R. Pego. Front migration in the nonlinear Cahn–Hilliard equation. *P. Roy. Soc. A*, 422:261–278, Apr 1989.

O. Pierre-Louis, C. Misbah, Y. Saito, J. Krug, and P. Politi. New nonlinear evolution equation for steps during molecular beam epitaxy. *Phys. Rev. Lett.*, 80:4221, 1998.

O. Pierre-Louis, G. Danker, J. Chang, K. Kassner, and C. Misbah. Nonlinear dynamics of vicinal surfaces. *J. Cryst. Growth*, 275:56, 2005.

P. Politi, G. Grenet, A. Marty, A. Ponchet, and J. Villain. Instabilities in crystal growth by atomic or molecular beams. *Phys. Rep.*, 324:271, 2000.

C. Polop, S. Bleikamp, and T. Michely. I. Phys. Institut, RWTH Aachen.

A. Schmidt and K. Siebert. *Design of adaptive finite element software: The finite element toolbox ALBERTA*. Number 42 in LNCSE. Springer, 2005.

R. L. Schwoebel and E. J. Shipsey. Step motion on crystal surfaces. *J. Appl. Phys.*, 37:3682–3686, 1966.

A. Voigt, editor. *Multiscale modeling in epitaxial growth*, volume 149 of *International Series of Numerical Mathematics*. Birkhäuser Verlag, Basel, 2005. ISBN 978-3-7643-7208-8; 3-7643-7208-7. Selected papers from the workshop held in Oberwolfach, January 18–24, 2004.

U. Weikard. *Numerische Lösungen der Cahn-Hilliard-Gleichung und der Cahn-Larché-Gleichung*. Dissertation, Rheinische Friedrich-Wilhelms-Universität Bonn, Oct 2002.

Micro Structures in Thin Coating Layers: Micro Structure Evolution and Macroscopic Contact Angle

Julia Dohmen[1], Natalie Grunewald[2], Felix Otto[2], and Martin Rumpf[1]

[1] Institut für Numerische Simulation, Universität Bonn Nussallee 15, 53115 Bonn, Germany, {julia.dohmen,martin.rumpf}@ins.uni-bonn.de
[2] Institut für Angewandte Mathematik, Universität Bonn, Wegelerstr. 10, 53115 Bonn, Germany, {grunewald,otto}@iam.uni-bonn.de

Summary. Micro structures of coating surfaces lead to new industrial applications. They allow to steer the wetting and dewetting behaviour of surfaces and in particular to enhance hydrophobicity. Here, we discuss the formation of micro structures in the drying process of a coating. Furthermore, for a given micro structured surface we show how to predict the effective contact angle of drops on the surface. At first, we derive a new approach for the simulation of micro structure evolution based on a gradient flow perspective for thin liquid films. This formulation includes a solvent dependent surface tension, viscosity and evaporation rate. In each time step of the resulting algorithm a semi implicit Rayleigh functional is minimized. The functional itself depends on the solution of a transport problem. We apply a finite difference discretization both for the functional and the transport process. As in PDE optimization a duality argument allows the efficient computation of descent directions. Next, given a certain micro structured coating we mathematically describe effective contact angles in different configurations and their impact on the macroscopic hydrophilic or hydrophobic surface properties. On periodic surfaces we aim at the computation of effective contact angles. This involves a geometric free boundary problem on the fundamental cell. Its solution describes vapor inclusions on the wetted surface. The free boundary problem is solved by a suitable composite finite element approach. Furthermore, we introduce a new model for the influence of micro structures on contact angle hysteresis. This model is adapted from elasto–plasticity and dry friction. It identifies stable contact angles not only as global or local energy minimizers but as configurations at which the energy landscape is not too steep.

1 Introduction

Micro structures in coatings are of great industrial relevance. They can be desirable and undesirable. On the one hand they might lead to rupture of a paint. On the other hand they can enhance hydrophobicity of the surface. Here we discuss two different aspects of these phenomena.

In Sect. 2 we consider a model for the formation of micro structures in a drying coating. These strucutures can for instance evolve from a non homogeneous solvent distribution in an originally flat coating. We model the coating by an adapted thin film model. It is based on a gradient flow model with solvent dependent viscosity, surface tension and evaporation rate, see Sect. 2.1. This introduces Marangoni effects to the film which can lead to a structured film height but also counteract rupture. It also takes into account the solvent evaporation in a coating, which is fast at low film heights, due to a faster heating up. A third effect considered is the hardening, i.e. the temporal change of the viscosity of the coating. In Sect. 2.2 and 2.3 we introduce a numerical algorithm based on a semi implicit time discretization, which takes advantage of the gradient flow structure. In each time step a corresponding Rayleigh functional is minimized in Sect. 2.5 we show numerical results.

In the second part in Sect. 3 we discuss the implications of a structured surface to contact angles of macroscopic drops sitting on the surface. The micro structures highly influence the contact angle and thereby the sticking of the drop to the surface. One governing effect is the formation of vapor inclusions on the surface at a micro scale. This reduces the contact of the drop to the surface – hence, it rolls off easily. We introduce an algorithm in Sect. 3.1, which simulates the vapor inclusions in a periodic setup. The corresponding liquid vapor interface is a minimal surface with prescribed microscopic contact angle of the triple contact line. In the limit of small scale periodicity of the surface this enables the calculation of effective contact angles.

Finally, in Sect. 3.2 we consider the stability of drop configurations on the micro structured surface. A new model is introduced which determines the stability of effective contact angles. Their stability depends on the micro configuration of the drop, i.e. on the possible vapor inclusions. The model allows for intervals of stable contact angles (contact angle hysteresis). It is adapted from elasto–plasticity and dry friction, and assumes a configuration not only to be stable if it minimizes (locally) the relevant surface energy but also if the energy landscape at this configuration is not too steep. This leads to different hysteresis intervals for configurations with and without vapor inclusions. A change in the vapor configuration at the surface can explain the highly non monotone dependence of the hysteresis on the surface roughness, known since the sixties, [JD64], as well as more recent experiments.

2 Modeling and Simulation of the Micro Structure Formation in Thin Coatings

2.1 Modeling Thin Coatings as a Gradient Flow

We propose a simple model for coatings similar to the one considered in [HMO97], which in spite of its simplicity reproduces many of the interesting features known for a drying paint. We assume the paint to consist of two components, the non–volatile resin and the volatile solvent, whose concentration is given by s. Together they form a well-mixed fluid with height h. In the simulations we plot both the height (on the left) and the solvent concentration (on the right), see Fig. 1. These are the two parameters describing the physical properties of the fluid:

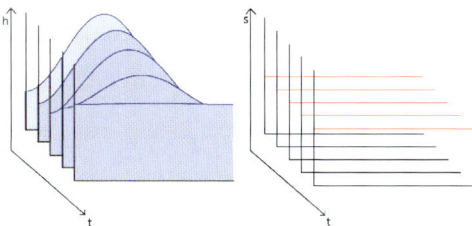

Fig. 1. A time evolution (*back to front*) of a coating is described by its height (*on the left*) and solvent concentration (*on the right*). Here the trivial case with constant Solvent Concentration is depicted

The solvent concentration influences the viscosity μ (the drying coating becomes harder with descreasing solvent concentration) as well as the surface tension σ (the surface tension increases with decreasing solvent concentration) and the evaporation rate e. The evaporation rate also depends on solvent concentration and on the height of the film, as a thin film dries fast due to its closeness to the warm substrate. We assume a well–mixed coating, where both components are transported by the same horizontal fluid velocity u.

This model can introduce micro structures even on an initially flat coating. Indeed, they may be originated in a inhomogeneous distribution of solvent. Local areas on the coating where the solvent concentration is high have less surface tension. This induces a Marangoni flow in the direction from high to low solvent concentration. This flow reduces the surface energy as the interface with less surface tension is strechted in comparison to the interface with high surface tension, which is condensed. Hence, fluctuation in the solvent concentration lead to a structured film height. On the other hand, surface tension primarily induces a flow which reduces the area of the interface. It therefore drives the fluid to a flat film. These two forces can in the absence of evaporation compensate each other leading to an inhomogeneous structured but stable film, c.f.[W93]. Figure 2 shows a Marangoni induced stable micro structure.

Furthermore, the combination of a height dependent evaporation rate e of the solvent and of Marangoni effects (i.e. the solvent dependent surface tension) counteracts film rupture at points, where the height of the film tends to zero. In fact, due to their closeness to the warm surface the film dries quickly at low film heights. This reduces the solvent concentration at these points, which again induces a Marangoni flow to the valleys on the film surface due to a higher surface tension in case of a low solvent concentration. This flow counteracts rupture. Indeed our simulations (Figs. 5 and 4) do not show a critical deepening of the film leading to rupture.

Gradient Flow Structure. For our model we firstly assume a balance of viscous and capillary forces but neglect the momentum of the fluid. We assume an over-damped limit in which the quasi stationary Stokes equations for an incompressible fluid are appropriate. By the well known lubrication approximation [BDO97] they can be reduced to the thin film equations, which are of gradient flow structure (cf. [GO03]). The height of the film h performs a steepest descent of an energy functional E:

$$\dot{h} = -\text{grad}E\big|_h. \tag{1}$$

To make sense of the gradient of the energy one has to identify the metric structure of the manifold $\mathcal{M}$ on which the gradient flow takes place. In this case, this is the manifold of all heights of the film with prescribed volume. The metric is described by its metric tensor $g_h(\delta h, \delta h)$ on the tangent spaces, which consist of the infinitesimal height variations δh. Denoting $\text{diff} E\big|_h .\delta h = \lim_{\varepsilon \to 0} \frac{1}{\varepsilon}(E(h + \varepsilon \delta h) - E(h))$ turns (1) into

$$g_h(\dot{h}, \delta h) = -\text{diff} E\big|_h .\delta h \quad \forall \, \delta h \in T_h \mathcal{M}. \tag{2}$$

Equation (2) can be seen as the Euler–Lagrange equation of

$$\mathcal{F}(\delta h) = \frac{1}{2} g_h(\delta h, \delta h) + \text{diff} E\big|_h .\delta h \tag{3}$$

with respect to δh. Indeed, the actual rate of change $\dot{h}$ minimizes $\mathcal{F}$ under all possible infinitesimal variations δh. We will use such a gradient flow structure to model thin coatings, inspired by the gradient flow model for thin films, which we will explained first.

Thin Films as a Gradient Flow. Thin fluid films are described by the well known thin film equation

$$\dot{h} = -\frac{\sigma}{3\mu} \text{div}(h^3 \nabla \Delta h), \tag{4}$$

for the height of the film [BDO97]. Here, we might impose either periodic or natural boundary conditions. This evolution is a gradient flow, as introduced in [O01]. The relevant energy is the linearized surface energy:

$$E(h) := \int_\Omega \sigma \left(1 + \frac{1}{2}|\nabla h|^2\right) dx.$$

The metric tensor is given by the minimal energy dissipated by viscous friction, i. e.

$$g_h(\delta h, \delta h) = \inf_u \left\{ \int_\Omega \frac{3\mu}{h} u^2 dx \right\},$$

where Ω is the underlying domain. Note that the metric tensor is base point dependent. The infimum is taken over any velocity profile u that realizes the given change in film height δh described by the transport equation

$$\delta h + \text{div}\,(h\,u) = 0. \tag{5}$$

On the first sight the metric tensor seems to be a complicated object, as it involves the minimization of the viscous friction. Therefore finding the minimizer of the functional $\mathcal{F}$ in (3) requires to solve a nested minimization problem. This can be avoided, if one describes the tangent space, i.e. all infinitesimal changes in film height h, directly by an admissible velocity fields u via (5) (of course the same δh may be described by many u's). In this sense the metric tensor can be lifted onto the space of admissible velocities u:

$$g_h(u, u) = \int_\Omega \frac{3\mu}{h} u^2 dx. \tag{6}$$

Rewriting (3) leads to a formulation of the gradient flow as the evolution

$$\dot{h} + \text{div}\,(h\,u^*) = 0, \tag{7}$$

where u^* minimizes the Rayleigh functional

$$\mathcal{F}(u) = \frac{1}{2} g_h(u, u) + \text{diff}\,E\big|_h . u \tag{8}$$

over all fluid velocities u. Here $\text{diff}\,E\big|_h . u$ is defined as $\text{diff}\,E\big|_h . \delta h$ with δh satisfying (5). It is now easy to see that the gradient flow given by (6)–(8) coincides with the evolution of the thin film equation (4). Indeed, we observe that u^* solves the Euler–Lagrange equation corresponding to the Rayleigh functional (8):

$$0 = g_h(u^*, u) + \text{diff}\,E\big|_h . u = \int_\Omega \frac{3\mu}{h} u^* \cdot u\,dx - \int_\Omega \sigma \nabla h \nabla \text{div}(h\,u)\,dx$$

for all test velocities u. For periodic or natural boundary conditions this immediately implies

$$u^* = \frac{\sigma h^2}{3\mu} \nabla \Delta h.$$

Finally, plugging u^* into (7) yields the thin film equation (4). The thin film is a special case of a thin coating, i.e. the one with constant solvent concentration. Numerical results for the spreading of a thin film are shown in Fig. 1.

Thin Coatings as a Gradient Flow. The model for thin coatings is more difficult, as the state of the paint is not only described by its film height h but also by the solvent concentration s in the film. We assume a thin film model, which is inspired by the gradient flow described above. Here, we adopt a point of view developed in [GP05]: The gradient flow evolves on the manifold of all possible film heights. The solvent will be transported along with the fluid and is taken into account as a vector bundle on the manifold. At any given film height, there is a vector space of possible solvent concentrations, the fiber. They are not part of the manifold. The tangent spaces therefore consist only of the infinitesimal changes in film height δh. These are induced by a velocity u (as explained above):

$$\delta h + \operatorname{div}(h\,u) = 0 \qquad (9)$$

The solvent concentration is transported by parallel transport. That is, we assume a mixed fluid, where the solvent is transported by the same velocity. As s is the concentration of solvent, the actual amount of solvent is given by $h\,s$. Therefore

$$\delta(hs) + \operatorname{div}(hs\,u) = 0. \qquad (10)$$

This vector bundle construction to model an extra component slaved to the transport of the fluid was introduced in [GP05] for a thin film with surfactant.

The gradient flow is now given by the reduced energy and the metric on the manifold. As in the thin film case, the relevant energy is the linearized surface energy:

$$E(h,s) := \int_\Omega \sigma(s)\left(1 + \frac{1}{2}|\nabla h|^2\right) dx. \qquad (11)$$

The surface tension σ depends on the solvent concentration s. This introduces Marangoni effects to the model, which we see in a drying coating. The metric is given by the minimal energy dissipated by viscous friction, where the viscosity μ depends on the solvent concentration. The drying coating becomes hard. One has the metric tensor

$$g_{h,s}(u,u) = \int_\Omega \frac{3\mu(s)}{h} u^2 dx. \qquad (12)$$

The gradient flow is (9) and (10) with the velocity field $u = u^*$, where u^* minimizes the Rayleigh functional

$$\mathcal{F}(u) = \frac{1}{2} g_{h,s}(u,u) + \operatorname{diff} E\big|_{h,s}.u \qquad (13)$$

over all velocities u. This model is similar to the thin film model, but has included the solvent features of a thin coating. On the one hand it tries to minimize the (linearized) surface energy (11) by mean surface tension and Marangoni flows. They reduce the energy by elongating the surface with low surface tension. One the other hand the flow is hindered by viscous friction (12). The viscous friction increases as the evaporation continues (as $\mu(s)$ is an increasing function). The only effect not yet modeled is the evaporation. On a continuous level this would include the modeling of the full vapor phase. On the discrete level the evaporation is included as a second step in an operator splitting method, see below.

2.2 Natural Time Discretization

Any gradient flow has a natural time discretization. It involves the natural distance function dist on the manifold $\mathcal{M}$ defined via

$$\text{dist}^2(h_0, h_1) := \inf_\gamma \left\{ \left(\int_0^1 \sqrt{g_{\gamma(t)}(\dot\gamma, \dot\gamma)}\, dt \right)^2 \right\},$$

with γ any smooth curve with $\gamma(0) = h_0$ and $\gamma(1) = h_1$. If $\mathcal{M}$ is actually Euclidean instead of genuinely Riemannian as in our case

$$\text{dist}^2(h_0, h_1) = |h_0 - h_1|^2. \tag{14}$$

If τ denotes the time step size, the solution h^{k+1} at step $k+1$ can be inferred from the state h^k at step k via the variational problem:

$$h^{k+1} = \operatorname{argmin}_h \left\{ \frac{1}{2\tau} \text{dist}^2\left(h, h^k\right) + E(h) \right\}. \tag{15}$$

As a motivation consider the Euclidean case (14). Here the Euler–Lagrange equation for (15) turns into the implicit Euler scheme

$$\frac{1}{\tau}\left(h^{k+1} - h^k\right) = -\nabla E\big|_{h^{k+1}}.$$

We want to use (15) as a starting point to construct a natural and stable discretization. The drawback of (15) is, it is fully nonlinear and it involves two nested minimizations.

One natural idea to overcome this drawback, which is also used for epitaxial growth, see the corresponding chapter in this book, is the following: We approximate the functional by its quadratic at h^k and then lift the variational problem on the level of possible velocities u in the spirit of (7) and (8). We first turn to the quadratic approximation: Writing $h = h^k + \tau \delta h$, we have

$$\frac{1}{2\tau} \text{dist}^2\left(h, h^k\right) + E(h) \approx$$
$$\frac{\tau}{2} g_{h^k}(\delta h, \delta h) + E(h^k) + \tau \operatorname{diff} E\big|_{h^k} . \delta h + \frac{\tau^2}{2} g_{h^k}\left(\delta h, \operatorname{Hess} E\big|_{h^k} \delta h\right), \tag{16}$$

where $\operatorname{Hess} E\big|_{h^k}$ denotes the Hessian of E in h^k. Hence we can solve

$$\delta h^* = \operatorname{argmin}_{\delta h} \left\{ \frac{1}{2} g_{h^k}(\delta h, \delta h) + \operatorname{diff} E\big|_{h^k} . \delta h + \frac{\tau}{2} g_{h^k}\left(\delta h, \operatorname{Hess} E\big|_{h^k} \delta h\right) \right\} \tag{17}$$

and then set $h^{k+1} = h^k + \tau \delta h^*$, cf. (3). However, as in (3), (17) still involves two nested minimizations. Therefore, using (5) we may lift (17) on the level of possible velocities u as before. This yields

$$u^{k+1} = \operatorname{argmin}_u \left\{ \frac{1}{2} g_{h^k}(u,u) + \operatorname{diff} E\big|_{h^k}.u + \frac{\tau}{2} g_{h^k}\left(u, \operatorname{Hess} E\big|_{h^k} u\right) \right\} \quad (18)$$

and then set $h^{k+1} = h^k + \tau \operatorname{div}\left(h^k u^{k+1}\right)$. Compare (18) to (7) and (8). This is the basis for the gradient flow algorithm used for epitaxial growth.

For our algorithm we use an alternative approach. We consider a semi implicit time discretization. For this we only approximate the squared distance dist^2 in (15) by its metric based approximation and keep E fully nonlinear. We use the following notation: For given velocity field u varying in space and fixed in time define the transport operator $\mathbf{h}(\cdot, \cdot)$, which maps a height field h^k at time t^k onto a height field $\mathbf{h}(h^k, u) = h(t^{k+1})$, where h solves the transport equation $\partial_t h + \operatorname{div}(h\,u) = 0$ with initial data $h(t^k) = h^k$. Given this operator, we again apply a linearization of the distance map dist in (15) and evaluate the energy on $\mathbf{h}[h_k, u]$. This energy is again implicitly defined via the velocity field u, which minimizes a corresponding functional. Thus, we define

$$u^{k+1} = \operatorname{argmin}_u \left\{ \frac{\tau}{2} g_{h^k}(u,u) + E\left(\mathbf{h}\left(h^k, u\right)\right) \right\}, \quad (19)$$

which can be considered as a semi-implicit alternative to the time discretization in (18). The new height field is then given by $h^{k+1} = \mathbf{h}(h^k, u^{k+1})$. Here, we still use the metric for the linearization of the distance map and evaluate this at the height field h^k at the old time t^k.

This gradient flow model for the thin film equation can easily be generalized for the thin coating model. To simplify the presentation let us introduce the vector $q = (h, hs)$ consisting of the two conservative quantities film height h and amount of solvent hs. Furthermore, we again define a transport operator $\mathbf{q}(\cdot, \cdot)$, which maps $q^k = (h^k, h^k s^k)$ at time t^k onto $\mathbf{q}(q^k, u) = q(t^{k+1})$, where q is a the solution of the system of transport equations

$$\partial_t h + \operatorname{div}(h\,u) = 0 \quad (20)$$
$$\partial_t (hs) + \operatorname{div}(hs\,u) = 0 \quad (21)$$

with initial data $q(t^k) = q^k = (h^k, h^k s^k)$. In analogy to (19), we consider an implicit variational definition of the motion field

$$u^{k+1} = \operatorname{argmin}_u \left\{ \frac{\tau}{2} g_{q^k}(u,u) + E\left(\mathbf{q}\left(h^k, u\right)\right) \right\}, \quad (22)$$

where $E[q]$ is given by (11). Hence, in every time step we ask for the minimizer of a functional whose integrand depends on the solution of a hyperbolic initial value problem. Indeed this is a PDE constrained optimization problem. In the next section we will solve this problem numerically based on a suitable space discretization and duality techniques.

2.3 Space Discretization for the Gradient Flow

Let us consider a discretization of (22) in one and two space dimensions and for simplicity restrict to a domain $\Omega = [0,1]^d$, where $d \in \{1,2\}$, and impose periodic boundary conditions. We suppose Ω to be regularly subdivided into N interval of width $\Delta := \frac{1}{N}$ ($d = 1$) or squares of edge length Δ ($d = 2$). By $Q = (Q_i)_{i \in I} = (H_i, H_i S_i)_{i \in I}$ and $U = (U_i)_{i \in I}$ we denote nodal vectors of discrete q and u quantities, respectively, where the ith component corresponds to a grid nodes x_i. Here I is supposed to be the lexicographically ordered index set of nodes (for $d = 2$ these indices are 2-valued, i. e. $i = (i_1, i_2)$, where the two components indicate the integer coordinates on the grid lattice). Spatial periodicity can be expressed by the notational assumption $Q_i = Q_{i+Ne}$ and $V_i = V_{i+Ne}$, where $e = 1$ for $d = 1$ and $e = (1,0)$ or $(0,1)$ for $d = 2$. Now, we define in a straightforward way a discrete energy value $E[Q]$ on $\mathbb{R}^{2\sharp I}$ and a discrete metric $G_Q[U,U]$ on $\mathbb{R}^{d\sharp I} \times \mathbb{R}^{d\sharp I}$:

$$E[Q] = \sum_{i \in I} \Delta^d \sigma(\tilde{S}_i) \left[1 + \frac{1}{2}(\nabla_i H)^2\right], \qquad (23)$$

$$G_Q(U,U) = \sum_{i \in I} \Delta^d \frac{3\mu(S_i)}{H_i} |U_i|^2, \qquad (24)$$

where $\tilde{S} = \frac{1}{2}(S_i + S_{i+1})$ ($d = 1$) or $\tilde{S} = \frac{1}{4}(S_i + S_{i+(0,1)} + S_{i+(1,0)} + S_{i+(1,1)})$ ($d = 2$) are interpolated values for the solvent concentration at cell centers, and $\nabla_i H = \frac{1}{\Delta}(H_{i+1} - H_i)$ ($d = 1$) or $\nabla_i H = \frac{1}{2\Delta}(H_{i+(1,0)} + H_{i+(1,1)} - H_i - H_{i+(0,1)}, H_{i+(0,1)} + H_{i+(1,1)} - H_i - H_{i+(1,0)})$ ($d = 2$) is the difference quotient approximation of the gradient of the height field. Next, we define an operator $\mathbf{Q}$, which computes $\mathbf{Q}(Q^k, U) = Q^{k+1} = (H_i^k, H_i^k S_i^k)_{i \in I}$ as the solution of an implicit Lax–Friedrich scheme for the associated transport problem for given data Q^k at time t^k and a discrete velocity vector U. Let us detail this here in the one dimensional case, where we obtain the following system of equations

$$\frac{Q_i^{k+1} - Q_i^k}{\tau} = \frac{U_{i+1} Q_{i+1}^{k+1} - U_{i-1} Q_{i-1}^{k+1}}{2\Delta} + \epsilon \frac{Q_{i+1}^{k+1} - 2 Q_i^{k+1} + Q_{i-1}^{k+1}}{\Delta^2}$$

for all $i \in I$ and a small positive constant ϵ. The two dimensional case is completely analogous. This scheme can be rewritten in matrix vector notation

$$Q^k = A(U) \mathbf{Q}(Q^k, U) \qquad (25)$$

where $A(U) \in \mathbb{R}^{2\sharp I \times 2\sharp I}$ is a matrix depending on the discrete vector field U, which can easily be extracted from the Lax-Friedrich scheme. For $\epsilon > 0$ this matrix is invertible. Thus, we obtain the explicit representation $\mathbf{Q}(Q^k, U) = A(U)^{-1} Q^k$ for the discrete transport operator. With these ingredients at hand, one obtains a discrete counterpart of the variational problem (22)

$$U^{k+1} = \mathrm{argmin}_{U \in \mathbb{R}^{d\sharp I}} \left\{ \frac{\tau}{2} G_{Q^k}(U,U) + E\left(\mathbf{Q}\left(Q^k, U\right)\right) \right\}. \qquad (26)$$

Finally, we define $Q^{k+1} = \mathbf{Q}(Q^k, U^{k+1})$. In each time step we aim at computing the discrete minimizer U^{k+1} via a gradient descent scheme on $\mathbb{R}^{d\sharp I}$. Hence, besides the energy on the right hand side of (26) we have to compute the gradient vector on $\mathbb{R}^{d\sharp I}$. For the variation of the energy $E(\mathbf{Q}(Q^k, U))$ in a direction $W \in \mathbb{R}^{d\sharp I}$ we get $\partial_U E(\mathbf{Q}(Q^k, U))(W) = \partial_Q E(\mathbf{Q}(Q^k, U))(\partial_U \mathbf{Q}(Q^k, U)(W))$. A direct application of this formula for the evaluation of the gradient of the energy E would require the computation of

$$\partial_U \mathbf{Q}(Q^k, U)(W) = -A^{-1}(U)(\partial_U A(U)(W))A^{-1}(U)Q^k$$

for every nodal vector W in $\mathbb{R}^{d\sharp I}$. To avoid this, let us introduce the dual solution $P = P(Q^k, U) \in \mathbb{R}^{2\sharp I}$ which solves

$$A(U)^T P = -\partial_Q E(\mathbf{Q}(Q^k, U)).$$

Computing the variation of the linear system (25) with respect to U we achieve

$$0 = (\partial_U A(U)(W))\mathbf{Q}\left(Q^k, U\right) + A(U)\left(\partial_U \mathbf{Q}\left(Q^k, U\right)(W)\right),$$

from which we then derive

$$\begin{aligned}
\partial_U E\left(\mathbf{Q}\left(Q^k, U\right)\right)(W) &= \partial_Q E\left(\mathbf{Q}\left(Q^k, U\right)\right)\left(\partial_U \mathbf{Q}\left(Q^k, U\right)(W)\right) \\
&= -A(U)^T P(Q^k, U) \cdot \left(\partial_U \mathbf{Q}\left(Q^k, U\right)(W)\right) \\
&= -P\left(Q^k, U\right) \cdot A(U)\left(\partial_U \mathbf{Q}\left(Q^k, U\right)(W)\right) \\
&= P\left(Q^k, U\right) \cdot (\partial_U A(U)(W))\mathbf{Q}\left(Q^k, U\right).
\end{aligned}$$

This representation of the variation of the energy can be evaluated without solving $d\sharp I$ linear systems of equations. In our implementation we consider the Armijo rule as a step size control in the descent algorithm on $\mathbb{R}^{d\sharp I}$.

2.4 Evolution of Thin Coatings with Solvent Evaporation

So far the model for the evolution of a thin film consisting of resin and solvent is considered as a closed system and formulated as a gradient flow. Evaporation of the solvent from the liquid into the gas phase – the major effect in the drying of the coating – still has to be taken into account. As already mentioned, incorporating this in a gradient flow formulation would require to model the gas phase as well. To avoid this we use an operator splitting approach and consider the evaporation separately as a right hand side in the transport equations. Thus, we consider the modified transport equations

$$\begin{aligned}
\partial_t h + \text{div}(h\, u) &= e(h, s), \\
\partial_t (hs) + \text{div}(hs\, u) &= e(h, s),
\end{aligned}$$

where $e(h, s) = -\frac{C}{c+hs}$ is the usual model for the evaporation [BDO97], where $C, c > 0$ are evaporation parameters. In the time discretization we now alternate the descent step of the gradient flow and an explicit time integration of the evaporation. In the first step, the velocity u^{k+1} is computed based on (22). Solving the corresponding transport equations (20) and (21) we obtain updated solutions for the height and the solvent concentration at time t^{k+1}, which we denote by $\tilde{h}^{k+1}$ and $\tilde{s}^{k+1}$, respectively. In the second step, applying an explicit integration scheme for the evaporation we finally compute

$$h^{k+1} = \tilde{h}^{k+1} + \tau e\left(\tilde{h}^{k+1}, \tilde{s}^{k+1}\right),$$
$$s^{k+1} = \left(h^{k+1}\right)^{-1}\left(\tilde{h}^{k+1}\tilde{s}^{k+1} + \tau e\left(\tilde{h}^{k+1}, \tilde{s}^{k+1}\right)\right).$$

For the fully discrete scheme, we proceed analogously and update the nodal values Q^{k+1} in each time step. In fact, given U^{k+1} as the minimizer of (26) we compute $\tilde{Q}^{k+1} = (\tilde{H}^{k+1}, \tilde{S}^{k+1}) = A(U^{k+1})^{-1}Q^k$ and then update pointwise $Q_i^{k+1} = \tilde{Q}_i^{k+1} + \tau e(\tilde{H}_i^{k+1}, \tilde{S}_i^{k+1})$.

2.5 Numerical Results

The numerical results show the features of thin coatings introduced by Marangoni and surface tension effects combined with evaporation and hardening. We will discuss them separately. A first test of our algorithm was to run it with constant solvent concentration, which turns the model for thin coatings into the simpler thin film model described above. Numerical results are already shown in Fig. 1. They are numerically consistent with results obtained by a finite volume scheme for the thin film equation [GLR02], where thin films with (and without) surfactant are simulated. Figure 2 shows the effects introduced by Marangoni forces. In particular an inhomogeneous solvent concentration can lead to a structure formation in the film height. In the absence of evaporation this structure becomes stable as the Marangoni forces are opposed by mean surface tension forces, which want to reduce the length of the film surface.

An inhomogeneous solvent concentration also introduces a structured film height via evaporation, Fig. 3. This leads – as only solvent evaporates – to valleys in the film located at positions with a high amount of solvent. Still the coating is by no means close to rupture, as this is opposed by Marangoni forces. Figure 5 shows that the combination of these effects leads to a micro

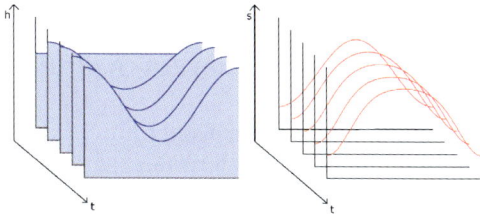

Fig. 2. Evolution of a coating with a marangoni flow introduced by an inhomogeneous solvent concentration

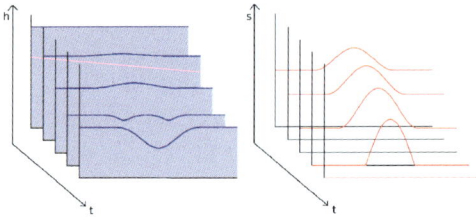

Fig. 3. Evolution of a coating with evaporating solvent

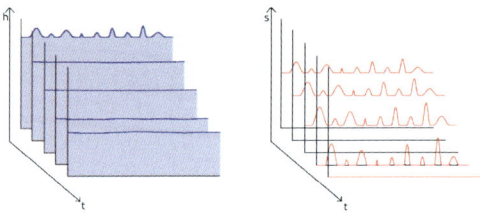

Fig. 4. The drying of a coating with (artifically) constant viscosity with a vanishing of micro structures

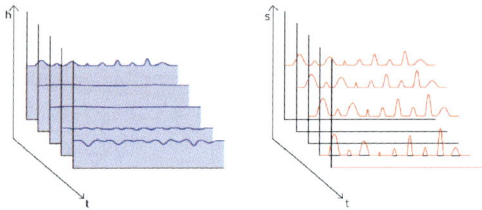

Fig. 5. The evolution of a coating with hardening, where micro structures persist

structure. This micro structures turns into a stable pattern of the dry coating. This is due to a solvent dependent viscosity, which leads to hardening during the drying process. Figure 4 shows that in a coating with constant viscosity the mean surface tension forces dominate the evolution at later times. This finally leads to a flat coating similar to the thin film case. Micro structures occur only at intermediate times.

3 Micro Structured Coatings and Effective Macroscopic Contact Angle

Micro structures in thin coatings are not only an unwanted feature, like the rupture of a coating. They also can be desirable, as micro structures enhance water repellent properties of a surface. This feature is known as the lotus effect. Among other plants, the lotus plant makes use of this [BN97], to let water roll off their leaves. One can also spot it at the back of a duck. The duck will stay dry while the water rolls off in pearls, as the feathers have a micro structure whose cavities are not filled with the water. To analyze this effect one has to understand how the form of the drops especially the contact angles are determined by the surface energy, which is the relevant energy in the quasi static case we are considering here.

The surface energy E is the sum of the energies of the three different interfaces in our problem. That is, the liquid/vapor interface Σ_{LV}, the solid/liquid interface Σ_{SL} and the solid/vapor interface Σ_{SV}. Each of these interfaces is weigthened with its surface tension:

$$E = |\Sigma_{SL}| \cdot \sigma_{sl} + |\Sigma_{LV}| \cdot \sigma_{lv} + |\Sigma_{SV}| \cdot \sigma_{sv}.$$

The shape of the drop is the one with the least energy given the volume of the drop. This also determines the contact angle, which is important to understand the lotus effect. Drops with large contact angles take a nearly pearl like form and roll of easily. Drops with small contact angles are flatter and stick more to the surface.

For a flat surface the contact angle θ^Y can be calculated using Young's law, which can be derived from minimizing property with respect to the surface energy (see below):

$$\cos \theta^Y = \frac{\sigma_{sv} - \sigma_{sl}}{\sigma_{lv}}. \qquad (27)$$

Drops on surfaces with micro structures are more complicated. They can either fill the micro structure with water, a situation described by Wenzel in [W36] (Fig. 6), or they can sit on air bubbles situated in the surface cavities, as considered by Cassie and Baxter in [CB44], see Fig. 7. For a nice review on this effect see either [Q02] or the book [GBQ04].

On a periodic surface it is possible to calculate effective contact angles. These are contact angles that would be attained in the limit of small scale periodicity. These contact angles determine the shape of the drop, see Figs. 6 and 7. The micro structure is much smaller than the size of the drop. It therefore makes sense to think of an effective surface tension of the micro

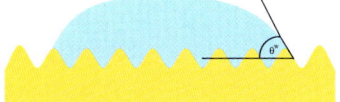

Fig. 6. A Wenzel type drop

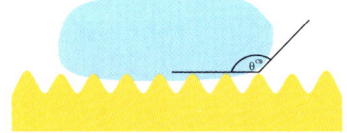

Fig. 7. A Cassie–Baxter type drop

structured surface. The justification for this is given in [AS05], where it is shown that the energy minimizing drops behave in the limit of small surface periodicity like the drops with the corresponding effective surface tensions. This is a mathematically rigorous argument using the Γ-convergence of the energies.

The effective surface tensions are the ones assigned to a macroscopically flat surface with a small scale micro structure. In the Wenzel situation the solid surface and thereby the solid/liquid interface as well as the solid/vapor interface are enlarged by the roughness r. (r equals the area of the surface on the unit square.) The effective surface tensions σ_{sl}^* and σ_{sv}^* are:

$$\sigma_{sl}^* = r \cdot \sigma_{sl} \quad \text{and} \quad \sigma_{sv}^* = r \cdot \sigma_{sv},$$

The effective contact angle θ^W is then determined by an adapted Young's law, cf. (27):

$$\cos \theta^W = \frac{\sigma_{sv}^* - \sigma_{sl}^*}{\sigma_{lv}} = r \cdot \frac{\sigma_{sv} - \sigma_{sl}}{\sigma_{lv}}.$$

Therefore a Wenzel type situation enlarges large contact angles and shrinks small ones in comparison to the flat surface case. Thus it enhances water repellent properties of a surface (with pearl like drops and large contact angles), as well as hydrophilic properties (with flat drops and low contact angles).

In the Cassie–Baxter situation the calculation of the effective surface tension is more difficult as it involves a determination of the size of the vapor bubbles at the micro scale, see Fig. 7. In a periodic set up this leads to a free boundary problem to be solved on the periodicity cell. The solution may be a configuration with or without vapor inclusions. At the triple line the contact angle for a flat surface θ^Y is attained. Below, we developed an algorithm which solves the free boundary problem and thereby determines the shape of the vapor inclusions.

The solution of the cell problem provides the area α of the liquid/vapor interface in one periodicity cell, the area β of the solid/liquid interface and the area of the solid/vapor interface, which is $r - \beta$, see Fig. 8. The effective

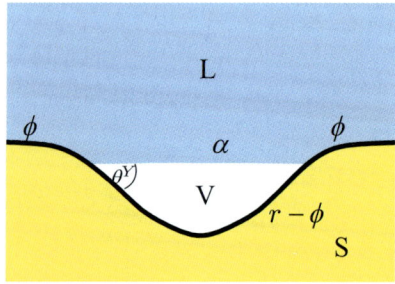

Fig. 8. The Configuration of a cell problem in the Cassie–Baxter regime

surface tension $\sigma^\star_{sl}$ is the sum of the surface tensions of the interfaces:
$$\sigma^\star_{sl} = \alpha \cdot \sigma_{lv} + \beta \cdot \sigma_{sl} + (r - \beta) \cdot \sigma_{sv}.$$
We obtain a modified Young's law (cf. (27)) for the effective solid/vapor surface tension $\sigma^\star_{sv} = r \cdot \sigma_{sv}$ and thereby determine the effective Cassie–Baxter contact angle:
$$\cos\theta^{CB} = \frac{\sigma^\star_{sv} - \sigma^\star_{sl}}{\sigma_{lv}} = -\alpha + \beta \cdot \cos\theta^Y.$$
For $\alpha \to 1$ and $\beta \to 0$ the Cassie–Baxter contact angle tends to 180°. This is the situation when the drop hardly touches the surface but rests mostly on the air pockets. The drop takes a nearly spherical shape and rolls off easily.

The effective contact angles calculated above are derived under the assumption of periodicity of the surface. An assumption typically not satisfied by natural surfaces. Theses surfaces show a highly inhomogeneous structure with both sizes and shape of the micro structure varying over several orders of magnitude, see Fig. 9.

A future perspective is to derive a mathematical model which captures these inhomogeneities. It should be based on a stochastical model where one asks for the expectation of the effective contact angle.

There is a second drawback of Young's law which describes the the absolut minimizer of the energy. In fact, drops on surface can have many different stable contact angles. Rain drops on a window sheet demonstrate this in our daily life. They stick to the window and do not roll off, in spite of the window being inclined. These drops are not spherical caps but take an non symmetric shape, see Fig. 10.

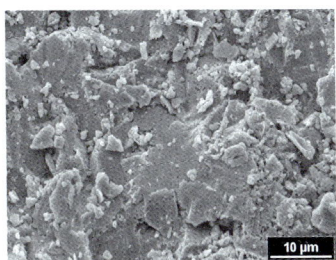
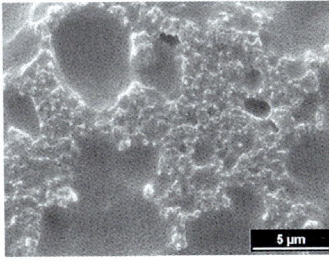

Fig. 9. Natural surfaces with micro structure (copyright: Bayer Material Science)

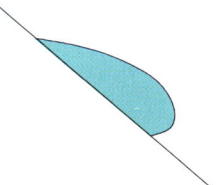

Fig. 10. A drop sticking to a tilted plane

The contact angles at the upward point part of the contact line are much smaller than those at the downward pointing part. Nevertheless all contact angles are stable, as the drop does not move. We developed a new model to understand which drops are stable, [SGO07], see Sect. 3.2. This model is adapted from models used in dry friction and elasto–plasticity. It mainly states that a drop should by stable, if the energy landscape is not to steep at its configuration.

3.1 Computing the Effective Contact Angle

In this section we will discuss how to compute the effective contact angle on a rough coating surface in the regime of the Cassie–Baxter model. Thus, we consider a periodic surface micro structure described by a graph on a rectangular fundamental cell Ω (cf. Fig. 11). The surface itself is supposed to be given as a graph $f : \Omega \to \mathbb{R}$, whereas the graph of a second function $u : \Omega \to \mathbb{R}$ represents the gas/liquid interface between a vapor inclusion on the surface and the covering liquid. In fact, we suppose $\{(x, y) \in \Omega \times \mathbb{R} \mid f(x) < y < u(x)\}$ to be the enclosed gas volume. Following [SGO07] we take into acount the total (linearized) surface energy on the cell Ω given by

$$E(u,f) = \int_{[u>f]} \sigma_{sv}\sqrt{1+|\nabla f|^2} + \sigma_{lv}\sqrt{1+|\nabla u|^2}\,dx + \int_{[u<f]} \sigma_{sl}\sqrt{1+|\nabla f|^2}\,dx$$

$$= \int_{[u>f]} (\sigma_{sv} - \sigma_{sl})\sqrt{1+|\nabla f|^2} + \sigma_{lv}\sqrt{1+|\nabla u|^2}\,dx$$

$$+ \int_{\Omega} \sigma_{sl}\sqrt{1+|\nabla f|^2}\,dx$$

Here, $[u > f] = \{x \in \Omega \mid f(x) < u(x)\}$ represents the non wetted domain of the vapor inclusion, also denoted by Ω_{sv}, and $[u < f] = \{x \in \Omega \mid f(x) > u(x)\}$ the wetted domain, respectively (cf. Figs. 7, 11). Let us emphasize that for fixed f the energy effectively depends only on $u|_{[u>f]}$. In the energy minimization we have to compensate for this by a suitable extension of u outside $[u > f]$. The variation of the energy E with respect to u in a direction w is given by

$$\partial_u E(u,f)(w) = \int_{\partial[u>f]} (v \cdot \nu)\left((\sigma_{sv} - \sigma_{sl})\sqrt{1+|\nabla f|^2} + \sigma_{lv}\sqrt{1+|\nabla u|^2}\right)d\mathcal{H}^1$$

$$+ \int_{[u>f]} \sigma_{lv}\frac{\nabla u \cdot \nabla w}{\sqrt{1+|\nabla u|^2}}\,dx,$$

where ν denotes the outer normal at the triple line $\partial[u > f]$ and v is the normal velocity field of this interface induced by the variation w of the height function

u. The relation between $v\cdot\nu$ and w is given by $(v\cdot\nu)(\nabla f\cdot\nu - \nabla u\cdot\nu) = w$. A minimizer u of $E(\cdot, f)$ describing the local vapor inclusion attached to the surface is described by the necessary condition $\partial_u E(u, f)(w) = 0$ for all smooth variations w. Applying integration by parts we deduce the minimal surface equation $-\mathrm{div}\frac{\nabla u}{\sqrt{1+|\nabla u|^2}} = 0$ for u on $[u > f]$ and the boundary condition

$$0 = \frac{(\sigma_{sv} - \sigma_{sl})\sqrt{1+|\nabla f|^2} + \sigma_{lv}\sqrt{1+|\nabla u|^2}}{\nabla f \cdot \nu - \nabla u \cdot \nu} + \frac{\sigma_{lv}\nabla u \cdot \nu}{\sqrt{1+|\nabla u|^2}}$$

on $\partial[u > f]$. The energy is invariant under rigid body motions. Hence, for a point x on $\partial[u > f]$ we may assume $\nabla f(x) = 0$. In this case $\nu(x) = -\frac{\nabla u(x)}{|\nabla u(x)|}$ and thus $\frac{\sigma_{ls} - \sigma_{sv}}{\sigma_{lv}} = \sqrt{1+|\nabla u(x)|^2} - \frac{|\nabla u(x)|^2}{\sqrt{1+|\nabla u(x)|^2}} = \frac{1}{\sqrt{1+|\nabla u(x)|^2}} = \cos(\theta)$, where θ is the contact angle between the solid–liquid and the liquid vapor interface. Hence, we have recovered Young's law on the micro scale of the cell problem.

Finally we end up with the following free boundary problem to be solved: Find a domain Ω_{sv} and a function u, such that the graph of u on Ω_{sv} is a minimal surface with Dirichlet boundary condition $u = f$ and prescribed

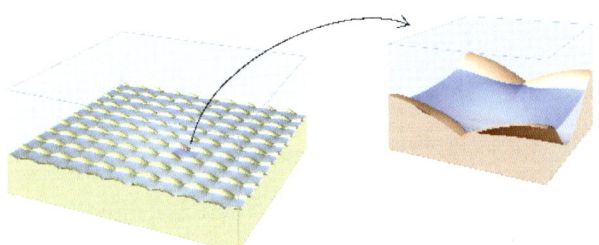

Fig. 11. The effective contact angle on a rough surface is calculated based on the numerical solution of a free boundary problem on a fundamental cell. The liquid vapor interface of the vapor inclusion on the surface forms a minimal surface with a contact angle on the surface of the solid determined by Young's law

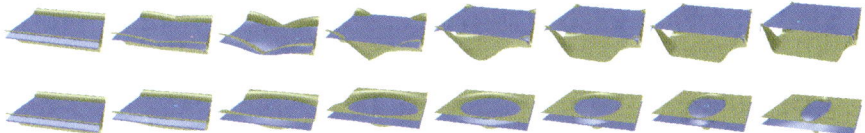

Fig. 12. Each row shows on the periodic cell a family of coating surfaces together with the liquid vapor interfaces of the corresponding vapor inclusions in the wetting regime of the Cassie–Baxter model. In the *first row* the transition in the surface configuration from a wavelike pattern in one axial direction to more spike type structures is depicted *from left to right*, whereas in the *second row* the transition from the same wave pattern to elongated troughs is shown

contact angle θ on $\partial\Omega_{sv}$, and this graph should be periodically extendable as a continuous graph on $\mathbb{R}^2$ (cf. Fig. 11 and Fig. 12).

The numerical solution of this free boundary problem is based on a time discrete gradient descent approach for a suitable spatially discrete version of the above variational problem. Let us denote by $\mathcal{V}_h$ the space of piecewise affine, continuous functions (with a continuous periodic extension on $\mathbb{R}^2$ on some underlying simplicial mesh of grid size h covering the rectangular fundamental cell Ω. For a discrete graph $F \in \mathcal{V}_h$ of the coating surface we start from some initial guess $U^0 \in \mathcal{V}_h$ for the (extended) discrete graph of the liquid vapor interface on top of the vapor inclusions and successively compute a family $(U^k)_{k\geq 0}$ with decreasing Energy $E(\cdot, F)$. For given U^k we first solve the discrete Dirichlet problem for a minimal surface on $\Omega_{sv}^k := [U^k > F]$ in a composite finite element space $\mathcal{V}_h^k$ [HS97, HS98] and based on that compute the next iterate U^{k+1}. In fact, following [HS97a] we define $\mathcal{V}_h^k$ as a suitable subspace of functions $W \in \mathcal{V}_h$ with $W = 0$ on $\partial\Omega_{sv}^k$. Thereby, the degrees of freedom are nodal values on the original grid contained in Ω_{sv}^k whose distance from $\partial\Omega_{sv}^k$ is larger than some $\epsilon = \epsilon(h) > 0$. Then, a constructive extension operation defines nodal values on all grid nodes of cells intersected by Ω_{sv}^k (for details we refer to [HS97a]). Hence, we compute a solution $\tilde{U}^{k+1}$ with $\tilde{U}^{k+1} - F \in \mathcal{V}_h^k$, such that

$$0 = \int_{\Omega_{sv}^k} \frac{\nabla \tilde{U}^{k+1} \cdot \nabla \Phi}{\sqrt{1 + |\nabla U^k|^2}} \mathrm{d}x$$

for all test functions $\Phi \in \mathcal{V}_h^k$. Next, based on $\tilde{U}^{k+1}$ data on $\partial\Omega_{sv}^k$ we compute a discrete descent direction $V^k \in \mathcal{V}_h$ as the solution of

$$G^k\left(V^{k+1}, \Phi\right) = -\partial_u E\left(\tilde{U}^k, F\right)(\Phi)$$

for all $\Phi \in \mathcal{V}_h$. Here, with the intention of a proper preconditioning of the gradient descent, we take into account the metric $G^k(\Psi, \Phi) = \sigma_{lv} \int_{\Omega_{sv}^k} \frac{\nabla \Psi \cdot \nabla \Phi}{\sqrt{1+|\nabla U^k|^2}}$. Given V^{k+1} we finally determine the actual descent step applying Amijo? step size control rule and compute $U^{k+1} = \tilde{U}^{k+1} + \tau^{k+1} V^{k+1}$ for a suitable timestep τ^{k+1}. Here, we implicitly assume that the built–in extension of $\tilde{U}^{k+1}$ on whole Ω is sufficiently smooth.

3.2 A New Model for Contact Angle Hysteresis

We consider a drop on a micro structured plane. Experiments show that there is an hysteresis interval $[\theta^r, \theta^a]$ of stable contact angles. It is bounded by the receding contact angle θ^r and the advancing contact angle θ^a. The dependence of this interval on the surface roughness is badly understood. We introduced a new model for contact angle hysteresis [SGO07] to understand the experimental evidence of a complicated dependence of the hysteresis interval on the roughness:

Well known experiments from the sixties [JD64] show that the width as well as the position of the hysteresis interval depend in a nonlinear way on the surface roughness, see Fig. 13.

Especially the receding contact angle shows a jump like behavior at a certain surface roughness.

Furthermore, recent experiments [QL03] show that the receding contact angle not only depends on the surface roughness, but also on the way the drop is put on the surface. In Fig. 14 we show how the receding contact angles depends on a pressure applied to press the drop into surface cavities.

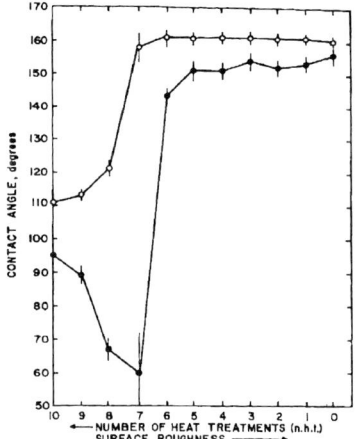

Figure 1. Water contact angles on TFE-methanol telomer wax surface as a function of roughness

○ Advancing angle
● Receding angle

Fig. 13. Experimental Dependence of Advancing and Receding Contact Angles on the Surface Roughness. Reprinted with Permission from [JD64]. Copyright (1964) American Chemical Society

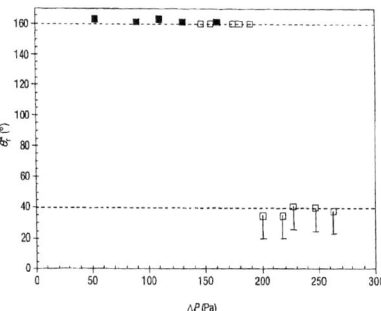

Fig. 14. Experimental Dependence of Receding Contact Angles on the Pressure Pushing the Drop onto the Surface. Reprinted from [QL03] with Permission

The pressure is then released and the contact angle is measured. Figure 14 again shows a jump like behavior of the receding contact angle.

We introduce a new model to capture these phenomena. It is similar to models used in dry friction [MT04] and elasto–plasticity [MSM06]. The main idea of our model is that stability of drops is primarily not related to global or local minimality of its interfacial energy, but rather to the fact that the local energy-landscape seen by the drop should not be too steep such that dissipation energy pays off the modify the configuration. To be be more precise, if the energy that would be gained moving the drop (i.e. controlled up to first order by the slope of the energy landscape) is smaller than the energy that would be dissipated while moving, then the drop will not move. In order to implement these concept, we use the derivative-free framework proposed in [MM05] (see also the review [M05]).

That is, we assume a drop L_0 (with its contact angle) to be stable if

$$E(L_0) - E(\tilde{L}) \leq dist(L_0, \tilde{L})$$

for all $\tilde{L}$ with the same volume. Here we have modeled the distance of two drops to be the area of the coating surface wetted by only one of them. This seems reasonable, as we know that the most energy is dissipated around the moving triple line. Therefore a drop which has significantly changed its bottom interface on the coating surface is far apart from its initial configuration.

Our new model implies two different diagrams of stable contact angles, depending on the type of drop (Wenzel or Cassie–Baxter type). These are shown in Figs. 15 resp. 16 in the case of a surface with flat plateau and vallees, separated by steep edges. The roughness of this type of surface can be increased by deepening the asperities without changing the size of the wetted surface plateau.

The hysteresis interval for Cassie–Baxter drops is much narrower than the one for Wenzel drops. This can explain qualitatively both the downward

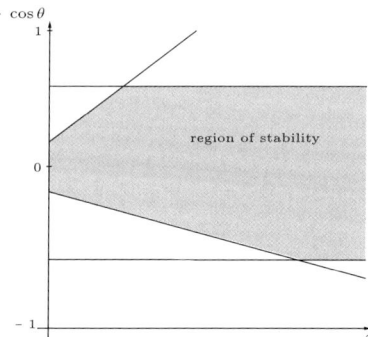

Fig. 15. Stable contact angles for Wenzel type drops

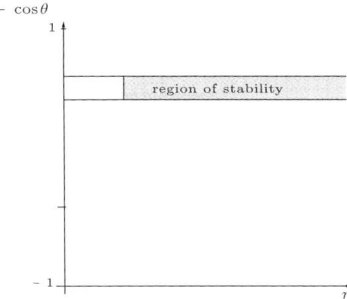

Fig. 16. Stable contact angles for Cassie–Baxter type drops

jump at large pressures of the receding contact angles in Fig. 14, and the jump behavior in Fig. 13.

The latter can be understood as a superposition of the two stability diagrams. The jump in the width of the hysteresis interval results from a transition from Wenzel type drops to Cassie–Baxter type drops. At low surface roughnesses Wenzel type drops are stable. They exhibit a wide hysteresis interval. At higher roughness, the stable configurations in the experiment are instead Cassie–Baxter. They display a much narrower hysteresis interval. The stable contact angles resulting from the transition from Wenzel to Cassie–Baxter drops are shown schematically in Fig. 17, where they are superposed on the experimental results of Johnson and Dettre. The comparison is only qualitative, because experimentally roughness is measured only indirectly, through the number of heat treatments undergone by the solid surface in the sample

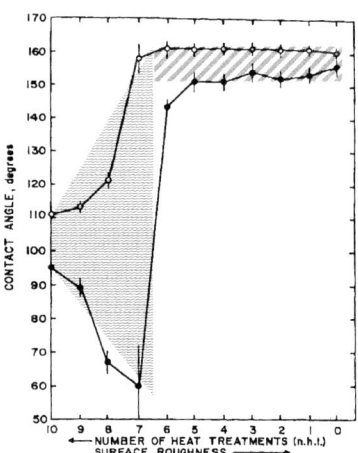

Fig. 17. A schematic sketch of the stable contact angles is given according to our model. The shaded regions represents the set of stable angles for varying surface roughness, superposed on experimental data from Fig. 13

preparation. The figure shows a transition from a regime in which the difference between advancing and receding contact angles increases monotonically with roughness, to one in which such a difference is smaller, and nonsensitive to roughness.

Figure 14 reflects the fact that the stability interval depends on the type of drop. Assuming that the corresponding surface has sufficiently large roughness, we see from Figs. 15 and 16 that forcing a transition from a Cassie–Baxter to a Wenzel type drop (by applying a large enough pressure) may reduce the lower end of the stability interval (i.e., the receding contact angle) from well above to well below 90°.

References

[AS05] Alberti, G., DeSimone, A.: Wetting of rough surfaces: a homogenization approach. Proc. R. Soc. London Ser. A Math. Phys. Eng. Sci., **461**, 79–97, (2005)

[BN97] Barthlott, W., Neinhuis, C.: Characterization and Distribution of Water-repellent, Self-cleaning Plant Surfaces. Annals of Botany, **79**, 667–677, (1997)

[BGLR02] Becker, J., Grün, G., Lenz, M., Rumpf, M.: Numerical Methods for Fourth Order Nonlinear Degenerate Diffusion Problems. Applications of Mathematics, **47**, 517–543, (2002)

[CB44] Cassie, A.B.D., Baxter, S.: Wettability of Porous Surfaces. Trans. Faraday Soc., **40**, 546–551, (1944)

[MSM06] Dal Maso, G., DeSimone, A., Mora, M.G.: Quasistatic evolution problems for linearly elastic-perfectly plastic materials. Arch. Rat. Mech. Anal., **180**, 237–291, (2006)

[GBQ04] de Gennes, P.G., Brochard-Wyart, F., Quéré, D.: Capillarity and Wetting Phenomena. Springer (2004)

[SGO07] DeSimone, A., Grunewald, N., Otto, F.: A new model for contact angle hysteresis. Networks and Heterogeneous Media, **2**, 2, 211–225, (2007)

[JD64] Dettre, R.H., Johnson, R.E.: Contact Angle Hysteresis II. Contact Angle Measurements on Rough Surfaces. Contact Angle, Wettability and Adhesion, Advances in Chemistry Series, **43**, 136–144, (1964)

[GO03] Giacomelli, L., Otto, F.: Rigorous lubrication approximation. Interfaces and Free boundaries, **5**, No. 4, 481–529, (2003)

[GP05] Günther, M., Prokert, G.: On a Hele–Shaw–type Domain Evolution with convected Surface Energy Density. SIAM J. Math. Anal., **37**, No. 2, 372–410, (2005)

[GLR02] Grün, G.,Lenz, M., Rumpf: A Finite Volume Scheme for Surfactant Driven Thin Film Flow. in: Finite volumes for complex applications III, M. Herbin, R., Kröner, D. (ed.), Hermes Penton Sciences, 2002, 567–574

[HS97] Hackbusch, W. and Sauter, S. A.: Composite Finite Elements for Problems Containing Small Geometric Details. Part II: Implementation and Numerical Results. Comput. Visual. Sci., **1** 15–25, (1997)

[HS97a] Hackbusch, W. and Sauter, S. A.: Composite Finite Elements for problems with complicated boundary. Part III: Essential boundary conditions. technical report, Universität Kiel, (1997)

[HS98]   Hackbusch, W. and Sauter, S. A.: A New Finite Element Approach for Problems Containing Small Geometric Details. Archivum Mathematicum, **34** 105-117, (1998)
[HMO97]   Howison, S.D., Moriarty, J.A., Ockendon, J.R., Terril, E.L., Wilson, S.K.: A mathematical model for drying paint layers. J. of Eng. Math., **32**, 377–394, (1997)
[QL03]   Lafuma, A., Quéré, D.: Superhydrophobic States. Nature Materials, **2**, 457–460, (2003)
[MM05]   Mainik, A., Mielke, M.: Existence results for energetic models for rate-independent systems. Calc.Var., **22**, 73–99, (2005)
[M05]   Mielke, M.: Evolution of rate-independent systems. Handbook of Differential Equations. Evolutionary Equations, **2**, 461–559, (2005)
[MT04]   Mielke, M., Theil, F.: On rate independent hysteresis models. NoDEA **11**, 151–189, (2004)
[BDO97]   Bankoff, S.G, Davis, S.H., Oron, A.: Long-scale evolution of thin liquid films. Rev. of modern Physics, **69**, No. 3, 931–980, (1997)
[O01]   Otto, F.: The geometry of dissipative evolution equations: the porous medium equation. Comm. Partial Differential Equations, **26**, No.1–2, 101–174, (2001)
[Q02]   Quéré, D.: Rough ideas on wetting. Physica A, **313**, 32–46, (2002)
[W36]   Wenzel, R.N.: Resistance of Solid Surfaces to Wetting by Water. Ind. Eng. Chem., **28**, 988–994, (1936)
[W93]   Wilson, S.K.: The levelling of paint films. Journal of Applied Mathematics, **50**, 149–166, (1993)

Part III

Biochemical Reactions and Transport

Modeling and Simulation of Hairy Root Growth

Peter Bastian[1], Jenny Bauer[4], Andrés Chavarría-Krauser[2], Christian Engwer[1], Willi Jäger[2], Sven Marnach[3], Mariya Ptashnyk[2], and Bernhard Wetterauer[4]

[1] IWR, Universität Heidelberg, INF 368, 69120 Heidelberg, Germany (now at[3])
{peter.bastian,christian.engwer}@ipvs.uni-stuttgart.de
[2] IAM, Universität Heidelberg, INF 294, 69120 Heidelberg, Germany
{andres.chavarria.krauser,jaeger,
mariya.ptashnyk}@iwr.uni-heidelberg.de
[3] IPVS, Universität Stuttgart, Universitätsstraße 38, 70569 Stuttgart, Germany
sven.marnach@ipvs.uni-stuttgart.de
[4] IPMB, Universität Heidelberg, INF 364, 69120 Heidelberg, Germany
jenny_bauerbiol@yahoo.de, bernhard.wetterauer@urz.uni-heidelberg.de

Summary. A multiscale approach is presented to model growth of hairy roots. On the macroscopic scale, a continuous model is derived, which includes growth and nutrient transport. Water transport is considered on the microscopic scale. A Discontinuous Galerkin scheme for complex geometries is used to compute the permeability of root bulks. This permeability constitutes the linkage between micro- and macroscopic scale. The models are applied then to describe shaker cultures of hairy roots and simulations are compared to measurements.

1 Introduction

Plants remain a major source of pharmaceuticals and biochemicals. Many of these commercially valuable phytochemicals are secondary metabolites that are not essential to plant growth. Hairy roots, obtained from plants through transformation by *Agrobacterium rhizogenes*, produce many of the same important secondary metabolites and can be grown in relatively cheap hormone-free medium. Thus they may provide an alternative to agricultural processes to produce phytochemicals on a large scale [9, 11]. Hairy roots can be cultivated under sterile conditions either in a bioreactor or in shake flasks. The fast growing hairy roots are unique in their genetic and biosynthetic stability and are able to regenerate whole viable plants for further subculturing [6]. The yield of secondary metabolites is determined by biomass accumulation and by the level of secondary metabolite produced per unit biomass. Therefore

Fig. 1. *Ophiorrhiza mungos* and hairy root of *O. mungos*

a number of biological studies have focused on the growth process, growth dynamic, and production of secondary metabolites in bioreactors of different design [17, 10, 9].

Hairy roots of *Ophiorrhiza mungos* Linn., the Chinese camptotheca tree, are currently gaining the interest of pharmacologists, since a secondary metabolite, *camptothecin*, can be used to treat cancer diseases [26]. Camptothecin is a modified monoterpene indole alcaloid produced by *Camptotheca acuminata*, *Nothapodytes foetida*, some species of the genus *Ophiorrhiza*, *Ervatamia heyneana*, and *Merrilliodendron megacarpum* [24, 27]. In order to produce camptothecin efficiently, it is necessary to optimize the biological processes behind its biosynthesis (either in bioreactors or shaker cultures). However, to achieve this, it is essential to understand metabolism, growth and transport processes of and in root networks.

The aim of the project was to derive a mathematical model which describes growth of root networks and nutrient transport through root tissues. To describe the biological system a multiscale approach was used. The processes on macroscopic and microscopic scale are linked. Numerical solutions were compared to experimental data obtained from *O. mungos* hairy roots grown as shaker cultures. The model and numerical algorithms are general enough to describe growth and transport processes in bioreactors.

2 Biological Processes

The processes observed in a bioreactor are water transport, diffusion and transport of nutrients in medium and roots, and growth of roots. These processes are taking place on different scales, each of which contributes to the global system.

On the macroscopic level roots form a dense bulk which resembles a porous medium. This allows to use well known modeling approaches to describe porous media. The root bulk is hence treated as a continuous medium of varying porosity, and all processes are defined on this continuum. Growth and nutrient transport are observed on the macroscopic scale and described through distributions. Growth is assumed to depend on nutrient concentration in the medium and inside the roots [20, 10, 21]. Three processes are responsible for changes in the mediums nutrient concentration: uptake on the root surface, convection due to pressure gradients and diffusion arising from concentration gradients. The macroscopic diffusion coefficient and the uptake kinetics depend on the density of the root network and are defined phenomenologically.

On the microscopic scale the root structure influences flow and transport processes around the root network, which has a complex highly ramified structure. The surface of a single root is covered with fine hairs, reducing conductivity [16]. Here it becomes clear that the microscopic structure determines substantially the macroscopic properties, in particular porosity and permeability.

Nutrient transport inside the roots is also a microscopic process. Since transport inside the root network is substantially faster in comparison to growth and branching, it is legitimate to consider only the average internal nutrient concentration and use a macroscopic internal nutrient concentration.

3 Macroscopic Model

Two densities are used to describe growth of hairy root networks: the root volume per unit volume ρ ($0 \leq \rho(\mathbf{x}, t) \leq 1$) and the cross section area of tips per unit volume n ($n(\mathbf{x}, t) \geq 0$). Growth can then be assumed to occur due to tip movement (elongation), tip formation (branching), and secondary thickening. Thus the change of density n is defined by a transport equation with growth velocity $\mathbf{v}$ and a branching term f. A similar approach has been used to model growth of fungi mycelia [7, 4]. The change of root density ρ is determined by the root volume produced due to tip movement. Secondary thickening is defined phenomenologically as a production term in the equation for ρ. Growth velocity and branching kinetics depend on the concentration of nutrients in the medium (denoted by $c(\mathbf{x}, t)$) and within the roots (denoted by $s(\mathbf{x}, t)$).

The transport of nutrients in the medium is defined by a convection-diffusion equation with a reaction term describing the active and passive nutrient uptake on the roots surface. Active uptake is assumed to be unidirectional (into the root network) and dependent only on the local medium nutrient concentration c. Passive uptake depends on the nutrient gradient between medium and roots, given by the difference $c - s$. Four processes which change the total internal nutrients $S = s V_r$, where $V_r(t)$ is the root volume, are considered here: uptake, growth, ramification and metabolism.

The precise formulation of the macroscopic model of hairy root growth reads

$$\partial_t n + \nabla \cdot (n\,\mathbf{v}) = f \qquad \text{in } (0,T) \times \Omega,$$
$$\partial_t \rho = n\,\|\mathbf{v}\| + q \qquad \text{in } (0,T) \times \Omega,$$
$$\partial_t \left((1-\rho)c\right) + \nabla \cdot (\mathbf{u}\,c) - \nabla \cdot (D_c(1-\rho)\nabla c) = -g \qquad \text{in } (0,T) \times \Omega, \qquad (1)$$
$$\tfrac{d}{dt} S = \int_\Omega g\,dx - \gamma_g \int_\Omega (n\,\|\mathbf{v}\| + q)\,dx - \gamma_r \int_\Omega f\,dx - \gamma_m S \qquad \text{in } (0,T),$$

with
$$\mathbf{v} = R\,s\,(\rho_{max} - \rho)\,(\nabla \mu + \alpha_\tau \boldsymbol{\tau}),$$
$$\nabla \mu = \alpha_c \nabla c - \alpha_\rho \nabla \rho - \alpha_n \nabla n,$$
$$q = \chi\,s\,\rho\,(\rho_{max} - \rho),$$
$$f = \beta\,c\,s\,\rho\,(\rho_{max} - \rho),$$
$$g = \frac{2\lambda n}{r}\,\rho\,(K_m c + P(c - s)),$$

where R is a growth rate, ρ_{max} is the maximal root density, χ is a secondary thickening rate, β is a branching rate, λ is the characteristic length of the uptake-active tissue around a tip, K_m is a constant describing active uptake rate, P is a permeability characterizing passive uptake, $\mathbf{u}$ is the flow velocity of the medium, D_c is a diffusion constant, and γ_g, γ_r and γ_m are constants describing the proportion of metabolites used for growth, ramification and metabolism, respectively. Since hairy roots are agravitropic [14] growth velocity can be assumed to be independent of gravity. Growth can then be presumed to occur along nutrient gradients and away from dense tissue. Pure densification of the root system is modeled by the local rotation $\boldsymbol{\tau}$, which is a unit vector orthogonal to $\nabla \mu$ and ∇n. It does not affect the density distribution of tips, although mass is still produced and ρ changes. Here α_c, α_ρ, α_n, and α_τ are phenomenological constants, which relate the growth driving gradients to the resulting growth velocity.

Initial density distributions and nutrient concentrations are prescribed. For both the bioreactor and the shake flasks the side walls of the reactor vessel (Γ_{sw}) are impermeable to the medium. In the bioreactor we have inflow (Γ_{in}) and outflow (Γ_{out}) boundaries. On Γ_{in} the nutrient concentration is given and Dirichlet boundary condition can be posed. On Γ_{out} we have outflow boundary condition. In the case of the shaker cultures no-flux boundary condition can be posed (i.e. $\partial \Omega = \Gamma_{\text{sw}}$ and $\Gamma_{\text{in}} = \Gamma_{\text{out}} = \emptyset$). Since roots cannot extend beyond the vessel, the tip density fulfills also the no-flux boundary conditions.

$$n(0,\mathbf{x}) = n_0(\mathbf{x}),\ \rho(0,\mathbf{x}) = \rho_0(\mathbf{x}) \qquad \text{in } \Omega,$$
$$c(0,\mathbf{x}) = c_0,\ S(0) = S_0, \qquad \text{in } \Omega,$$
$$n\,\mathbf{v} \cdot \boldsymbol{\nu} = 0 \qquad \text{on } (0,T) \times \partial\Omega,$$

$$\begin{aligned}
c &= c_D & \text{on} \quad &(0,T) \times \Gamma_{in}, \\
D_c(1-\rho)\nabla c - \mathbf{u}\,c) \cdot \boldsymbol{\nu} &= 0 & \text{on} \quad &(0,T) \times \Gamma_{out}, \\
\nabla c \cdot \boldsymbol{\nu} &= 0 & \text{on} \quad &\Gamma_{sw}.
\end{aligned}$$

The water velocity $\mathbf{u}$ is defined by Darcy's law. We distinguish between Dirichlet condition given by pressure values on the boundary Γ_D and Neuman conditions defined by flux through the boundary Γ_N. Depending on the experimental setup, either Neumann or Dirichlet conditions are posed on Γ_{in} and Γ_{out}. For a shake flask and on the sidewalls (Γ_{sw}) of a bioreactor, no-flux (i.e. homogeneous Neumann) conditions need to be posed.

$$\begin{aligned}
\nabla \cdot \mathbf{u} &= 0 & \text{in} \quad &\Omega, \\
\mathbf{u} &= -K\nabla p & \text{in} \quad &\Omega, \\
p &= p_0 & \text{on} \quad &\Gamma_D \subset \partial\Omega, \\
\mathbf{u} \cdot \boldsymbol{\nu} &= j & \text{on} \quad &\Gamma_N = \partial\Omega \setminus \Gamma_D.
\end{aligned} \qquad (2)$$

Here K is the permeability function of the root network. On the macroscopic scale K changes with density, but derivation of the relation is cumbersome if not impossible for general geometries. It must be computed hence for a certain structure on the microscopic scale.

The effective permeability K is assumed to be of the form

$$K = K_0 \cdot K_{\text{rel}}, \qquad (3)$$

where K_0 is the average coefficient relating the flow velocity $\mathbf{u}$ to the pressure gradient ∇p in an empty reactor ($\rho = 0$). $K_{\text{rel}}(\rho)$ a dimensionless relative permeability which reflects the local root structure. K_0 can be obtained by determining the Hagen–Poiseuille flow in the reactor, while K_{rel} is computed using simulations of the microscopic model.

4 Microscopic Model

On the microscopic scale we consider water flow between single root branches. To simplify the problem, we assume an incompressible potential flow:

$$\begin{aligned}
\nabla \cdot \mathbf{u} &= 0 & \text{in} \quad &\Omega, \\
\mathbf{u} &= -\nabla p & \text{in} \quad &\Omega, \\
p &= p_0 & \text{on} \quad &\Gamma_D \subset \partial\Omega, \\
\nabla p \cdot \boldsymbol{\nu} &= j & \text{on} \quad &\Gamma_N = \partial\Omega \setminus \Gamma_D.
\end{aligned} \qquad (4)$$

The domain Ω has a complex geometry, given by the root structure. Water uptake by the roots (growth) is small compared to the water flow, therefore Neumann boundary conditions with $j = 0$ can be assumed on the root surfaces.

4.1 Numerical Methods for Microscale Simulations

The root structure on the microscopic scale exhibits a complex shape. Classical numerical methods require a grid resolving such complex geometries. Creating these grids is a very sophisticated process and generates a high number of unknowns. We developed a discretization scheme for complex geometries, based on a Discontinuous Galerkin (DG) discretization on a structured grid and a structured grid for the construction of trial and test functions [8]. This method offers a discretization where the number of unknowns is not directly determined by the possibly very complicated geometrical shapes, but still allows the provision of fine structures, even if their size is significantly smaller than the grid cell size.

Let $\Omega \subseteq \mathbb{R}^d$ be a domain. On a sub-domain $\Omega^* \subseteq \Omega$ we want to solve Eqn. (4). The shape of Ω^* is usually based on geometrical properties retrieved from experiments, like micro-CT images, or from computations. $\mathcal{T}(\Omega) = \{E_0, \ldots, E_{M-1}\}$ is a partitioning, where the mesh size $h = \min\{\operatorname{diam}(E_i) \mid E_i \in \mathcal{T}\}$ is not directly determined by the geometrical properties. Nevertheless error control on solution of the partial differential equation might require a smaller h due to the shape of $\partial \Omega$. For Ω^* a triangulation based on $\mathcal{T}(\Omega)$ is defined $\mathcal{T}(\Omega^*) = \{E_n^* \mid E_n^* = \Omega^* \cap E_n \wedge E_n^* \neq \emptyset\}$, see Fig. 2. As E_n^* is always a subset of E_n we will call E_n fundamental element of E_n^*. The internal skeleton Γ_{int} and external skeleton Γ_{ext} of the partitioning are denoted by

$$\Gamma_{\text{int}} = \left\{\gamma_{e,f} = \partial E_e^* \cap \partial E_f^* \mid E_e^*, E_f^* \subset \Omega^* \text{ and } E_e^* \neq E_f^* \text{ and } |\gamma_{e,f}| > 0\right\},$$
$$\Gamma_{\text{ext}} = \left\{\gamma_e = \partial E_e^* \cap \partial \Omega^* \mid E_e^* \subset \Omega^* \text{ and } |\gamma_{e,f}| > 0\right\}.$$

In the finite element mesh $\mathcal{T}(\Omega^*)$ each element E_n^* can be shaped arbitrarily. Using DG, unlike conforming methods, the shape functions can be chosen in-

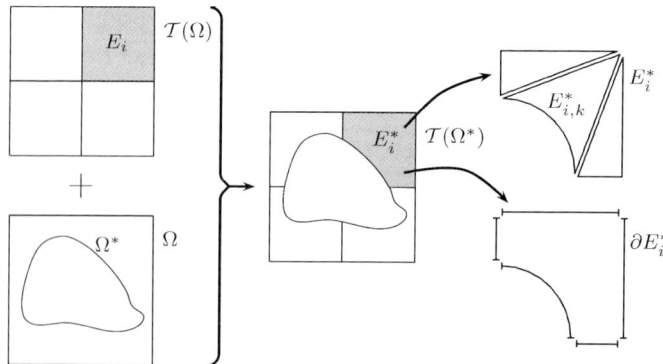

Fig. 2. Construction of the partitions $\mathcal{T}(\Omega^*)$ from the partitions $\mathcal{G}$ and $\mathcal{T}$ of the domain Ω and of E^* from its fundamental element and Ω^*. The local triangulation of E_i^* and ∂E_i^*

dependently from the shape of the element. Note that certain DG formulations are especially attractive, because they are element wise mass conservative and therefore able to accurately describe fluxes over element boundaries. We use a DG formulation with local base functions $\varphi^*_{n,j}$ being polynomial functions $\varphi_{n,j} \in P_k$ defined on the fundamental element E_n, with their support restricted to E^*_n:

$$\varphi^*_{n,j} = \begin{cases} \varphi_{n,j} & \text{inside of } E^*_n \\ 0 & \text{outside of } E^*_n \end{cases}, \qquad (5)$$

$P_k = \{\varphi : \mathbb{R}^d \to \mathbb{R} \mid \varphi(x) = \sum_{|\alpha| \le k} c_\alpha x^\alpha\}$ is the space of polynomial functions of degree k and α a multi–index. The resulting finite element space is defined by

$$V^*_k = \{v \in L_2(\Omega^*) \mid v|_{E^*_n} \in P_k\} \qquad (6)$$

and is discontinuous on the internal skeleton Γ_{int}. With each $\gamma_{e,f} \in \Gamma_{\text{int}}$ we associate a unit normal n. The orientation can be chosen arbitrarily, in this implementation we have chosen n oriented outwards E^*_e for $e > f$ and inwards otherwise. With every $\gamma_e \in \Gamma_{\text{ext}}$ we associate n oriented outwards Ω^*. The jump $[.]$ and the average $\langle . \rangle$ of a function $v \in V^*_k$ at $x \in \gamma \in \Gamma_{\text{int}}$ are defined as

$$[v] = v|_{E^*_e} - v|_{E^*_f} \qquad \text{and} \qquad \langle v \rangle = \frac{1}{2}\left(v|_{E^*_e} + v|_{E^*_f}\right).$$

Assembling the local stiffness matrix in DG requires integration over the volume of each element E^*_n and its surface ∂E^*_n. Integration is done using a local triangulation of E^*_n. E^*_n is subdivided into a disjoint set $\{E^*_{n,k}\}$ of simple geometric objects, i.e. simplices and hypercubes, with $\bar{E}^*_n = \bigcup_k \bar{E}^*_{n,k}$, see Fig. 2.

The integral over a function f on E^*_n can then be evaluated as

$$\int_{E^*_n} f(x)\, dx = \sum_k \int_{E^*_{n,k}} f(x)\, dx,$$

where $\int_{E^*_{n,k}} f\, dx$ is evaluated using standard quadrature rules.

4.2 Numerical Estimation of Macroscopic Parameters from Microscopic Simulations

Following the approach described in the previous subsection and applying this method to the microscopic problem in (4), the relative permeability, as introduced in (3), can be computed from direct simulation of flow through a root bulk:

$$K_{\text{rel}} = \left(\int_{\Gamma_{\text{in}}} \mathbf{u}\, dx\right) \cdot \left(\int_{\Gamma_{\text{in}}} \frac{p_2 - p_1}{h} dx\right)^{-1}, \qquad (7)$$

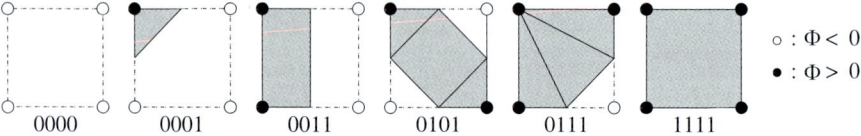

Fig. 3. The Marching Cube algorithm in $\mathbb{R}^2$ distinguishes six basic cases, depending on the value of a function Φ in the corners. The pictures show these six different cases, together with their key in the look-up table

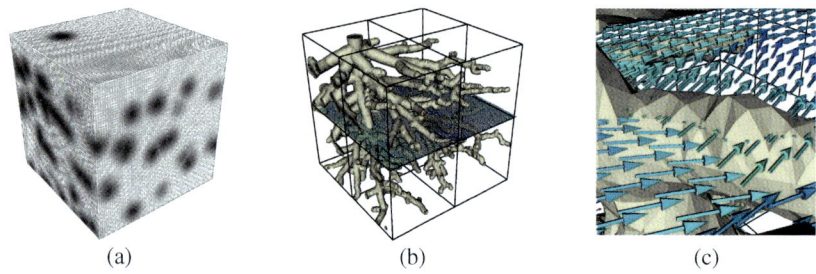

Fig. 4. (**a**) shows the scalar function describing the geometry, the marching simplex algorithm yields the geometry visible in (**b**). Pressure and velocity are computed on the given domain, using direct simulations with a Discontinuous Galerkin scheme. The resulting velocity can be seen in (**b**). (**c**) shows a closeup

where $\Gamma_{in} \subset \Gamma_D$ describes the inflow boundary. Dirichlet boundary conditions are posed both on the inflow and the outflow boundary.

The domain Ω^* is implicitly given by a scalar function Φ. This scalar function will usually be obtained through post processing of image data, i.e. from CT images. In these calculations we use artificially generated structures (Fig. 4.2a), based on structural parameters (using the PlantVR software [5]). The sub-domain boundary $\partial\Omega^*$ is given as an iso-surface $\Phi = 0$.

The local triangulation is based on the *Marching Cube/Simplex* Algorithm [13]. These algorithms give a surface reconstruction for an iso-surface. Each vertex of an element can be below or above the value of the iso-surface, read inside or outside the sub-domain. For a cube element in $\mathbb{R}^2$ this gives 16 different cases. Each of these cases corresponds to one of six basic cases and can be transformed using simple geometric operations (see Fig. 3). A look-up table maps each case to the appropriate surface reconstruction. The key for the look-up table is given by assigning the state of each corner (inside $\rightarrow$ 1, outside $\rightarrow$ 0) to one bit of an integer. The look-up table was extended to provide a surface and a volume reconstruction.

Using the formulation described in [18, 19], (4) reads: Find $p \in V_k^*$ such that
$$a_\epsilon(p,v) + J_{\sigma\beta}(p,v) = l_{\epsilon\sigma\beta}(v) \qquad \forall v \in V_k^*.$$

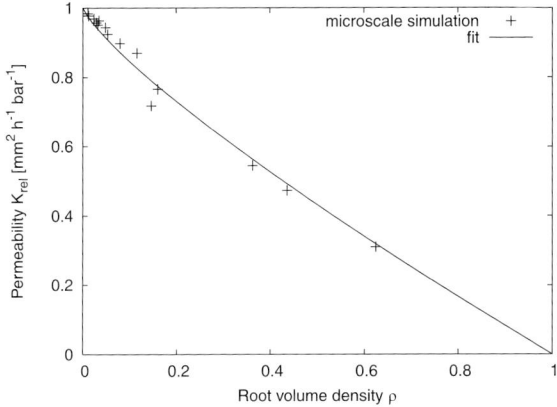

Fig. 5. Estimation of macroscale parameters from microscale simulations. Direct simulation of (4) for different root bulks with different densities yields correlation of K_{rel} and ρ. Model assumed: $K_{rel} = 1 - \rho^b$, where $b = 0.82 \pm 0.02$

The bilinear form

$$a_\epsilon(p,v) = \sum_{E_e^* \in \mathcal{T}^*} \int_{E_e^*} (K\nabla p) \cdot \nabla v \, dV$$

$$+ \sum_{\gamma_{ef} \in \Gamma_{int}} \int_{\gamma_{ef}} \epsilon \langle (K\nabla v) \cdot n \rangle [p] - \langle (K\nabla p) \cdot n \rangle [v] \, ds$$

$$+ \sum_{\gamma_e \in \Gamma_D} \int_{\gamma_e} \epsilon (K\nabla v) \cdot n \, p - (K\nabla p) \cdot n \, v \, ds$$

is parametrized by $\epsilon = \pm 1$. Choosing $\epsilon = 1$ we get a non–symmetric scheme introduced by Oden, Babuška and Baumann in [15]. For $\epsilon = -1$ we obtain the Symmetric Interior Penalty method which needs an additional stabilization term added to the bilinear form:

$$J_{\sigma\beta}(p,v) = \sum_{\gamma_{ef} \in \Gamma_{int}} \frac{\sigma}{|\gamma_{ef}|^\beta} \int_{\gamma_{ef}} [p][v] ds + \sum_{\gamma_e \in \Gamma_D} \frac{\sigma}{|\gamma_e|^\beta} \int_{\gamma_e} pv \, ds$$

with $\sigma > 0$ and β, where β depends on the dimension of Ω ($\beta = 1$ when dim = 2). Choosing $\epsilon = 1$ and $\sigma > 0$ results in the Non–Symmetric Interior Penalty method.

The right hand side is a linear form

$$l_{\epsilon\sigma\beta}(v) = \sum_{E_e^* \in \mathcal{T}^*} \int_{E_e^*} f \, v \, dV + \sum_{\gamma_e \in \Gamma_N} \int_{\gamma_e} J \, v \, ds$$

$$+ \sum_{\gamma_e \in \Gamma_D} \int_{\gamma_e} \epsilon (K\nabla v) \cdot n \, g \, ds + \sum_{\gamma_e \in \Gamma_D} \frac{\sigma}{|\gamma_{ef}|^\beta} \int_{\gamma_e} v \, g \, ds.$$

Direct simulation of Eq. (4) for different realizations of root bulks, yields K_{rel} and ρ. K_{rel} is assumed to be of the form $K_{\text{rel}}(\rho) = 1 - \rho^b$ and b is determined using a mean square fit.

Fig. 5 shows the dependence of K_{rel} on ρ and the fitted function $K_{\text{rel}}(\rho)$, where $b = 0.82 \pm 0.02$.

5 Application of the Macroscopic Model

There are two common ways of cultivating hairy roots, either in shake flasks or in bioreactors. Shaker cultures are used more often because of their simple assembly and usage in experiments and biological research. However, experiments in shaker cultures do not provide information about the spatial structure and distribution of roots. Bioreactors have rather industrial applications and are more complex to operate and to use as experimental set-ups. In the work presented here both cultivation methods were considered. In fact, equations (1) are able to describe both situations, as these differ only slightly in the method used to guarantee nutrient supply. While the principle of shake flask is based on permanent shaking of medium and culture, bioreactors use medium fluxes to ensure nutrient and oxygen supply.

5.1 Simulation of Shaker Cultures

For numerical simulation Eqs. (1) can be simplified to reduce the amount of free parameters. Uptake of nutrients can be considered to be purely of active nature, neglecting the passive transport ($P = 0$). Moreover, the energy cost for branching of new tips can be neglected ($\gamma_r = 0$). Since the root branches are very thin and variation in radius is small, root thickening can be neglected as well ($\chi = 0$). The main purpose of shaking is to supply oxygen and to ensure a homogeneous distribution of nutrients. This means that transport in the medium is non-limiting to uptake and growth. In the simulation this homogeneous distribution can be achieved via large diffusion. Active water transport is neglected.

A personal computer was used to simulate the macroscopic model (1), using a implementation of the numerical schemes based on the *DUNE* framework [3, 1]. For spatial discretization of the first and third equation in (1) a cell centered finite volume scheme on a structured grid was used [12]. The diffusive and convective/reactive part of the third equation in (1) were decoupled for discretization in time (second order operator splitting [23]). To prevent both instabilities in the transport term and effects from strong numerical diffusion, the convection equation was solved using an explicit second order Godunov upwind scheme with a minmod slope limiter [25, 12]. The diffusive part of the equation was solved implicitly. The ordinary differential equations for the

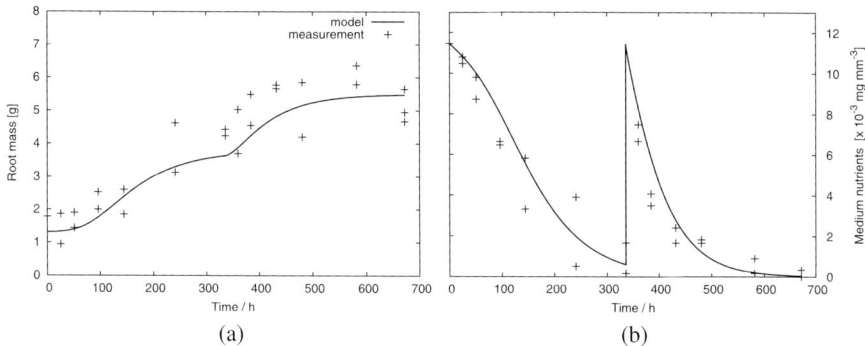

Fig. 6a,b. Comparison of simulation and experimental data from hairy roots grown as shaker cultures. The evolution in time of root mass (**a**) and concentration of sucrose in the medium (**b**) are compared to measurements. At 336 hours cultures were transferred into fresh medium

root density and the inner nutrient concentration S [second and fourth Eq. in (1)] were solved with Euler's method [22].

The parameters in the model are chosen such that the numerical results fit experimental data obtained from *O. mungos* hairy roots (Fig. 6). Gradients of nutrients and tip density, with moderate tissue compaction (local rotation), were chosen here as the driving force of growth. Very good agreement is found between measurements and simulation (mass increase: $R^2 = 0.85$, nutrient uptake: $R^2 = 0.93$). The model delivers spatial information on growth patterns as well. A simulation of a two dimensional flask is found in Fig. 7. The distributions are assumed to be constant in one of the the three dimensions. Simulation in three dimensions could however be easily implemented. Measurements deliver at the moment only data describing the kinetics of mass increase and nutrient uptake (compare Fig. 6), which do not include quantitative information on spatial patterns. Therefore, verification of the patterns in Fig. 7 is at the moment not possible. It is hence not clear which growth process dominates in the growth force $\nabla \mu + \alpha_\tau \, \tau$. Do hairy roots follow rather nutrient gradients than space gradients, or is diffusion of root tips more important? Or is mass increase a consequence of tissue compaction (local rotation)? It will probably be a mixture of all and other processes not accounted for. A detailed discussion regarding this issue can be found in Bastian et al. (2007b).

5.2 Hairy Roots Bioreactor

Root growth in bioreactors can be described by the full macroscopic model (1). Active water transport has to be considered. The flow velocity **u** can be calculated using Darcy's-Law (2), which requires the permeability $K(\rho)$. The relation between K and the root volume density ρ is determined by the microscopic model (compare Fig. 5). The model and numerical algorithms are

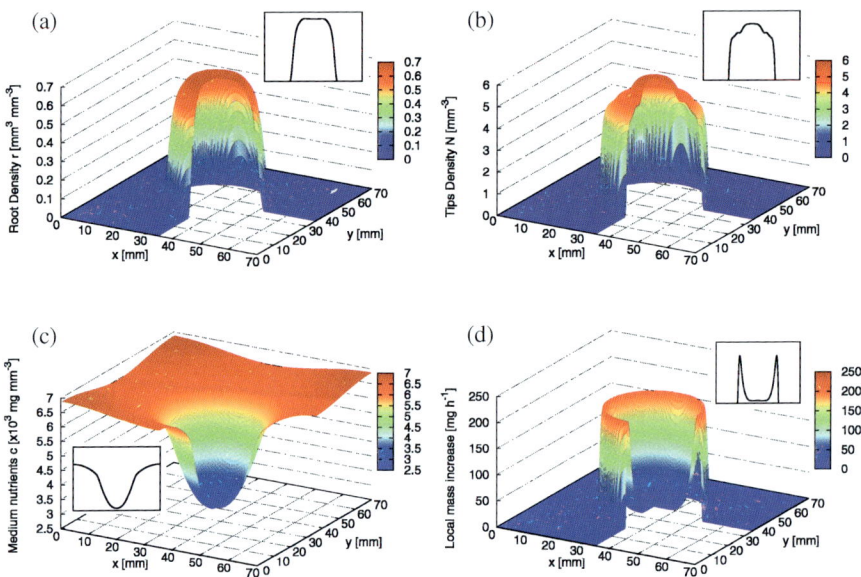

Fig. 7a–d. Simulated two dimensional spatial growth patterns of hairy roots grown as shaker cultures. (**a**) root volume density, (**b**) root tip density, (**c**) nutrient concentration in medium, and (**d**) local mass increase (all after 380 h of growth)

elaborated sufficiently to calculate and optimize the transport and growth processes in bioreactors. For meaningful simulation, parameters related to the three dimensional structure are required for (1). However these are not available from shaker culture experiments.

In order to study three dimensional growth patterns, a small experimental bioreactor (*WEWA I*) for root cultures was constructed in the group of Prof. Wink (IPMB, Universität Heidelberg) to deliver the information needed. "WEWA I"s construction is based on the *Low Cost Mist Bioreactor* (LCMB) of the company ROOTec GmbH (Heidelberg, Germany; Patentnumber: US 2003/0129743A1 / EP 1268743B1). The bioreactor system consists of a 3 l reactor reservoir (Fig. 8), a medium reservoir, a gas reservoir, and a mist chamber for the distribution of medium over the fixed root bed. Every 15 min the root inoculum is sprayed with 20 ml medium. Due to spraying, a pressure difference of 0.3 bar arises, which is used to return the surplus medium into the medium reservoir. The pressure in the reactor is controlled by a pneumatic relief valve (0.5 bar opening pressure). Instead of the B5 medium (1% sucrose; ROOTec GmbH, Heidelberg, Germany) used in shaker cultures, a modified version with 0.5% sucrose is used to obtain an optimal medium osmolarity. In order to prevent contaminations in the reactors a bacteriostatic (0.5 g/l Claforan) and a fungicide (40 mg/l Nystain) were added to the medium.

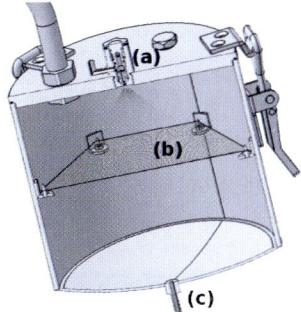

Fig. 8a–c. Scheme of the *WEWA I* mist chamber: injection valve of medium (**a**), mesh on which cultures are fixed (**b**), drain for surplus medium (**c**)

"WEWA I" is being put into operation at the moment. Several test runs have been done so far. However, these test measurements have not supplied reliable enough experimental data to identify the model parameters. Therefore meaningful simulations of the processes in bioreactors were not possible until now.

6 Conclusion

The work presented here has a general interest for modeling and simulation of complex growing networks like root systems and growth processes in bioreactors. A macroscopic model describing water and nutrient transport through a growing root network was derived. The growth of roots in a water solution and the dense growth habit of hairy roots give the possibility to define growth as a change of tissue volume density. This allows further to expand a one dimensional growth model, namely pure elongation of single roots, to a continuous three dimensional model, which delivers information on spatial growth patterns. Linkage of the microscopic and the macroscopic scale is accomplished by the elaborate and novel numerical algorithms developed for solution of elliptic equations on complex shaped domains. Comparison of numerical solutions and measurements of hairy roots shaker cultures, showed that the model is able to describe very well the kinetics of growth and nutrient uptake. Moreover, model and numerical algorithms are general enough to describe growth in bioreactors. Optimization of camptothecin production is still an open task due to the lack of experimental data. The model presented here is a good step forward towards achieving this and represents a sound base for further research. The numerical methods developed for microscale simulations are of interest for a wide range of applications, ranging from pore scale problems for water resources to cell biology. In future work these methods will be extended to time-dependent problems.

References

[1] P. Bastian, M. Blatt, A. Dedner, C. Engwer, R. Klöfkorn, R. Kornhuber, M. Ohlberger, and O. Sander. A generic grid interface for parallel and adaptive scientific computing. part ii: Implementation and tests in dune. Preprint, 2007. Submitted to *Computing*.

[2] P. Bastian, A. Chavarría-Krauser, C. Engwer, W. Jäger, S. Marnach, and M. Ptashnyk. Modelling *in vitro growth* of dense root networks. In preparation, 2007.

[3] P. Bastian, M. Droske, C. Engwer, R. Klöfkorn, T. Neubauer, M. Ohlberger, and M. Rumpf. Towards a unified framework for scientific computing. In R. Kornhuber, R.H.W. Hoppe, D.E. Keyes, J. Périaux, O. Pironneau, and J. Xu, editors, *Domain Decomposition Methods in Science and Engineering*, number 40 in Lecture Notes in Computational Science and Engineering, pages 167–174. Springer-Verlag, 2004.

[4] G. P. Boswell, H. Jacobs, F. A. Davidson, G. M. Gadd, and K. Ritz. Growth and function of fungal mycelia in heterogeneous environments. *Bulletin of Mathematical Biology*, 65:447–477, 2003.

[5] S. Chuai-Aree, S. Siripant, W. Jäger, and H.G. Bock. PlantVR: Software for Simulation and Visualization of Plant Growth Model. In *Proceedings of PMA06: The Second International Symposium on Plant growth Modeling, Simulation, Visualization and Applications*. Beijing, 2006.

[6] P. M. Doran. *Hairy Roots: Culture and Applications*. Harwood Academic Publishers, Amsterdam, 1997.

[7] L. Edelstein-Keshet. *Mathematical models in biology*. The Random House/ Birkhäuser Mathematics Series, New York, 1988.

[8] C. Engwer and P. Bastian. A Discontinuous Galerkin Method for Simulations in Complex Domains. Technical Report 5707, IWR , Universität Heidelberg, http://www.ub.uni-heidelberg.de/archiv/5707/, 2005.

[9] Y. J. Kim, P. J. Weathers, and B. E. Wyslouzil. Growth of *Artemisia annua* hairy roots in liquid and gas-phase reactors. *Biotechnology and Bioengineering*, 80:454–464, 2002.

[10] Y. J. Kim, P. J. Weathers, and B. E. Wyslouzil. Growth dynamics of *Artemisia annua* hairy roots in three culture systems. *Journal of Theoretical Biology*, 83:428–443, 2003.

[11] Y. J. Kim, B. E. Wyslouzil, and P. J. Weathers. Invited review: Secondary metabolism of hairy root cultures in bioreactors. *In Vitro Cell. Dev. Biol. Plant*, 38:1–10, 2002.

[12] Randall J. LeVeque. *Finite Volume Methods for Hyperbolic Problems*. Cambridge University Press, 2002.

[13] William E. Lorensen and Harvey E. Cline. Marching cubes: A high resolution 3d surface construction algorithm. In *SIGGRAPH '87: Proceedings of the 14th annual conference on Computer graphics and interactive techniques*, pages 163–169, New York, NY, USA, 1987. ACM Press.

[14] E. Odegaard, K. M. Nielsen, T. Beisvag, K. Evjen, A. Johnsson, O. Rasmussen, and T. H. Iversen. Agravitropic behaviour of roots of rapeseed (*Brassica napus* L.) transformed by *Agrobacterium rhizogenes*. *J Gravit Physiol.*, 4(3):5–14, 1997.

[15] J. T. Oden, I. Babuška, and C. E. Baumann. A discontinuous hp-finite element method for diffusion problems. *Journal of Computational Physics*, 146:491–519, 1998.
[16] M. Ptashnyk. Derivation of a macroscopic model for nutrient uptake by a single branch of hairy-roots. University Heidelberg, Preprint, 2007.
[17] D. Ramakrishnan and R. W. Curtis. Trickle-bed root culture bioreactor design and scale-up: Growth, fluid-dynamics, and oxygen mass transfer. *Biotechnology and Bioengineering*, 88(2):248–260, 2004.
[18] B. Rivière and M.F. Wheeler. Discontinuous galerkin methods for flow and transport problems in porous media. *Communications in Numerical Methods in Engineering*, 18:63–68, 2002.
[19] B. Rivière and M.F. Wheeler. A posteriori error estimates and mesh adaptation strategy for discontinuous galerkin methods applied to diffusion problems. *Computers & Mathematics with Applications*, 46(1):141–163, 2003.
[20] D. Robinson. Tansley review no. 73. The responses of plants to non-uniform supplies of nutrients. *The New Phytologist*, 127:635–674, 1994.
[21] S. R. Schnapp, W. R. Curtis, R. A. Bressan, and P. M. Hasegawa. Growth yields and maintenance coefficients of unadapted and NaCI-adapted tobacco cells grown in semicontinuous culture. *Plant Physiol*, 96:1289–1293, 1991.
[22] J. Stoer and R. Burlisch. *Numerische Mathematik 2*. Springer, 4 edition, 2000.
[23] Gilbert Strang. On the construction and comparison of difference schemes. *SIAM J. Numer. Anal.*, 5:506–517, 1968.
[24] H. Sudo, T. Yamakawa, M. Yamazaki, N. Aimi, and K. Saito. Bioreactor production of camptothecin by hairy root cultures of *Ophiorrhiza pumila*. *Biotechnology Letters*, 24:359–363, 2002.
[25] P. K. Sweby. High resolution schemes using flux-limiters for hyperbolic conservation laws. *SIAM J. Num. Anal.*, 21:995–1011, 1984.
[26] C. H. Takimoto, J. Wright, and S. G. Arbuck. Clinical applications of the camptothecins. *Biochim Biophys Acta*, 1400:107–119, 1998.
[27] M. Wink, A. W. Alfermann, R. Franke, B. Wetterauer, M. Distl, J. Windhövel, O. Krohn, E. Fuss, H. Garden, A. Mohagheghzadeh, E. Wildi, and P. Ripplinger. Sustainable bioproduction of phytochemicals by plant in vitro cultures: anticancer agents. *Plant Genetic Resources. NIAB.*, 3(2):90–100, 2005.

Simulation and Optimization of Bio-Chemical Microreactors

Rolf Rannacher and Michael Schmich

Institut für Angewandte Mathematik, Universität Heidelberg,
Im Neuenheimer Feld 293/294, 69120 Heidelberg, Germany
{rolf.rannacher,michael.schmich}@iwr.uni-heidelberg.de

Summary. We describe the main steps in the development of a tool for the numerical simulation and optimization of a bio-chemical microreactor. Our approach comprises a 2D/3D grid generator for the prototypical Lilliput® chip, a stabilized finite element discretization in space coupled with an implicit time-stepping scheme and level-set techniques.

1 Description of the Project

Aim of this project is the development of a numerical tool for the simulation and optimization of so-called microfluidic diagnosis chips which are used for instance in clinical microbiology. In their channels and cavities, fluid dynamical and chemical processes take place. Because of the large surface-to-volume ratio, microfluidic aspects like capillary pressure, adhesion and viscosity, in contrast to macroscopic flow problems, dominate the overall behaviour of such chips.

1.1 The Lilliput® Analysis Laboratory

An example of such a "lab-on-a-chip" is the Lilliput® analysis laboratory, see Fig. 1, which was developed by Boehringer Ingelheim microParts GmbH for Merlin Diagnostika GmbH. This $20 \times 37 \times 3$ mm plastic chip has 96 reaction arrays with a volume of only $1.8\,\mu l$ each. This corresponds to one hundredth of the volume of conventional titer plates. Hence, much more tests can be performed with the same amount of sample.

1.2 The Mathematical Model

The flow on the diagnosis chip is modelled as a two phase flow. Here the two phases are the liquid and the air that is replaced by it. The flow is governed

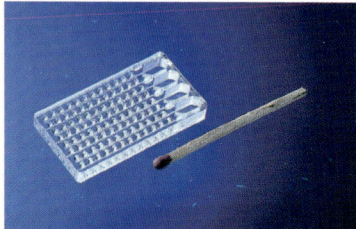 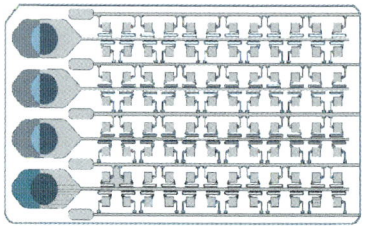

Fig. 1. "Lab-on-a-chip" Lilliput®

by the incompressible Navier–Stokes equations. It is driven by a geometry dependent pressure difference – as a simplified model for capillary pressure – as well as other volume forces like gravitation. The whole system of equations written in primitive variables reads

$$\rho \partial_t v - \mu \Delta v + \rho(v \cdot \nabla)v + \nabla p = \rho f,$$
$$\nabla \cdot v = 0,$$
$$\partial_t \rho + v \cdot \nabla \rho = 0.$$

This set of equations is completed by appropriate initial and boundary conditions. For the tracking of the interface between both phases we use a level set approach, see e. g. [OF03]. To this end, we introduce the so-called level set function ϕ which is positive in the fluid and negative in the air. We then have

$$\partial_t \phi + v \cdot \nabla \phi = 0.$$

Let ρ_{fl} and ρ_{gas} denote the densities of the fluid and the air respectively and let μ_{fl} and μ_{gas} denote the corresponding viscosities. Introducing the Heaviside function

$$H(x) = \begin{cases} 0 & x < 0, \\ 1 & x > 0, \end{cases}$$

we may write

$$\rho = \rho_{\text{gas}} + (\rho_{\text{fl}} - \rho_{\text{gas}})H(\phi) \quad \text{and} \quad \mu = \mu_{\text{gas}} + (\mu_{\text{fl}} - \mu_{\text{gas}})H(\phi).$$

1.3 Goal of the Numerical Simulation

Goal of the numerical simulation is to analyse the process of the filling of the chip with liquid. This is essential for the subsequent optimization of the geometry of the chip such that an almost uniform filling of the reaction arrays under all relevant operating conditions is achieved. This requires a careful balance between driving forces like capillary pressure and antagonistic forces like gravitation or friction.

To achieve this goal, many difficulties have to be overcome:

- very complex three dimensional geometry (see Fig. 1),
- highly anisotropic meshes,
- small time steps to resolve the whole filling process,
- locally refined meshes for capturing the interface which strongly influences the velocity of propagation.

These complications result in high computational costs.

2 The Solution Approach

Our solution approach is based on a Galerkin finite element method in space coupled, in the first implementation, with the backward Euler time stepping scheme for the temporal discretization. The finite element method is based on a variational formulation of the governing equations.

Let $(\cdot, \cdot)$ denote the inner product of L^2 on Ω. Furthermore, $H^1(\Omega)$ denotes the space of L^2 functions with generalized (distributional) first-order derivatives in $L^2(\Omega)$.

2.1 Temporal and Spatial Discretization

Choosing discrete time points $0 = t_0 < \cdots < t_m < \cdots < t_M = T$ and applying the backward Euler scheme to the system of equations yields

$$\rho k_m^{-1}(v^m - v^{m-1}) - \mu \Delta v^m + \rho(v^m \cdot \nabla)v^m + \nabla p^m = \rho f^m, \qquad (1)$$

$$\nabla \cdot v^m = 0, \qquad (2)$$

$$k_m^{-1}(\phi^m - \phi^{m-1}) + v^m \cdot \nabla \phi^m = 0, \qquad (3)$$

for $m = 1, \ldots, M$, where $k_m := t_m - t_{m-1}$. This system has to be complemented by appropriate initial and boundary conditions. Furthermore, we replace H by a regularized version H_ε defined as

$$H_\varepsilon(x) := \begin{cases} 0 & \text{if } x < -\varepsilon, \\ \frac{1}{2}\left(1 + \frac{x}{\varepsilon} + \frac{\sin(\frac{\pi x}{\varepsilon})}{\pi}\right) & \text{if } |x| \leq \varepsilon, \\ 1 & \text{if } x > \varepsilon, \end{cases}$$

and set

$$\rho_\varepsilon := \rho_{\text{gas}} + (\rho_{\text{fl}} - \rho_{\text{gas}})H_\varepsilon(\phi), \qquad \mu_\varepsilon := \mu_{\text{gas}} + (\mu_{\text{fl}} - \mu_{\text{gas}})H_\varepsilon(\phi).$$

After introducing the spatial discretization, we will choose $\varepsilon = O(h)$ varying with the local mesh size h.

The variational formulation of the problem is obtained by multiplying equations (1)–(3) by appropriate test functions $\varphi := (\psi, \xi, \chi)$ and integrating over the domain Ω. This leads us to define the semi-linear form $a(\cdot)(\cdot)$:

$$a(u^m)(\varphi) := \left(\rho_\varepsilon v^m + k_m \rho_\varepsilon (v^m \cdot \nabla) v^m, \psi\right) + k_m(\mu_\varepsilon \nabla v^m, \nabla \psi)$$
$$- k_m(p^m, \nabla \cdot \psi) + k_m(\nabla \cdot v^m, \xi) + (\phi^m + k_m v^m \cdot \nabla \phi^m, \chi)$$

with $u^m := (v^m, p^m, \phi^m)$. The variational form of system (1)–(3) then reads: For $m = 1, \ldots, M$ find $u^m \in V + \hat{u}$ such that

$$a(u^m)(\varphi) = F(\varphi) \quad \forall \varphi \in V,$$

where

$$F(\varphi) = (\rho_\varepsilon v^{m-1} + k_m \rho_\varepsilon f^m, \psi) + (\phi^{m-1}, \chi) - k_m P \int_{\Gamma_{\text{in}}} \psi \cdot n \, ds$$

and $\hat{u}$ represents prescribed Dirichlet boundary conditions while P is the prescribed mean value of the pressure on the inflow boundary which drives the flow. The function space V is the tensor product of certain subspaces of $H^1(\Omega)$ for the velocity and the level set function while the space for the pressure is $L^2(\Omega)$.

The spatial discretization via the Galerkin finite element method is defined on quadrilateral/hexahedral meshes $\mathcal{T}_h = \{K\}$ covering the domain $\overline{\Omega}$. The trial and test spaces $V_h \subset V$ consist of continuous, piecewise polynomial functions. The polynomial space is Q_1, the space of (isoparametric) tensor-product polynomials of degree 1. For a detailed description of this standard construction process, we refer to [BS02]. In order to facilitate local mesh refinement, we allow the cells to have nodes which lie on midpoints of faces of neighboring cells, see Fig. 2. There are no degrees of freedom associated to such "hanging nodes". The value of a finite element function at such points is determined by interpolation to enforce global continuity.

Since we use equal-order trial functions for v and p, our scheme requires pressure stabilization because the Babuška–Brezzi inf-sup-stability condition is not fulfilled. In addition, the convective terms need stabilization. This is done via the so-called *Local Projection Stabilization* (LPS), see e.g. [BB06]. To this end, we introduce the interpolation operator $i_{2h} : V_h \to V_{2h}$ from the current grid to a coarser grid in which four/eight adjacent cells of $\mathcal{T}_h$ form a macro cell in $\mathcal{T}_{2h}$. This can always be done if the mesh has a patch

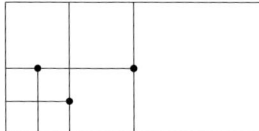

Fig. 2. Mesh with hanging nodes

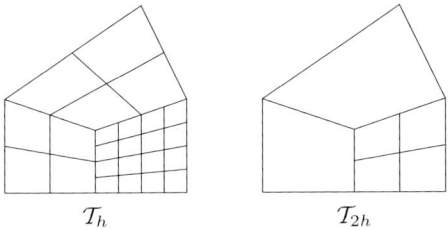

Fig. 3. Mesh with patch structure and corresponding coarsened mesh

structure, see Fig. 3. Let then the fluctuation operator $\pi : V_h \to V_h$ be defined as $\pi := \mathrm{id} - i_{2h}$. The stabilization process comprises in the modification of the original semi-linear form by a mesh-dependent semi-linear form:

$$a_h(u^m)(\varphi) := a(u^m)(\varphi) + s_h(u^m)(\varphi)$$

where $s_h(u^m)(\varphi)$ is given by

$$s_h(u^m)(\varphi) := \sum_{K \in \mathcal{T}_h} k_m \Big\{ \alpha_K (\nabla \pi p^m, \nabla \pi \xi)_K \\ + \delta_K^{(v)}(\rho_\varepsilon(v^m \cdot \nabla)\pi v^m, \rho_\varepsilon(v^m \cdot \nabla)\pi \psi)_K \\ + \delta_K^{(\phi)}(v^m \cdot \nabla \pi \phi^m, v^m \cdot \nabla \pi \chi)_K \Big\}.$$

The first term accounts for the pressure stabilization whereas the second and third term serve as stabilization of the convective term for the velocity and the level set function respectively. The cellwise parameters α_K, $\delta_K^{(v)}$, and $\delta_K^{(\phi)}$ are chosen as

$$\alpha_K := \alpha_0 \frac{h_K^2}{6\mu_\varepsilon + h_K \|\rho_\varepsilon v^m\|_K},$$

$$\delta_K^{(v)} := \delta_0^{(v)} \frac{h_K^2}{6\mu_\varepsilon + h_K \|\rho_\varepsilon v^m\|_K + k_m^{-1} h_K^2},$$

$$\delta_K^{(\phi)} := \delta_0^{(\phi)} \frac{h_K^2}{h_K \|v^m\|_K + k_m^{-1} h_K^2}$$

with $\alpha_0, \delta_0^{(v)}, \delta_0^{(\phi)} \approx 0.3$. Hence, the discrete system of equations to be solved reads: For $m = 1, \ldots, M$ find $u_h^m \in V_h + \hat{u}_h$ such that

$$a_h(u_h^m)(\varphi_h) = F(\varphi_h) \quad \forall \varphi_h \in V_h. \tag{4}$$

2.2 Solution of the Algebraic Systems

In each time step, we have to solve a nonlinear system of equations. This is done via Newton's method. If we denote by u_n the n-th iterate in computing

the discrete solution u_h^m, one Newton step for (4) consists of solving the linear system
$$a'_h(u_n)(\delta u_n, \varphi_h) = F(\varphi_h) - a_h(u_n)(\varphi_h) \quad \forall \varphi_h \in V_h, \tag{5}$$
for the correction $\delta u_n := u_{n+1} - u_n$. The directional derivative of the semi-linear form $a_h(\cdot)(\cdot)$ at u_n in direction δu_n, given by
$$a'_h(u_n)(\delta u_n, \varphi_h) := \lim_{s \to 0} \frac{1}{s} \Big\{ a_h(u_n + s\delta u_n)(\varphi_h) - a_h(u_n)(\varphi_h) \Big\},$$
is derived analytically on the continuous level and then discretized. The linear subproblems (5) are solved by the GMRES method with preconditioning by a multigrid iteration. The multigrid component uses a block ILU smoother in which those unknowns corresponding to the same node of the mesh are blocked. For a detailed description of this approach, we refer to [BR99].

3 Numerical Results

The first step was to implement a grid generator for (a part of) the geometry of the chip. The specific features of this geometry lead to anisotropic cells with aspect ratios of approximately 1:10. A picture of the resulting coarse grid which we used in our computations can be seen in Fig. 4.

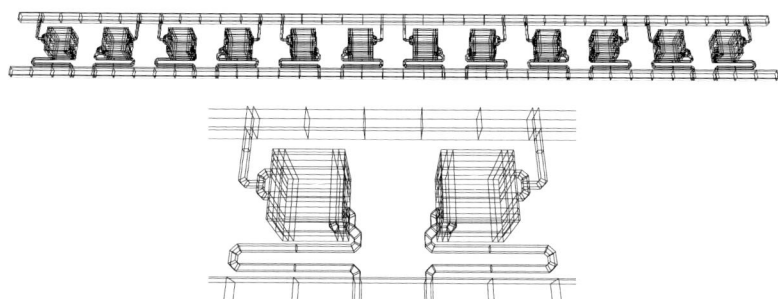

Fig. 4. Three dimensional coarse grid used for computation (*bottom:* zoom into coarse grid)

3.1 Simulation

In the first computation we simulated a flow with two identical phases. At the beginning, the whole chip is filled with fluid 1 while fluid 2 is entering at the lower left part of the computational domain. This computation was done without respecting gravitational forces, i.e., we set $f = 0$. Figure 5 shows the propagation of fluid 2 (red) at different times.

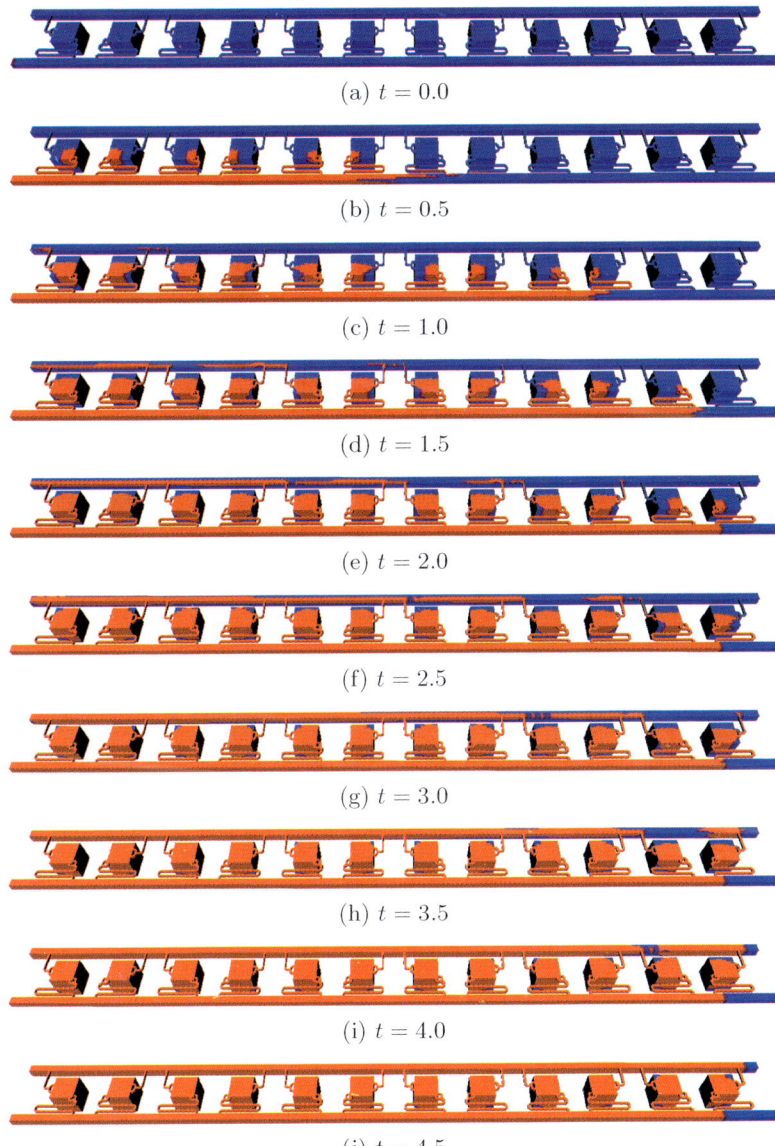

Fig. 5a–j. Computation 1: Filling process (phase 1 *blue*, phase 2 *red*)

The next computation, which was done on only a part of the chip's domain, simulated a real two phase flow under the influence of gravitation. Figure 6 shows the corresponding sequence of snapshots. One can nicely see how the liquid flows through the channels down into the reaction arrays and fills them before leaving them.

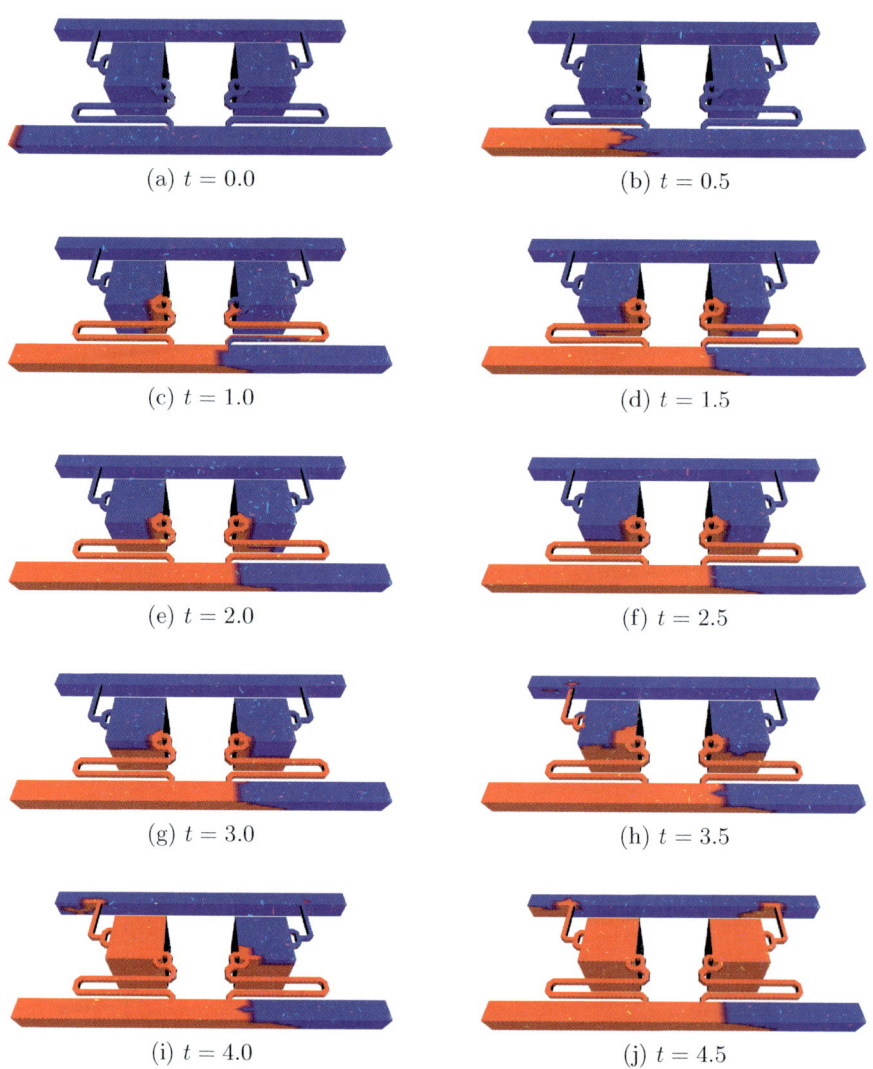

Fig. 6a–j. Computation 2: Filling process (phase 1 *blue*, phase 2 *red*)

For a better tracking of the interface between both phases, we like to apply adaptive mesh refinement. The selection of the cells to be refined here is still a quite heuristic one: We mark a cell K to be refined if the level-set function ϕ has positive as well as negative values in K which corresponds to the fact that the interface which is given by points x with $\phi(x) = 0$ lies in this cell. This technique has been applied to a 2D version of the chip. Figure 7 shows by a sequence of locally refined meshes that this method works quite well. We see that the interface between both phases is always lo-

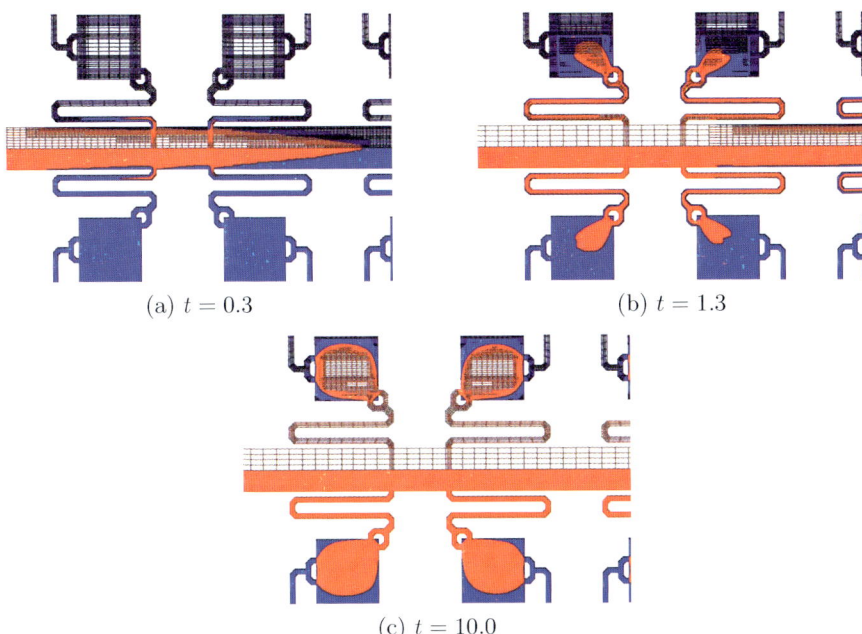

Fig. 7a–c. Computation 3: Adaptive mesh refinement for tracking of the interface

cated in the most refined zone. In the next step, the mesh refinement will be done automatically, driven by residual and sensitivity information, in order to obtain the correct propagation speed of the interface. For further information on this so-called *Dual Weighted Residual* (DWR) method, we refer to [BR01].

3.2 Optimization

On the 2D version of the chip, we also performed a first simple optimization. By slightly modifying the chip's geometry, we were able to achieve a much more efficient filling of the reaction arrays. The result of this optimization process can be seen in Fig. 8 where the corresponding filling states at different times are compared.

4 Conclusion and Outlook

We have presented a tool for the numerical simulation of chemical microreactors. The method is based on the incompressible Navier–Stokes equations with non-homogeneous density and uses equal-order finite elements together with local projection stabilization as well as level-set techniques for the tracking

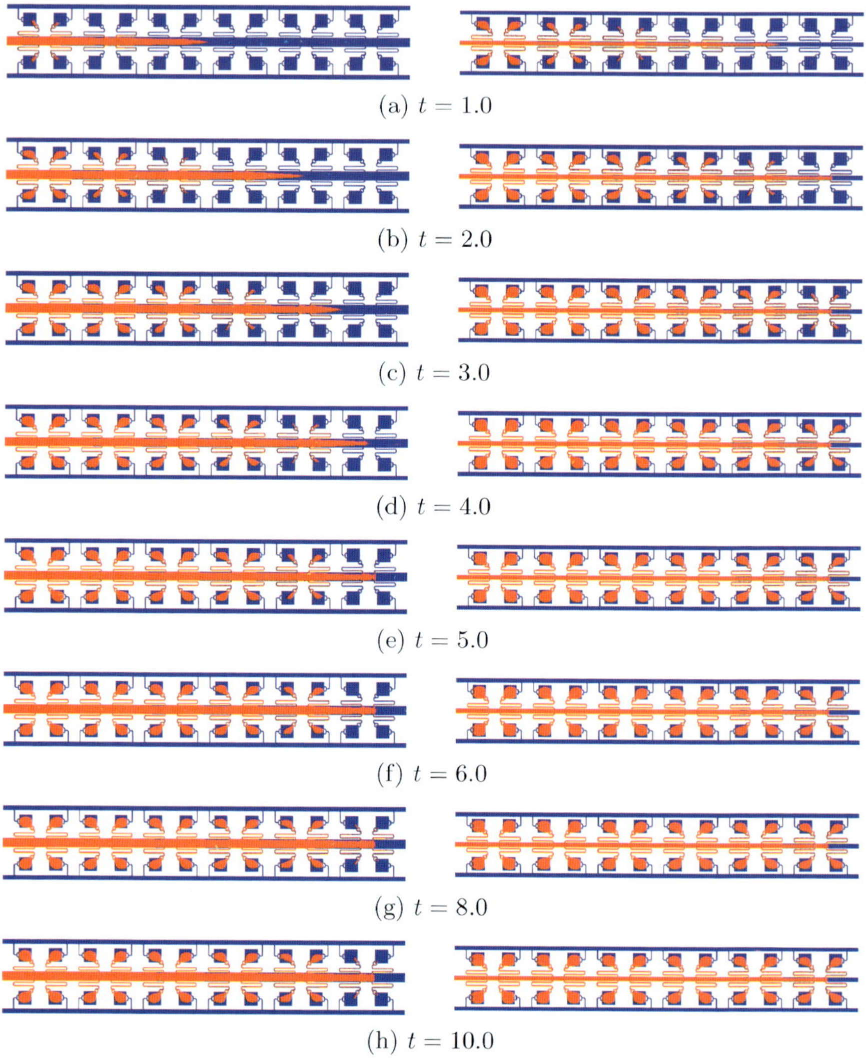

(a) $t = 1.0$

(b) $t = 2.0$

(c) $t = 3.0$

(d) $t = 4.0$

(e) $t = 5.0$

(f) $t = 6.0$

(g) $t = 8.0$

(h) $t = 10.0$

Fig. 8a–h. Solution before (*left*) and after (*right*) geometry "optimization"

of the interface. At first, computations with two identical phases and without gravitation have been performed. In a second step, we simulated a two phase flow under the influence of gravitation. In this context, adaptive mesh refinement for tracking the interface between both phases has been tested for a 2D version of the chip. The next step will be to combine the 3D solver components with adaptivity based on the DWR method for a better interface tracking. Finally, we will incorporate effects by surface tension as well as capillary forces.

References

[BS02] Brenner, S. C., Scott, L. R.: The Mathematical Theory of Finite Element Methods. Springer, New York, Berlin, Heidelberg (2002)

[OF03] Osher, S., Fedkiw, R.: Level Set Methods and Dynamic Implicit Surfaces. Springer, New York, Berlin, Heidelberg (2003)

[BR01] Becker, R., Rannacher, R.: An optimal control approach to a posteriori error estimation in finite element methods. In: Iserles, A. (ed) Acta Numerica 2001. Cambridge University Press (2001)

[BR99] Braack, M., Rannacher, R.: Adaptive finite element methods for low-Mach-number flows with chemical reactions. In: Deconick, H. (ed) 30th Computational Fluid Dynamics. The von Karman Institute for Fluid Dynamics, Belgium (1999)

[BB06] Braack, M., Burman, E.: Local projection stabilization for the Oseen problem and its interpretation as a variational multiscale method. SIAM J. Numer. Anal., **43**(6), 2544–2566 (2006)

Part IV

Computeraided Medicine

Modeling and Optimization of Correction Measures for Human Extremities

René Brandenberg, Tobias Gerken, Peter Gritzmann, and Lucia Roth

Zentrum Mathematik, Technische Universität München, Boltzmannstr. 3, 85747 Garching bei München, Germany
{brandenb,gerken,gritzman,roth}@ma.tum.de

Summary. Deformities of the lower extremities can be congenital or post-traumatic. Recent progress in their treatment arises from a new technique based on fully implantable intramedullary limb lengtheners. This allows patient friendly and effective treatment of complex length and rotation deformities. The advancement in technology and growing demands on the precision of a correction require a revision of common operation planning schemes. In this paper, we discuss the questions from computational geometry that arise in semi-automated operation planning based on three-dimensional CT data. We present algorithms for these problems that have been implemented and integrated into a newly developed 3D planning device.

1 Introduction

Deformities of the lower extremities can be congenital or post-traumatic. They can be treated according to the principle of callus distraction which allows for bone growth among adults. This was first studied systematically by Siberian orthopedist G. Ilizarov [33], [34]. Classical treatment of such deformities relies on the Ilizarov apparatus, and standard operation planning schemes [44] are based on 2D X-rays (see Fig. 1).

Recent progress arises from a new technique based on fully implantable intramedullary limb lengtheners (see Fig. 1) that were designed by R. Baumgart and his team at the Limb Lengthening Center Munich [6]. Treatment with the aid of this device is much gentler on the patient. Moreover, the risk of infections is minimized as the apparatus, once implemented, is fully secluded within the bone, with no parts penetrating the skin. In contrast to the Ilizarov apparatus, however, the lengthening direction is completely determined with the implantation of the device and non-invasive corrections during the distraction phase are impossible. This fact results in increasing demands on the precision of common operation planning procedures. Therefore, in cooperation with the Limb Lengthening Center, a new planning scheme based on three-dimensional computer tomography (CT) data was developed.

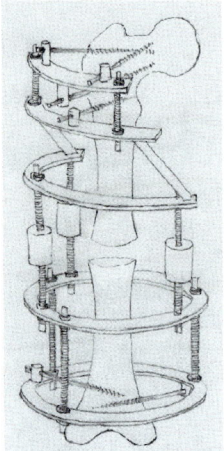 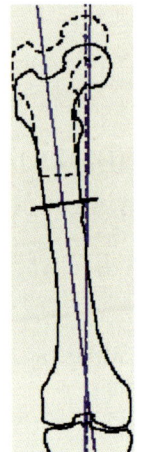 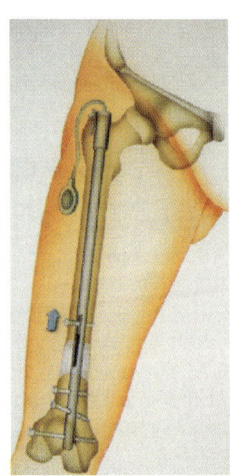

Fig. 1. The Ilizarov apparatus, classical operation planning in 2D, and an implanted intramedullary nail

The paper is organized as follows: In Sect. 2, we briefly describe the treatment of deformities. Additionally, we identify the basic tasks that are specific to the planning procedure for the new technique. In Sect. 3, we introduce the underlying mathematical problems that arise from the previous section. Algorithmic solutions to these problems are discussed in Sect. 4. Finally, in Sect. 5, we describe the resulting software tool.

2 Medical Treatment of Deformities

In this section, we briefly discuss the basic tasks of deformity correction. For more detailed information on the medical background see [6], [44] and the references there. The general procedure in deformity correction can be characterized by the following steps (compare Fig. 2):

A. Analysis of the state of the patient's musculoskeletal system.
B. Identification of the site of the osteotomy and the positioning of the limb lengthening device in order to achieve an optimal post-treatment status.
C. Implantation of the intramedullary nail and callus distraction.
D. Consolidation of newly formed bone tissue and removal of the device.

The present paper is concerned with the planning procedure (Step B), where mathematical models come into play. Based on the current state of the patient's musculoskeletal system and relying on biomechanical considerations, a goal state has to be defined. Then, the site of the osteotomy and the position of the limb lengthening device must be determined accordingly. The planning process relies on critical points and axes. These characterize the

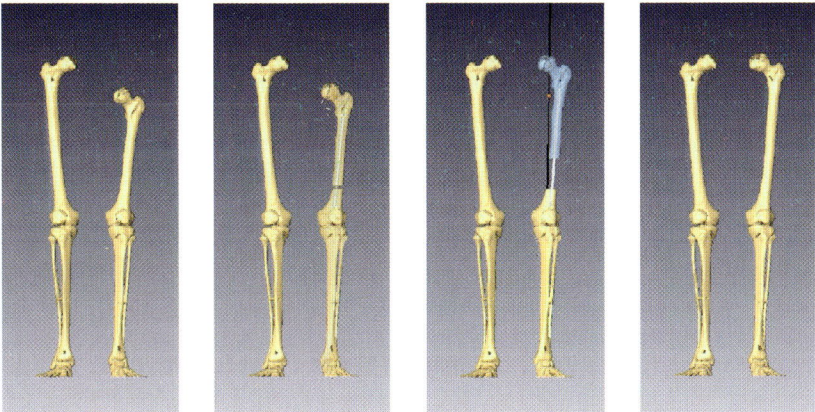

Fig. 2. Schematic illustration of a deformity correction procedure: initial state, post-operational state, projected post-treatment state, and consolidated state

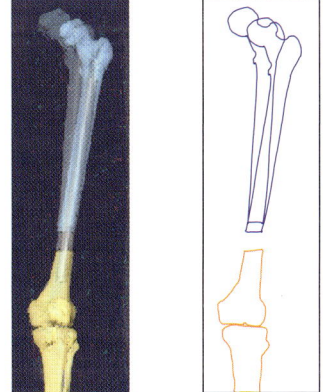

Fig. 3. Slight inaccuracies cause significant deviation after lengthening

anatomical and the (idealized) mechanical components of the musculoskeletal system. Deformities are identified by comparison of the axes and angles to those of normal alignments. Ideally, hip, knee and ankle should be collinear in the frontal view [44]. Naturally, the details of Step B depend on the specific device. In particular, limb lengthening with intramedullary nails has to be carried out along the anatomical axis of the corresponding bone.

We stress again the critical fact that once placed inside the bone, no post-operational corrections of the implanted device are possible. Lengthening with a slightly improperly placed nail may result in serious malalignment that may necessitate further operations (compare Fig. 3). Hence, the intramedullary technique demands high precision in operation planning.

3 Basic Geometric Problems in Operation Planning

3.1 Medical Modeling

In this section, we discuss the mathematical questions that result from modeling the planning procedure. Naturally, our mathematical model must reflect the actual medical situation as exactly as necessary but must also meet the demands on computation times for an interactive software tool executable on standard devices. We do not aim at a fully automated planning process but at a tool supporting the physician's work, while still allowing comprehensive use of expertise and experience.

We presume that a voxel based model of the patient's skeleton has been extracted from three-dimensional CT data via standard image processing techniques. More precisely, the skeleton may be given as a finite point set $P \subset \mathbb{R}^3$. Registration of limb images in the context of orthopedic interventions is practical and, 'since the bone contrast [in CT or X-ray images] is very high, most methods, including segmentation tasks, can be automated' [39, p. 20]. For details about medical imaging, see e.g. [5], [47] and the references therein.

The primal task is now to identify different centers and axes of P, which represent the geometry of the extremity, including the center of the femur head or the knee joint, the site of the deformity, and the anatomical axis of the thigh bone. In addition, the placement of the intramedullary nail has to be determined.

The center of a the femur head may be found via approximating the concerned part of the bone by a Euclidean ball. First models for the central part of a long bone may be circular or elliptical cylinders, others may involve several cylindrical segments. This coarse approximation is appropriate for the specific application as the key feature for a successful operation is the anatomically correct axis direction.

A special geometric approximation problem in operation planning is finding the site for an optimal osteotomy. A deformity involving twisting and bending of the bone axis can be corrected via a single cut, followed by a suitable rotation and translation of the two resulting bone segments [29]. The affected part of the bone can be modeled by two cylindrical segments with intersecting axes. The site of the osteotomy is therefore determined by the point where the axes meet (the center of the deformity) and their directions.

When specifying the approximation strategy, one should be aware of the fact that the data points are usually distributed non-uniformly on the bone surface. In fact, in order to minimize radiation exposure, high-precision CT is limited to the parts of the skeleton which are most relevant for the operation (e.g. joints). A lower precision is used for less relevant parts. Since the planning procedure relies in an essential way on the mechanical and anatomical axes, we are interested in the geometric structure underlying the data points. The planning procedure also includes the device positioning. This step is carried out as follows in the 3D model: the part of the bone cut by the osteotomy is virtually set to its position in the goal state and the extracted nail is placed so as to appropriately connect the two parts. The post-operational state is now determined by simply contracting the lengthening device and translating the cut part accordingly (compare Fig. 2).

The intramedullary nail can be seen as a (circular) finite cylinder (compare Fig. 1). In order to locate a proper placement, we need an appropriate geometric model for the interior of the bone. A simple set of data points does not provide this directly. Therefore, we represent the bone by a structure built from blocks that are the convex hulls of two 'adjacent' circles or ellipses respectively. The ellipses are approximations of the original CT-slices of bone layers. Note that these layers need not be parallel any more, since, in the goal state, bone parts have been cut, rotated and translated.

Consequently, we are lead to a set of basic geometric problems, all involving the *containment* of objects, which can be classified within the more general settings of computational geometry or computational convexity. While our main application 'lives' in 3D, we will formally introduce the relevant problems in general dimensions in order to give a more complete account of their structure. Later, we will specialize again to the specific medical application.

3.2 Mathematical Description

The first geometric problem occurs when approximating parts of the thigh bone with basic geometric shapes (Fig. 4). These shapes are obtained from simple geometric objects via homothety or similarity: Two sets C, C^* are *similar* if there exists a rotation $\Phi : \mathbb{R}^d \to \mathbb{R}^d$, $c \in \mathbb{R}^d$ and $\rho \geq 0$, such that $C^* = \rho\Phi(C) + c$. If Φ can be chosen as the identity, C and C^* are *homothetic*.

Problem 1 (Basic k-Containment). For a finite point set $P \subset \mathbb{R}^d$ and closed convex sets $C_1, \ldots, C_k \subset \mathbb{R}^d$, find $C^* = \bigcup_i \rho_i \Phi_i(C_i) + c_i$ such that $P \subset C^*$, minimizing $\max_i \rho_i$.

Important examples for basic 1-containment are the computation of the smallest enclosing Euclidean ball ($C_1 = \mathbb{B}^d$, the Euclidean unit ball) or a smallest enclosing cylinder ($C_1 = l + \mathbb{B}^d$, where l denotes a 1-dimensional linear subspace). Of course, as we have seen, more complicated container shapes and higher numbers of different containers occur naturally. Other objective functions may be chosen in the above problem definition, such as the sum of the dilatation factors. The above choice is justified in our application since the proportions of the bone parts (and the geometric objects representing them) are roughly constant.

Some of our medical tasks involve more general containments. For instance, the problem of finding the center of deformity leads to a restricted version of the double-ray center problem, a 2-containment problem where the transformations of the two container sets are not independent (which is the reason for the 'double' instead of a 2 in the name). The double-ray center problem

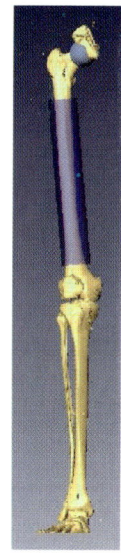
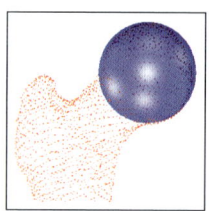

Fig. 4. Approximating parts of the thigh bone with basic geometric shapes: complete leg and detail

is a well known mathematical problem. For our medical application, an additional constraint about the position of the center c of the deformity needs to be enforced (see Sect. 4.3 for details).

Problem 2 (Restricted Double-Ray Center). For a finite set of points $P \subset \mathbb{R}^d$, an approximate center $c' \in \mathbb{R}^d$, and a maximal distance δ, determine two rays $r_1 = \{c + \lambda v_1 : \lambda \geq 0\}$ and $r_2 = \{c + \lambda v_2 : \lambda \geq 0\}$, with $v_1, v_2 \in \mathbb{R}^d \setminus \{0\}$, emanating from the same center $c \in \mathbb{R}^d$ such that $P \subset (r_1 \cup r_2) + \rho \mathbb{B}^d$, $\|c - c'\| \leq \delta$, minimizing ρ.

As we saw above, the device positioning problem involves non-convex containers.

Problem 3 (Cylinder Inclusion). For given length λ, radius ρ, and a sequence $E_1, \ldots, E_m$ of $(d-1)$-dimensional ellipsoids, find a line segment L of length λ such that for the cylinder $Z(L, \rho) = L + (\rho \mathbb{B} \cap L^\perp)$

$$Z(L, \rho) \subset \bigcup_{i=1}^{m-1} \mathrm{conv}\,(E_i \cup E_{i+1})$$

holds or decide that no such cylinder exists.

4 Mathematical Treatment

In this section, we give an overview of the state of the art for the mathematical problems raised in Sect. 3 and their relatives as well as our basic algorithmic approaches. First, a general framework is provided through the notion of *Optimal Containment*. Second, the specific problems from Sect. 3 are discussed in detail.

4.1 Optimal Containment

The notion of containment problems generically summarizes all problems of computing, approximating or measuring the sets which among a given class are the largest contained in or the smallest containing a given set [26].

Optimal Containment. *Let $\mathcal{B}$ and $\mathcal{C}$ be families of closed sets in $\mathbb{R}^d$ and $\omega : \mathcal{B} \times \mathcal{C} \to \mathbb{R}_0^+ \cup \{\infty\}$ a functional. The task is to find $B^* \in \mathcal{B}$ and $C^* \in \mathcal{C}$ such that $B^* \subset C^*$ and $\omega(B^*, C^*)$ is minimal or to decide that no such pair exists.*

In order to cope with optimal containment problems algorithmically, specialized problem classes have to be considered. Since set inclusion provides a partial order on the objects in $\mathcal{B}$ and $\mathcal{C}$, it is natural to ask for ω to be monotonically decreasing in the first argument and increasing in the second with respect to this order.

In case of the outer containment problems considered here, the inner object is a specific point set and $\mathcal{C}$ is a family of sets obtained from one or more reference containers. The function ω therefore depends only on the containers. In an instance of Problem 1, k basic convex containers C_i are fixed and $\mathcal{C}$ represents all feasible transformations of the containers; i.e., $\mathcal{C}$ is the set of all unions $\bigcup_{i=1}^{k} \rho_i \Phi_i(C_i) + c_i$ with $\rho_i \geq 0$, Φ_i a rotation, and $c_i \in \mathbb{R}^n$ for all $i = 1, \ldots, k$. In the double-ray center problem, the transformations of the two rays are not independent. Hence, $\mathcal{C}$ consists of the Minkowski sum of unions of two rays with common origin c and a ball of radius ρ there.

In Problem 3, an instance is defined by a specific outer container formed by the given ellipsoids and $\mathcal{B}$ is the set of all cylinders of a given length and radius. All feasible inner objects can therefore be obtained by rotations and translations of one representative cylinder. (Of course, in theory, this kind of problem can also be formulated as an outer containment problem since all involved transformations are invertible.)

One should recognize that in all three problems discussed here, $\omega(C^*)$ is obtained from the dilatation factors of C^*. Other objectives may be considered in many variants, some of which are relevant to the medical tasks described here. Ongoing research is concerned with containment under affinity, where the objective may for instance involve the volume of C^*.

In the survey [26], the complexity of some convex containment problems in general dimensions is addressed, classifying the problems according to the representation of the container, the enclosed object and the functional ω. Due to the present application, our focus is the three-dimensional case. However, we will point out NP-hardness results for specific problems in varying dimensions as they can, to a certain extent, explain jumps in the computational costs from dimension two to three.

4.2 Problem 1: Basic k-Containment

Theoretical background: First, let us consider the task of covering a point set with a (Euclidean) ball of minimal radius as it occurs when determining the center of the femur head. Since all Euclidean balls are translations and dilatations of the unit ball, this is a case of basic 1-containment under homothety with $C_1 = \mathbb{B}^d$, see e.g. [14], [20], [25], [37], [52], or [46] for an overview. The computational complexity of 1-containment under homothety in general is addressed in [17] and [26].

While the smallest enclosing ball can be regarded as containment under homothety, the smallest enclosing cylinder is an example of containment under similarity and is known to be NP-complete in general dimensions [40] (a result derived from the close relation to line
stabbing problems, see also [1], [27], [31], and compare Sect. 4.4). In fixed dimensions polynomial time solvability is shown in [19], but neither this result nor the polynomial time approximation schemes in [30] and [31] provide practical algorithms in 3D.

Both smallest enclosing balls and cylinders are special cases of outer radii of convex bodies [24], [25], [50]. Apart from shape approximation, the problem of computing inner and outer radii arises in various applications in computer science, optimization, statistics, and other fields [25].

In general, containment under similarity is substantially harder than containment under homothety. Here is another example: while deciding whether a point set can be covered by an axis aligned unit cube is trivial, the complexity of the corresponding task allowing rotations is unknown, though conjectured to be NP-complete in general dimensions [40]. Note that a regular simplex in dimension d can be covered by a rotated unit cube scaled by a dilatation factor of $\sqrt{d}$ if and only if a Hadamard matrix of order $d+1$ exists.

The medical application requires also approximations by several objects. Regarding basic k-containment problems under homothety with $k \geq 2$, as in the case of $k = 1$, most attention has been placed on the Euclidean k-center problem which asks for the minimal radius to cover a given point set with k balls. For $k \geq 2$, it is NP-complete in general dimensions [40]. The k-center problem for axis aligned cubes can still be solved in polynomial time for $k = 2$ but is NP-complete for $k \geq 3$ [40]. Many practical approaches focus on the planar 2-center problem, e.g. [15], [18]. However, if k is part of the input, approximation to a certain accuracy is NP-complete for both, circles and squares, even in the plane [41].

Very few papers deal with k-containment problems under similarity as in higher dimensions it combines the difficulties of both rotation with $k = 1$

and homothety for $k > 1$. A special case is the k-line center problem, that is covering a point set with k cylinders of equal radius. While in [42] it is shown that already the planar case is $\mathbb{NP}$-complete if k is part of the input, a linear time approximation scheme for every fixed k in 2D is given in [3].

Algorithms: 1-Containment under homothety is fundamental here since it occurs not only in many applications itself but also in subroutines for even harder containment problems. Already three-dimensional instances of 1-containment under similarity or basic k-containment under homothety require considerable computational effort. Usually, these problems are solved within branch-and-bound schemes that rely on the solution of many instances of 1-containment under homothety [10], [14], [16], [36]. Of course, basic 1-containment under homothety is a convex programming problem and can in principle (assuming well-boundedness) be solved by the ellipsoid method if a suitable separation oracle is known for C_1. Typically, however, the ellipsoid method is not competitive when more information about the container is available. For the special case of the smallest enclosing ball problem, for instance, one can choose from a considerable number of fast algorithms, see e.g. [20], [37] and [52] for implementations and experimental studies. In [11], different solution strategies for 1-containment under homothety are formulated and tested. Depending on the container shape, some instances can be formulated as Linear or Second-Order-Cone Programs. A cutting plane algorithm also based on a separation oracle but with good practical performance is provided as well.

So, while basic 1-containment under homothety is usually solvable in theory and in practice, a suitable approach to 1-containment under similarity is less straightforward. Even in the case of a smallest enclosing cylinder, the complexity increases significantly from planar to 3D instances. In [12], algebraic methods for smallest enclosing and circumscribing cylinders of simplices are addressed and a polynomial formulation with significantly reduced algebraic complexity is provided. Exact algorithms for smallest enclosing cylinders in 3D are proposed in [1] and [48]. In [31], it is shown that the radius can be approximated in linear time in both the dimension and the number of input points.

When $k \geq 2$, basic k-containment can be reduced to k basic 1-containment problems by finding an optimal partition of the point set P. A core set approach yields a linear time approximation scheme for the Euclidean k-center problem in general dimensions [14]. The algorithm uses the fact that the core set sizes depend neither on the dimension nor on the number of input points, but only on the error ϵ and therefore, for a fixed error bound, enumeration of all possible partitions of the core sets is feasible (in principle). A more practical algorithm based on a simple branch-and-bound scheme is proposed in [36]; however, computations for 3D point sets from practical applications are impossible when $k > 4$ or $\varepsilon < 0.01$. In [10], algorithms for the general basic k-containment problem under homothety are developed. One should be aware that these problems do not admit core sets of sizes independent of the dimension in general [11]. Moreover, the algorithms that do compute small core sets in the Euclidean case may construct dimension dependent core sets when the container is a cube, even though every diametral pair of points is a core set of size 2 in this case. See [7] for a fast algorithm when C is the cube and $k = 2$.

Treatment in practice: For the smallest enclosing ball, we use an approximation algorithm based on Second-Order-Cone Programming [37]. Basic 1-containment problems under homothety with other container shapes are solved by an approach suggested in [11].

The smallest enclosing cylinder computation is based on non-orthogonal projections along a-priorily fixed directions given by an adaptive discretization of the unit ball [16]. Though this algorithm has exponential running time in general dimensions, it is more practical for three-dimensional input than the polynomial time approximation scheme in [31] aiming at high dimensions. Nonetheless, a direct implementation of the algorithm is slow and additional effort has to be made in order to reduce running times in practice [45]. In general, computing the radius of the smallest enclosing cylinder is a non-convex problem which can have many local minima. However, the bone data is not a worst-case input but in fact rather cylindrical and therefore large regions for the cylinder axis can be discarded in advance. A first step towards practical computations is to derive a-priori bounds for the radius, e.g. via ellipsoidal approximation [45], semidefinite programming [50], or diameter approximation [31]. If one allows cylinders with elliptical (rather than just circular) cross-sections, ellipsoidal approximation even allows good a-priori bounds on the minimal volume of the elliptical cross-sections and almost optimal a-posteriori results in our application.

We use regular discretizations of the 3D sphere instead of techniques developed with a view towards general dimensions. The discretization is refined adaptively, the direction vectors are ranked according to their likelihood to improve the currently best solution and processed in a branch-and-bound scheme. Incrementally generated core sets [51] are used to reduce the processing time caused by the size of P.

In order to solve basic k-containment problems when $k \geq 2$, we use practical techniques described in [10]. A-priori upper and lower bounds (with provable quality even for different, not necessarily symmetric containers) are computed via methods based on diameter partitioning. Moreover, relaxations of a mixed integer Second-Order-Cone formulation of the problem are solved at each node of the branch-and-bound-tree instead of simply computing the dilatation factors for the current partition. Besides being capable of solving containment problems with general and possibly different containers, the described approach even improves the performance for the Euclidean k-center problem [10].

Besides the general methods discussed here, using a fast local search heuristic can be profitable when significant additional information about a basic containment problem is provided by the application.

4.3 Problem 2: Restricted Double-Ray Center

Theoretical background: The planar double-ray center problem is discussed in [23]. Here we focus on the 3D case, though in principle, the algorithm we formulate generalizes to higher dimensions. Related problems include the smallest enclosing cylinder (compare Problem 1), the anchored ray problem [21], and the 2-line center problem [2], [3]. Note that for finite point sets covering with two cylinders is the same as covering with two half-cylinders. However, the additional constraint in the double-ray center problem is forcing the two axes to have a common point (see Sect. 4.2) which adds further difficulties. In fact, a partition of the point set into two parts approximated by a single ray each does not yield a reduction to independent subproblems. Even when the partition of the point set is known, the position of one ray always depends on the other as the rays must issue from a common center. This makes approaches based on projections and algebraic formulations difficult in spite of the fact that the problem has fewer degrees of freedom than the 2-line center problem [22].

In case of the general double-ray center, a slight shift of just one of the input points may change the optimal configuration completely [22]. The same may happen in case of the smallest enclosing cylinder, but only for 'non-cylindrical' input data. In case of the double-ray center problem, however, even well-shaped input may yield unwanted optimal solutions. For instance, the optimal solution can degenerate to parallel rays meeting at infinity. In order to get meaningful solutions for the application, the additional constraint on the center c of the deformity has to be respected.

Algorithms: In [22] an approach based on testing all relevant partitions of the point set and solving the resulting optimization problems is proposed. At first, the fraction of the points close to the center is determined and afterwards the other points are split by separating hyperplanes into two subsets belonging to one ray each. The final optimization problem is still non-convex, but can be handled by discretizations of the direction space (similar to those considered for smallest enclosing cylinders in Sect. 4.2). The algorithm is polynomial in the number of input points in fixed dimensions, but a full computation is still impracticable for point sets of sizes relevant for our application.

Of course, when the specified ball containing c is very small, the problem character changes and it can be advantageous to consider alternate approaches, such as testing for the center position instead of the ray directions.

Treatment in practice: In the medical application, the ray directions are limited and possible sites for the center are in fact constrained to a small area, as, obviously, it should be located inside the bone (see Fig. 5). Constraints on the ray directions, e.g., on the angle between a ray and a given axis or between the two rays, can easily be taken into account in the algorithm. Since many possible ray directions can be disregarded for the medical application, the restricted version now actually allows for practical computations.

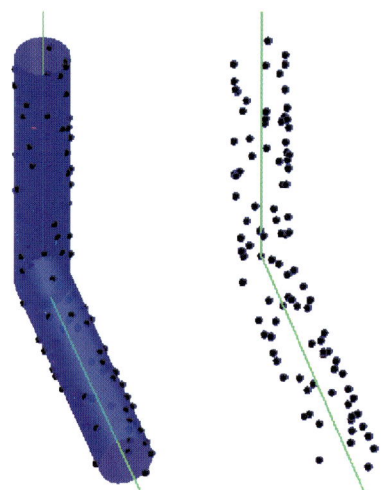

Fig. 5. The double-ray center problem in 3-space

An initial approximation of the double-ray center problem can easily be found directly allowing a local search strategy in the tool. The algorithm then starts from an approximate solution which is used to constrain the possible partitions of the point set, the position of the center and the directions of the emanating rays and optimizes the site of the cut for the restricted problem.

4.4 Problem 3: Cylinder Inclusion

Problem modification: For an optimal positioning of the nail, medical requirements have to be met. Criteria for a 'good' nail position are the stability of the arrangement and the expected healing process of the bone tissue. Therefore, the nail should leave enough margin on all sides and the cut surfaces of the bone parts divided by the osteotomy should have a large overlap in the post-operational state. In addition, the nail has to penetrate both bone parts by a sufficient depth to guarantee stability.

Our approach for the nail positioning is therefore to center the nail in a cylindrical corridor of maximal radius through the bone. This meets the demands from the application since the minimal margin between nail and bone boundary is maximized. Moreover, it guarantees that the cut surfaces in the post-operational state (or, more precisely, their projections in the plane perpendicular to the lengthening direction) overlap at least by the area of a disk of the computed maximal radius. When the cylindrical corridor traverses all the blocks forming the bone model, the nail can be placed in a way that maximizes the penetration of the two bone segments.

Hence, in the modified problem we are looking for optimal traversing cylinders.

Optimal Traversing Cylinder. Let $\mathcal{E} = \{E_1, \ldots, E_m\}$ be a finite set of $(d-1)$-dimensional ellipsoids in $\mathbb{R}^d$. The task is to find a line $l \subset \mathbb{R}^d$ and a maximal $\rho \geq 0$ such that $\emptyset \neq (l + \rho \mathbb{B}^d) \cap \operatorname{aff}(E) \subset E$ for all $E \in \mathcal{E}$, or decide that no such line exists.

Theoretical background: The associated feasibility problem reduces to the question whether a given arrangement of ellipsoids allows (at least) for a traversing (or stabbing) line. When the dimension is part of the input, $\mathbb{NP}$-completeness of this task can be shown using a modification of a proof from [40]. The original theorem shows the hardness of line stabbing for full-dimensional balls, yet the construction can be adopted to work also for carefully chosen $(d-1)$-dimensional balls (compare [27]). Stabbing problems are strongly related to visibility computations [28]. Applications from computer graphics are common to both. The structure of the solution space can be quite complex in general. In 3D for degenerate ellipses the results for transversals of line segments apply and the solution space may consist of up to m connected components [13]. Visually, two lines are in the same connected component if one can continuously be transformed into the other without leaving the solution space. Lines in the same connected component define the same geometric permutation of the sets to be traversed, i.e. the induced orders are either the same or reversed. However, when additional information about the input data is available, the structure of the solution space is often much simpler. For instance, when all ellipsoids are parallel, the solution space can actually be seen to form a convex set in $\mathbb{R}^{2(d-1)}$ [4]. For disjoint full-dimensional balls and a fixed geometric permutation, the set of directions of traversing lines is also convex [8].

In our application, there is only one reasonable ordering of the ellipses, yet this information is not explicitly included in the modified formulation. However, the information has not been lost, since for inputs from the application, $\operatorname{aff}(E_i)$ has dimension $d-1$ and $\operatorname{aff}(E_i) \cap E_j = \emptyset$ for all $E_i \neq E_j \in \mathcal{E}$. This suffices to show that the arrangement admits at most one geometric permutation [35] as any of the hyperplanes $\operatorname{aff}(E_i)$ determines a partition of $\mathcal{E} \setminus \{E_i\}$ into two disjoint subsets. It also implies $l \not\subset \operatorname{aff}(E_i)$ for any i and, consequently, ensures that the modified formulation allows no more than the desired solutions.

Algorithms and treatment in practice: For the algorithmic solution of the problem, we are only interested in directions permitting traversing cylinders of (not too small) positive radius. Assuming we know the direction of the axis, we can project the ellipsoids in $\mathcal{E}$ along the axis direction to obtain ellipsoids in $\mathbb{R}^{d-1}$ and then compute the largest ball in the intersection of the projected ellipsoids (again a containment problem under homothety) via semidefinite programming; see [9, pp. 45, 46]. An approximately optimal axis direction can be found by a discretization procedure of the sphere similar to that of Sect. 4.2 for computing smallest enclosing cylinders.

The size of the discretization can be kept suitably small. For any two ellipsoids E_i, E_j from $\mathcal{E}$, all feasible direction vectors are normalizations of points from the difference set $E_i - E_j$. The most promising strategy is therefore to pick the lowermost and uppermost ellipsoids E_1 and E_m to generate the directions.

Figure 6 shows an approximately optimal cylinder as computed by an implementation of the described algorithm.

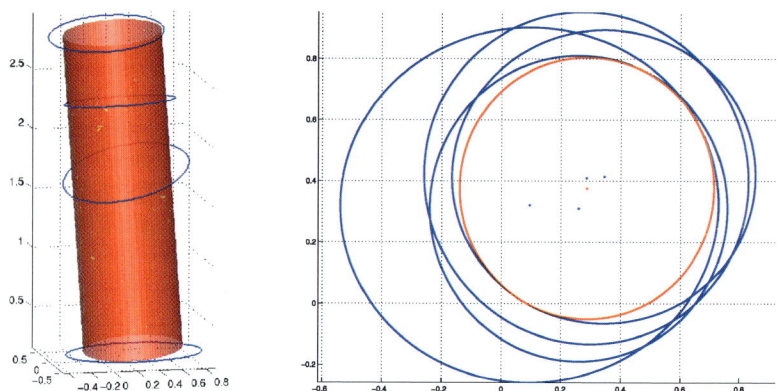

Fig. 6. Example of a traversing cylinder and its projection

5 Software

The algorithms described in Sect. 4 were implemented in **Matlab**© [49]. A new software tool (see Fig. 7, [32]) for operation planning based on three-dimensional CT data extending the **amira**© visualization software [43] was developed. Additional customized components were created in the C++ programming language.

The tool features a simulation of the osteotomy, the repositioning of the bone and the implantation of the intramedullary nail. It allows both for straight and hand-drawn cuts. The cut part of the bone can be positioned according to specified axes and points. When the position of the intramedullary nail is specified, an animation of the postoperative limb lengthening is possible. Semi-automated planning aids are provided for the determination of relevant axes and points of the musculoskeletal system via geometric approximations. Also, a test for the feasibility of a chosen nail position is included. Alternatively, the maximal corridor and an optimal nail positioning can be computed. Usually, additional medical factors like the local state of the bone tissue and the stability of the resulting bone parts affect the placement of the osteotomy and the implantation of the device. Therefore, at any stage of the

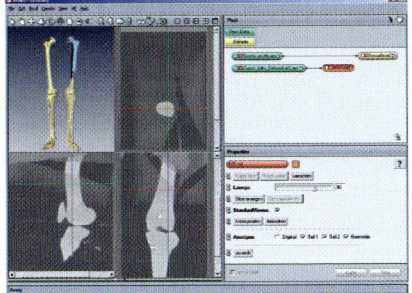 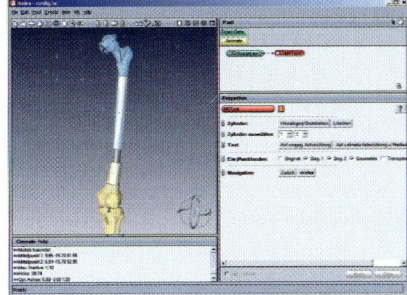

Fig. 7. 3D–Operation Planning Tool

planning process, the user is able to modify the geometric structure manually, save it, or reload it.

Acknowledgements

The project was carried out in cooperation with the Limb Lengthening Center Munich (ZEM), Germany [38]. It is a great pleasure to acknowledge stimulating discussions with Prof. Baumgart, Dr. Thaller and other members of ZEM. Funding from the German Federal Ministry of Education and Research (BMBF), grant number 03GRNGM1, is gratefully acknowledged.

References

1. P.K. Agarwal, B. Aronov, and M.Sharir. Line transversals of balls and smallest enclosing cylinders in three dimensions. *Discrete Comput. Geom.*, 21:473–388, 1999.
2. P.K. Agarwal, C.M. Procopiuc, and K.R. Varadarajan. A $(1 + \epsilon)$-approximation algorithm for 2-line-center. *Comput. Geom. Theory Appl.*, 26(2):119–128, 2003.
3. P.K. Agarwal, C.M. Procopiuc, and K.R. Varadarajan. Approximation algorithms for a k-line center. *Algorithmica*, 42:221–230, 2005.
4. N. Amenta. K-transversals of parallel convex sets. In *Proc. 8th Canadian Conf. Comput. Geom.*, pages 80–86, 1996.
5. I.N. Bankman, editor. *Handbook of Medical Imaging: Processing and Analysis*. Academic Press, Inc., Orlando, FL, USA, 2000.
6. R. Baumgart, P. Thaller, S. Hinterwimmer, M. Krammer, T. Hierl, and W. Mutschler. A fully implantable, programmable distraction nail (fitbone) – new perspectives for corrective and reconstructive limb surgery. In K.S. Leung, G. Taglang, and R. Schnettler, editors, *Practice of Intramedullary Locked Nails. New developments in Techniques and Applications*, pages 189–198. Springer Verlag Heidelberg, New York, 2006.
7. S. Bespamyatnikh and D. Kirkpatrick. Rectilinear 2-center problems. In *Proc. 11th Canadian Conf. Comput. Geom.*, pages 68–71, 1999.

8. C. Borcea, X. Goaoc, and S. Petitjean. Line transversals to disjoint balls. To appear in Discrete Comput. Geom.
9. S. Boyd, L. El Ghaoui, E. Feron, and V. Balakrishnan. *Linear matrix inequalities in system and control theory*, volume 15 of *SIAM studies in applied mathematics*. SIAM, Philadelphia, 1994.
10. R. Brandenberg and L. Roth. k-Containment. Manuscript.
11. R. Brandenberg and L. Roth. Minimal Containment under Homothetics – Solution Methods in Practice. Manuscript.
12. R. Brandenberg and T. Theobald. Algebraic methods for computing smallest enclosing and circumscribing cylinders of simplices. *Appl. Algebra Eng. Commun. Comput.*, 14(6):439–460, 2004.
13. H. Brönnimann, H. Everett, S. Lazard, F. Sottile, and S. Whitesides. Transversals to line segments in $\mathbb{R}^3$. *Discrete. Comput. Geom.*, 34(3):381–390, 2005.
14. M. Bădoiu, S. Har-Peled, and P. Indyk. Approximate clustering via core-sets. In *Proc. 34th Annu. ACM Sympos. Theory of Computing*, pages 250–257. ACM Press, 2002.
15. T.M. Chan. More planar two-center algorithms. *Comput. Geom.*, 13(3):189–198, 1999.
16. T.M. Chan. Approximating the diameter, width, smallest enclosing cylinder, and minimum-width annulus. In *Proc. 16th Annu. Sympos. Comput. Geom.*, pages 300–309, 2000.
17. B.C. Eaves and R.M. Freund. Optimal scaling of balls and polyhedra. *Math. Prog.*, 23:138–147, 1981.
18. D. Eppstein. Faster construction of planar two-centers. In *Proc. 8th Symp. Discrete Algorithms*, pages 131–138. ACM and SIAM, January 1997.
19. U. Faigle, W. Kern, and M. Streng. Note on the computational complexity of j-radii of polytopes in $\mathbb{R}^n$. *Math. Prog.*, 73(1):1–5, 1996.
20. K. Fischer, B. Gärtner, and M. Kutz. Fast smallest-enclosing-ball computation in high dimensions. In *Proc. 11th Annu. European Symp. Algorithms*, pages 630–641, 2003.
21. F. Follert. Maxmin location of an anchored ray in 3-space and related problems. In *Proc. 7th Canadian Conf. Comp. Geom.*, pages 7–12, 1995.
22. T. Gerken. On the Double-Ray Center Problem in 3-Space with an Application to Surgical Operation Planning. Diplomarbeit, Zentrum Mathematik, TU München, 2003.
23. A. Glozman, K. Kedem, and G. Shpitalnik. Computing a double-ray center for a planar point set. *J. Comput. Geom.*, 9(2):103–123, 1999.
24. P. Gritzmann and V. Klee. Inner and outer j-radii of convex bodies in finite-dimensional normed spaces. *Discrete Comput. Geom.*, 7:255–280, 1992.
25. P. Gritzmann and V. Klee. Computational complexity of inner and outer j-radii of polytopes in finite-dimensional normed spaces. *Math. Program.*, 59:163–213, 1993.
26. P. Gritzmann and V. Klee. On the complexity of some basic problems in computational convexity I: Containment problems. *Discrete Math.*, 136:129–174, 1994.
27. P. Gritzmann and T. Theobald. On algorithmic stabbing problems for polytopes. Manuscript.
28. P. Gritzmann and T. Theobald. On the computational complexity of visibility problems with moving viewpoints. In J.E. Goodman, J. Pach, and E. Welzl,

editors, *Combinatorial and Computational Geometry*, volume 52 of *MSRI Publications*, pages 377–397. Cambridge University Press, 2005.
29. L. Gürke, W. Strecker, and S. Martinoli. Korrektur mehrdimensionaler Deformationen durch eine einzige Osteotomie: graphische Analyse und Operationstechnik. *Der Unfallchirurg*, 102:684–690, 1999.
30. S. Har-Peled and K. Varadarajan. Projective clustering in high dimensions using core-sets. In *Proc. 18th Annu. Symp. Comput. Geom.*, pages 312–318. ACM Press, 2002.
31. S. Har-Peled and K.R. Varadarajan. High-dimensional shape fitting in linear time. *Discrete Comput. Geom.*, 32:269–288, 2004.
32. D. Häublein. Entwicklung von Planungssoftware für die Extremitätenchirurgie. Diplomarbeit, Fakultät für Informatik, TU München, 2007.
33. G. Ilizarov. Clinical application of the tension-stress effect for limb lengthening. *Cli. Orthop.*, 250:8–26, 1990.
34. G. Ilizarov. *Transosseous osteosynthesis: Theoretical and clinical aspects of the regeneration and growth of tissue*. Springer Verlag, Berlin, Heidelberg, New York, 1992.
35. M. Katchalski. Thin sets and common transversals. *J. Geometry*, 14(2):103–107, 1980.
36. P. Kumar. *Clustering and reconstructing large data sets*. PhD thesis, Department of Computer Science, Stony Brook University, 2004.
37. P. Kumar, J.S.B. Mitchell, and E.A. Yıldırım. Approximate minimum enclosing balls in high dimensions using core-sets. *J. Exp. Algorithmics*, 8, 2003.
38. Limb Lengthening Center Munich (ZEM), www.beinverlaengerung.de.
39. J. Maintz and M. Viergever. A survey of medical image registration. *Medical Image Analysis*, 2(1):1–36, 1998.
40. N. Megiddo. On the complexity of some geometric problems in unbounded dimension. *J. Symbolic Comp.*, 10:327–334, 1990.
41. N. Megiddo and K.J. Supowit. On the complexity of some common geometric location problems. *Siam J. Comput.*, 10:182–196, 1990.
42. N. Megiddo and A. Tamir. On the complexity of locating linear facilities in the plane. *Oper. Res. Lett.*, 1:194–197, 1982.
43. Mercury Computer Systems, Inc., www.mc.com.
44. D. Paley. *Principles of Deformity Correction*. Springer, 2002.
45. S. Rittsteiger. Shape Fitting Algorithmen–Theorie, Implementation und Anwendung in der chirurgischen Operationsplanung. Diplomarbeit, Zentrum Mathematik, TU München, 2006.
46. L. Roth. Exakte und ϵ-approximative Algorithmen zur Umkugelberechnung. Diplomarbeit, Zentrum Mathematik, TU München, 2005.
47. J.C. Russ. *The Image Processing Handbook*. CRC Press, 5th edition, 2006.
48. E. Schömer, J. Sellen, M. Teichmann, and C.-K. Yap. Smallest enclosing cylinders. *Algorithmica*, 27:170–186, 2000.
49. The MathWorks, www.mathworks.com.
50. K.R. Varadarajan, S. Venkatesh, Y. Ye, and J. Zhang. Approximating the radii of point sets. *SIAM J. Comput.*, 36(6):1764–1776, 2007. Preliminary versions in Proc. 43th IEEE Symp. Foundations of Computing, IEEE Comp. Soc., 2002, 561–569 and in APPROX 2003 + RANDOM 2003, LNCS, vol. 2764, Springer-Verlag, 2003, 178–187.

51. H. Yu, P.K. Agarwal, R. Poreddy, and K.R. Varadarajan. Practical methods for shape fitting and kinetic data structures using core sets. In *Proc. 20th Annu. ACM Sympos. Comput. Geom.*, pages 263–272, 2004.
52. G.L. Zhou, K.C. Toh, and J. Sun. Efficient algorithms for the smallest enclosing ball problem. *Comput. Optim. Appl.*, 30(2):147–160, 2005.

Image Segmentation for the Investigation of Scattered-Light Images when Laser-Optically Diagnosing Rheumatoid Arthritis

Herbert Gajewski[1], Jens A. Griepentrog[1], Alexander Mielke[1], Jürgen Beuthan[2], Urszula Zabarylo[2], and Olaf Minet[2]

[1] Forschungsverbund Berlin e.V., Weierstraß-Institut für Angewandte Analysis und Stochastik, Mohrenstraße 39, 10117 Berlin, Germany
{gajewski,griepent,mielke}@wias-berlin.de
[2] Charité Universitätsmedizin Berlin, Campus Benjamin Franklin, Institut für Medizinische Physik und Lasermedizin, Fabeckstraße 60–62, 14195 Berlin, Germany
{juergen.beuthan,urszula.zabarylo,olaf.minet}@charite.de

1 Introduction

With 1–2 % of the population affected, rheumatoid arthritis (RA) is the most frequent inflammatory arthropathy. In most cases, RA initially affects the small joints, only, especially the finger joints. The inflammations of the joints caused by this disease usually start with a synovitis. At the same time, there is a change in the filtration properties of the synovialis, which increases the enzyme rate within the synovia thus accelerating the progress of inflammation.

In a later stage, granulation and neovascularisation occur in the synovia (Figs. 1 and 2), which may finally lead to the destruction of cartilage and bone structures [1]. So, it is rather unsurprising that the optical parameters [2, 3] (Table 1) change in these early stages of the disease.

The examination of scattered-light images in laser-optical diagnostics opens up new possibilities in medicine on the basis of tissue optics [4, 5]. For quantitative prognoses on the successful application of lasers for diagnostics and therapy in humans, one has to understand how light propagates in biological tissues. Various scientific studies in this field used Monte-Carlo simulations (MCS) for this purpose [6].

Any information gained through the light from the interior of the body differs decisively from that acquired by radiological procedures like radiograms or NMR. The light is strongly scattered and reflected on the boundaries in almost any biological tissue which makes conventional projecting impossible. Compared to optical imaging in geometrical optics, both shape and size of

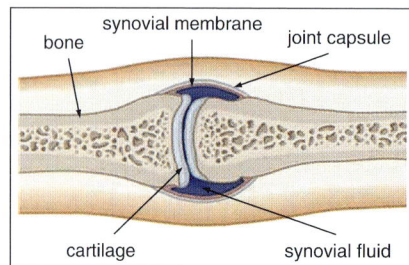 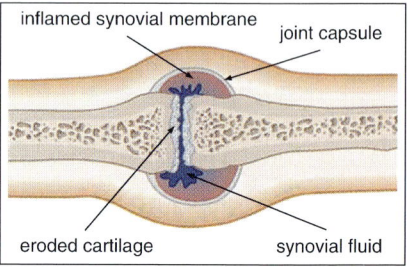

Fig. 1. Schematic of a small joint (*left*) healthy; (*right*) rheumatoid arthritis; compare [1]

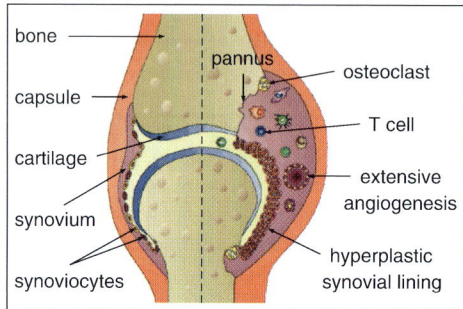

Fig. 2. Detail of a joint in an early stage of rheumatoid arthritis (according to Fig. 1)

Table 1. Optical parameters (ex vivo) of healthy and rheumatic human finger joints at 685 nm [3]; mean values of 14 samples

Tissue	healthy finger joint		rheumatic finger joint	
	μ_a (1/mm)	μ_s (1/mm)	μ_a (1/mm)	μ_s (1/mm)
skin	0,02	1,95	0,02	1,95
bone	0,08	2,1	0,08	2,1
cartilage	0,17	1,8	0,17	1,8
articular capsule	0,15	0,6	0,24	1,1
synovia	0,004	0,006	0,011	0,012

the image details are decisively influenced by scattering and absorption in the tissue. A medical application example is the excitation and detection of fluorescent light in superficial tumours for an exact evaluation and resection of the tumour at some safety margin for healthy tissue.

We investigate the finger joint screening by red laser light in order to diagnose early rheumatic inflammations on finger joints. Thanks to their small size, the joints are well suited for this non-invasive examination method [2, 7, 8, 9]. A first clinical study [10] showed that laser diaphanoscopy of finger joints, if complemented by adapted image processing, may be a valuable additional tool for follow-ups of inflamed joints. In this process dominant gray shades are assigned to the significant image details (bones, cartilage and synovia) which are interpreted as phases. Consequently, the scattered light appears as a mixture of particles of different phases which are subjected to a non-local separation with the aim of being segmented. On this basis a new algorithm for image segmentation and reconstruction was developed [11, 12].

2 Optical Imaging

2.1 Tissue Optics

Any material is optically characterized by the absorption coefficient μ_a, the scattering coefficient μ_s, and the anisotropy factor g (mean cosine of the scattering angle) at different wavelengths. Due to the low absorption and high scattering in the tissue, the light in the near infrared (NIR) can deeply penetrate into the tissue. The light propagation in homogeneous scattering media is determined by the radiative transfer equation [13]:

$$\frac{dI(r,\hat{s})}{ds} = -\mu_t I(r,\hat{s}) + \mu_s \int_S p(\hat{s},\hat{s}') I(r,\hat{s}') \, d\Omega',$$

where $d\Omega'$ is the differential solid angle in $\hat{s}'$ direction, while S stands for the unit sphere. The left term of the transfer equation determines the change rate of the intensity at point r in $\hat{s}$ direction. The right term describes the intensity loss due to the total interaction $\mu_t = \mu_a + \mu_s$ and the intensity gained by light scattering from all other directions in the direction $\hat{s}$, whereas μ_t merely indicates the attenuation of the so-called ballistic photons of the incident beam. The phase function $p(\hat{s},\hat{s}')$ describes the scattering of the light, which comes from the direction $\hat{s}$ and is diverged to $\hat{s}'$. The approximation of the phase function often used in tissue optics was proposed in another context by Henyey and Greenstein back in 1941 [14]:

$$p_{\mathrm{HG}}(\cos\theta) = \frac{1}{2} \frac{1-g^2}{(1+g^2-2g\cos\theta)^{3/2}}.$$

The anisotropy factor g is within the interval $[-1, 1]$ and indicates how strongly the scattering deviates from isotropic conditions ($g = 0$). In the case of biological applications the strongly forward-directed scattering predominates, i.e. $g > 0$. In most practical problems the transport equation cannot be analytically solved. The stochastic method of Monte-Carlo simulation has been used since 1983 [15] to model the interaction between light and tissue. Meanwhile, the MCS has become a standard method to calculate the laser light distribution in tissue, with a large number of photon trajectories being calculated based on the probability density functions for scattering and absorption; refer [16].

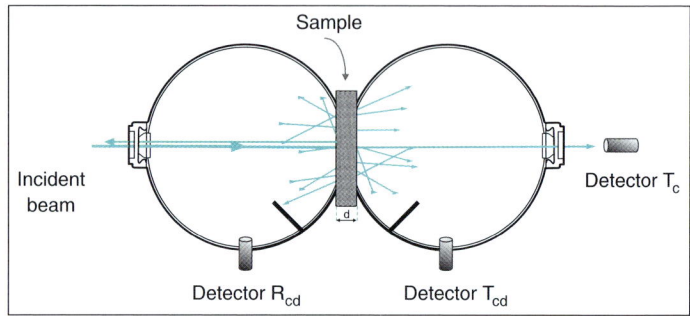

Fig. 3. Double integrating sphere for measuring the reflectance and transmittance of a tissue layer

The light distribution inside the material, the diffuse reflectance as well as the diffuse and the collimated transmittance of a substance can be calculated from these parameters using the MCS. In order to determine these microscopic coefficients, however, the inverse method is employed. The diffuse and collimated intensities are measured for optically thin samples (Fig. 3) and the parameters are adapted to the measured values using the MCS and the gradient method [17]. The data for various kinds of tissue affected by RA are listed in Table 1.

2.2 Experimental Set-Up

The experimental set-up for scattered-light measurements on finger joints is shown in Fig. 4; refer [8]. A laser diode (Lasiris, 670 nm, max. 5 mW cw)

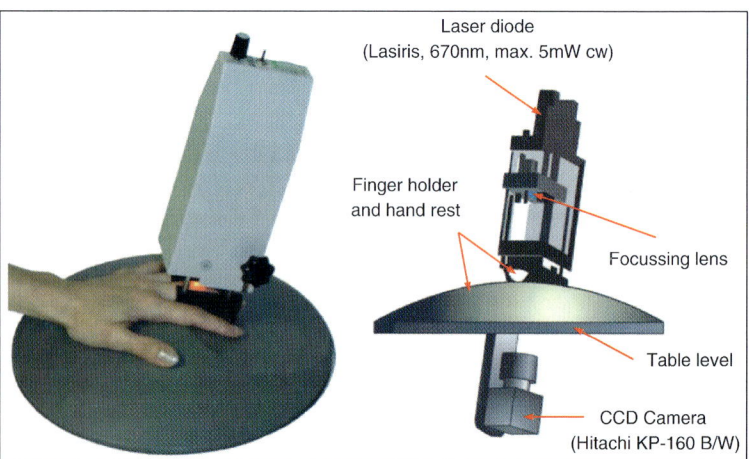

Fig. 4. Experimental set-up for measuring the scattered-light distribution on finger joints [8]: The system consists of a laser diode, a camera and an ergonomically adapted hand rest with finger holder

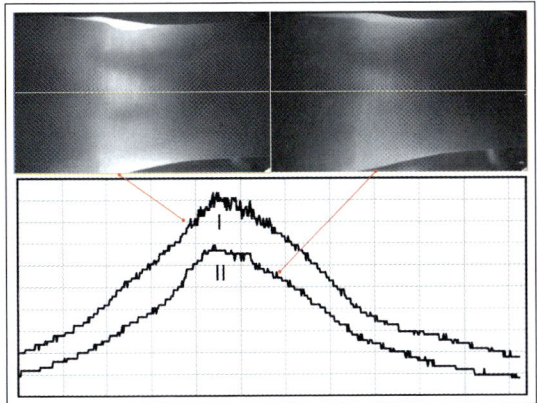

Fig. 5. Scattered-light distributions on a healthy (I) and a rheumatoid (II) finger joint

screens the finger joint. The scattered-light image is taken by a CCD camera (Hitachi KP-160 B/W) on the opposite side.

Figure 5 shows two examples for the gradient of the scattered-light intensities measured on a healthy finger joint (curve I) and a diseased one (curve II). Scheel et al. [10] have shown that laser diaphanoscopy of finger joints may be a valuable contribution to sensitive follow-ups of inflamed joints. For the classification of one-dimensional scattered-light distributions a neuronal net was used; refer Fig. 5.

The potential for medical diagnostics is, however, limited by the light scatter in the skin. This scatter caused by the skin does not contain any useful information and can be eliminated by deconvolution, which increases the diagnostic value of this non-invasive optical procedure. For known optical parameters the kernel of the deconvolution operator can be exactly constructed by means of the MCS [9].

3 Mathematical Algorithms for Image Segmentation

Contrary to segmentation procedures based on solving evolution equations (refer [18, 19, 20]), our algorithm for non-local image segmentation is based on a direct descent method for the free energy of the image, refer [11, 12].

We acquire the (already normalized) grayscale picture, the pixels of which cover a domain $\Omega \subset \mathbb{R}^n$, by a measurable function $c : \Omega \to [0, 1]$. In order to segment this picture c in respect of the given gray tones

$$a_0, \ldots, a_m \in [0, 1], \quad 0 = a_0 < \cdots < a_m = 1,$$

the following algorithm is used.

3.1 Decomposition into Phases

The aim is to decompose the picture c into several phases
$$u_0^0, \ldots, u_m^0 : \Omega \to [0, 1],$$
whereby the i-th component u_i^0, each, corresponds to the gray tone $a_i \in [0, 1]$:
$$0 \leq u_0^0, \ldots, u_m^0 \leq 1, \quad \sum_{i=0}^{m} u_i^0 = 1. \tag{1}$$

For this purpose we choose a continuous partition $\zeta_0, \ldots, \zeta_m : [0, 1] \to [0, 1]$ of unity and weights $b_0, \ldots, b_m \in \mathbb{R}$ in such a way that
$$0 \leq \zeta_i \leq 1, \quad \zeta_i(a_i) = 1, \quad \sum_{i=0}^{m} \zeta_i = 1, \tag{2}$$
$$0 < b_i < 1, \quad \sum_{i=0}^{m} b_i = 1, \quad \int_0^1 \zeta_i(s)\, ds = b_i \tag{3}$$
hold for each $i \in \{0, \ldots, m\}$ and define the transformation
$$c \mapsto u^0 = (u_0^0, \ldots, u_m^0) = (\zeta_0(c), \ldots, \zeta_m(c)). \tag{4}$$

Using the following partition of unity for concrete simulations, we choose weights $b_0, \ldots, b_m \in (0, 1)$ with the property
$$\sum_{j=0}^{m} b_j = 1, \quad b_i^* = \sum_{j=0}^{i-1} b_j \in (a_{i-1}, a_i) \quad \text{for all } i \in \{1, \ldots, m\},$$
and define exponents $\omega_i > 0$ and continuous functions $h_i : [a_{i-1}, a_i] \to \mathbb{R}$ by
$$\omega_i = \frac{a_i - b_i^*}{b_i^* - a_{i-1}}, \quad h_i(s) = \left(\frac{a_i - s}{a_i - a_{i-1}}\right)^{\omega_i} \quad \text{for } s \in [a_{i-1}, a_i].$$

Then, we get a partition $\zeta_0, \ldots, \zeta_m : [0, 1] \to [0, 1]$ of unity with the properties (2) and (3) by setting
$$\zeta_0(s) = \begin{cases} h_1(s) & \text{if } s \in [a_0, a_1], \\ 0 & \text{otherwise,} \end{cases}$$
and
$$\zeta_m(s) = \begin{cases} 1 - h_m(s) & \text{if } s \in [a_{m-1}, a_m], \\ 0 & \text{otherwise,} \end{cases}$$
for the cases $i = 0$ and $i = m$ as well as
$$\zeta_i(s) = \begin{cases} 1 - h_i(s) & \text{if } s \in [a_{i-1}, a_i], \\ h_{i+1}(s) & \text{if } s \in [a_i, a_{i+1}], \\ 0 & \text{otherwise,} \end{cases}$$
in the case $i \in \{1, \ldots, m-1\}$.

3.2 Non-Local Phase Separation

The multi-component distribution $u^0 = (u_0^0, \ldots, u_m^0)$ defined by (4) is subjected to a non-local phase separation process, which minimizes the free energy F of the system under the constraint of mass conservation. For this purpose, we developed a descent method delivering a sequence (u^k) of intermediate states and converging to an equilibrium state u^*, refer [11].

Equilibrium distributions $u^* = (u_0^*, \ldots, u_m^*) : \Omega \to [0,1]^{m+1}$ of the multi-component system are such states $u = (u_0, \ldots, u_m)$, in which the free energy $F(u) = \Phi(u) + \Psi(u)$, defined as the sum of the segmentation entropy

$$\Phi(u) = \int_\Omega \sum_{i=0}^m u_i \log(u_i)\, dx, \tag{5}$$

and the non-local interaction energy

$$\Psi(u) = \frac{1}{2} \int_\Omega \sum_{i,j=0}^m \left(u_i\, K_{ij} u_j + (u_i - u_i^0)\, L_{ij}(u_j - u_j^0) \right) dx, \tag{6}$$

has a critical point under the constraint of mass conservation

$$\int_\Omega u\, dx = \int_\Omega u^0\, dx. \tag{7}$$

To describe the non-local interactions between the type i and $j \in \{0, \ldots, m\}$ particles, we introduce compact linear operators K_{ij} and L_{ij} in (6), selecting the maps K_{ij} in such a way that particles of similar type drag on while particles of different type are repellent, which leads to the desired phase separation. At the same time, the suitable selection of the operators L_{ij} ensures that the final state u^* is kept close to the initial value u^0.

In our concrete applications the interaction operators K_{ij} and L_{ij} are defined as solution maps to elliptic boundary value problems with Neumann boundary conditions:

$$-\varrho_{ij}^2 \Delta(K_{ij}\psi) + K_{ij}\psi = \sigma_{ij}\psi \quad \text{in } \Omega, \quad \nu \cdot \nabla(K_{ij}\psi) = 0 \quad \text{on } \partial\Omega, \tag{8}$$

$$-r_{ij}^2 \Delta(L_{ij}\psi) + L_{ij}\psi = s_{ij}\psi \quad \text{in } \Omega, \quad \nu \cdot \nabla(L_{ij}\psi) = 0 \quad \text{on } \partial\Omega. \tag{9}$$

Here, we give the respective effective ranges $\varrho_{ij}, r_{ij} > 0$, and intensities $\sigma_{ij}, s_{ij} \in \mathbb{R}$ of the interactions between the type i and $j \in \{0, \ldots, m\}$ particles, with both matrices being assumed to be symmetric. The cases $\sigma_{ij} > 0$ and $\sigma_{ij} < 0$, respectively, correlate with the repellent and dragging interactions.

We minimize the free energy $F = \Phi + \Psi$ of the multi-component system under the constraint (7) by solving the appropriate Euler–Lagrange equations, with the following descent method being employed: Assuming $\tau \in (0, 1]$ is a suitably chosen relaxation parameter and u^0 is the initial distribution and knowing the old state u^k, the intermediate state v^k and the corresponding Lagrange multiplier $\lambda^k \in \mathbb{R}$ are uniquely determined by solving the auxiliary problem

$$\lambda^k = D\Phi\left(v^k\right) + D\Psi\left(u^k\right), \quad \int_\Omega v^k \, dx = \int_\Omega u^0 \, dx, \tag{10}$$

with a strongly monotone operator. As the new state, we define

$$u^{k+1} = \tau v^k + (1-\tau) u^k. \tag{11}$$

Our analytical investigations in [11] have shown that for $k \to \infty$ both the sequence $\left(u^k, \lambda^k\right)$ converges to a solution (u^*, λ^*) of the Euler–Lagrange equations

$$\lambda^* = D\Phi(u^*) + D\Psi(u^*), \quad \int_\Omega u^* \, dx = \int_\Omega u^0 \, dx, \tag{12}$$

and the sequence $\left(F\left(u^k\right)\right)$ of the free energies converges monotonically decreasing to the limit value $F(u^*)$.

3.3 Image Composition of the Segmented Phases

In the last step we calculate the segmented image $c^* : \Omega \to [0,1]$ as convex combination of the gray values $a_0, \ldots, a_m \in [0,1]$ in terms of the weight functions $u_0^*, \ldots, u_m^*$:

$$u^* \mapsto c^* = \sum_{i=0}^m u_i^* a_i.$$

3.4 Numerical Implementation

Basing on our analytical results we are in a position to carry out numerical simulations, which consequently implement right into the sphere of machine accuracy the descent method (5)–(11) earlier described herein, in particular its analytical convergence. For this purpose, we have developed a stand-alone, extendable and portable programme for the segmentation of gray tone pictures in respect of an arbitrary number of gray tones. Our programme is based on two widely used libraries, the sources of which are freely accessible. It can be used as a stand-alone programme, but also be integrated into existing image processing programmes. Data transfer, input and output processes are organized using the *netpbm* library for image files in *portable gray map* format. The descent method parameters can be imported through a separate control and configuration file. The rectangular geometry of the image file permits the discrete Fourier transformation to be used for effectively solving the linear elliptic boundary value problems (8) and (9) with the library *fftw* being utilized for that purpose. Due to the strong monotonicity of the related operator in the auxiliary problem (10) the Lagrange multipliers can be quickly and precisely calculated by means of the Newton method.

4 Simulation Results

Our image segmentation algorithm is simulated based on the reconstruction of a grainy test pattern as well as on the segmentation and evaluation of scattered-light images of finger joints, with two-dimensional images being represented naturally as rectangular domains $\Omega \subset \mathbb{R}^2$. The interaction range is given in the unit of length related to the problem, i.e., related to the edge length of a square pixel. Generally speaking, our method can also be used for the segmentation of image data which are available in multi-dimensional domains $\Omega \subset \mathbb{R}^n$.

4.1 Reconstruction of a Grainy Test Pattern

Subjecting our new method to a performance test, we start with reconstructing an artificially generated grainy test pattern in respect of three gray values ($m = 2$)

$$a_0 = 0, \ a_1 = \frac{49}{100}, \ a_2 = 1, \quad \text{of the weights} \quad b_0 = \frac{39}{100}, \ b_1 = \frac{22}{100}, \ b_2 = \frac{39}{100}$$

and the parameters

$$\varrho_{ij} = r_{ij} = 2, \quad (\sigma_{ij}) = \begin{pmatrix} -10 & +10 & +10 \\ +10 & -10 & +10 \\ +10 & +10 & -10 \end{pmatrix}, \quad (s_{ij}) = \begin{pmatrix} +12 & -12 & -12 \\ -12 & +12 & -12 \\ -12 & -12 & +12 \end{pmatrix}$$

according to (6), (8) and (9). Due to the selected matrix (σ_{ij}) particles of different type are strongly repellent while particles of similar type are dragging on at equal force. At the same time, the inverted sign structure of the matrix (s_{ij}) keeps the reconstructed image close to the initial value u^0. Figure 6 shows the numerical results of the grainy test pattern with 200×200 pixels. A satisfactory result is delivered, of course, not before all three components will have been reconstructed.

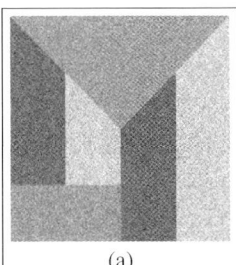 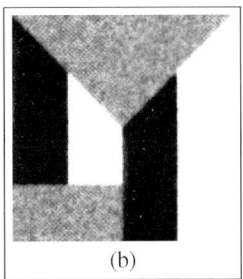 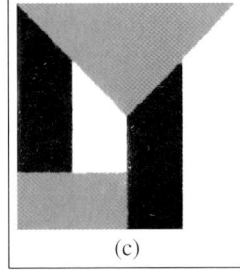

(a) (b) (c)

Fig. 6a–c. Image reconstruction: (**a**) Initial value (grainy pattern); (**b**) reconstruction result referring to the components black and white (gray region remains grainy); (**c**) reconstruction result referring to all three components, i.e. black, gray and white

4.2 Segmentation and Evaluation of Medical Scattered-Light Images

In our joint project we use the non-local image segmentation method for the investigation of scattered-light images for laser-optically diagnosing rheumatoid arthritis. This disease is the most frequent inflammatory arthropathy in humans. Initially, it typically affects the small joints, especially the finger joints. The disease starts with an inflammation of the capsular ligaments as a consequence of which the synovia opacifies. With the disease progressing granulation and neovascularisation develop in the capsular ligaments, and both cartilage and bone structures are destroyed.

We demonstrate our new method exemplarily by means of two finger joints of the right hand of a patient suffering from rheumatoid arthritis. So doing, we took one each scattered image of 385×209 pixels, of both joints at two times at an interval of circa six months. Figures 7 und 8 present not only the originals, but also the results of the image segmentation referring to the four gray values ($m = 3$) black, dark gray, light gray and white,

$$a_0 = 0, \; a_1 = \frac{1}{3}, \; a_2 = \frac{2}{3}, \; a_3 = 1, \quad \text{the weights} \quad b_0 = b_1 = b_2 = b_3 = \frac{1}{4}$$

and the parameters

$$\varrho_{ij} = r_{ij} = 4, \quad (\sigma_{ij}) = \begin{pmatrix} -4 & +4 & +4 & +4 \\ +4 & -4 & +4 & +4 \\ +4 & +4 & -4 & +4 \\ +4 & +4 & +4 & -4 \end{pmatrix}, \quad (s_{ij}) = \begin{pmatrix} +4 & -4 & -4 & -4 \\ -4 & +4 & -4 & -4 \\ -4 & -4 & +4 & -4 \\ -4 & -4 & -4 & +4 \end{pmatrix}$$

according to (6), (8) and (9). In both cases, the progress of the disease is already clearly visible by the increasing opacity of the synovia. To offer the rheumatologist an additional diagnostic option, our method calculates moreover a relative distance of the original scattered-light image u^0 to the result u^* of the respective image segmentation by means of squared expressions like

$$\int_\Omega \sum_{i=0}^m |u_i^* - u_i^0|^2 \, dx \quad \text{or} \quad \int_\Omega \sum_{i,j=0}^m (u_i^* - u_i^0) \, L_{ij} (u_j^* - u_j^0) \, dx,$$

refer (6). This distance gets larger for both the two finger joints in the course of time which we interpret as a progress of the disease. This is based on the observation that the high-definition and high-contrast image of a healthy joint seems to change less during image segmentation than the obliterated low-contrast image of a diseased joint.

Our figures merely show the results of two selected finger joints of the diseased patient. If all the investigated finger joints are considered, our assessment of the progress of the disease corresponds with the radiologist's diagnosis to 70–80 %.

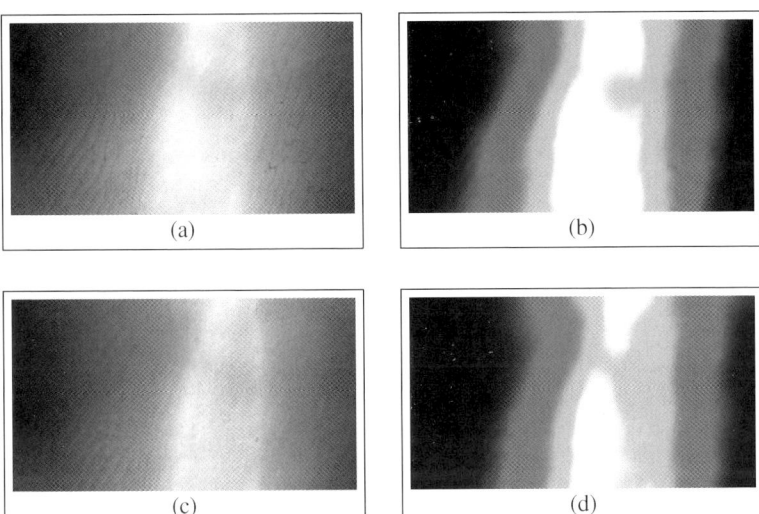

Fig. 7a–d. Progressing rheumatoid arthritis of a finger joint on the index of the patient's right hand: (**a**) Scattered-light image of an earlier date; (**b**) respective image segmentation result; (**c**) scattered-light image of a later date; (**d**) respective image segmentation result

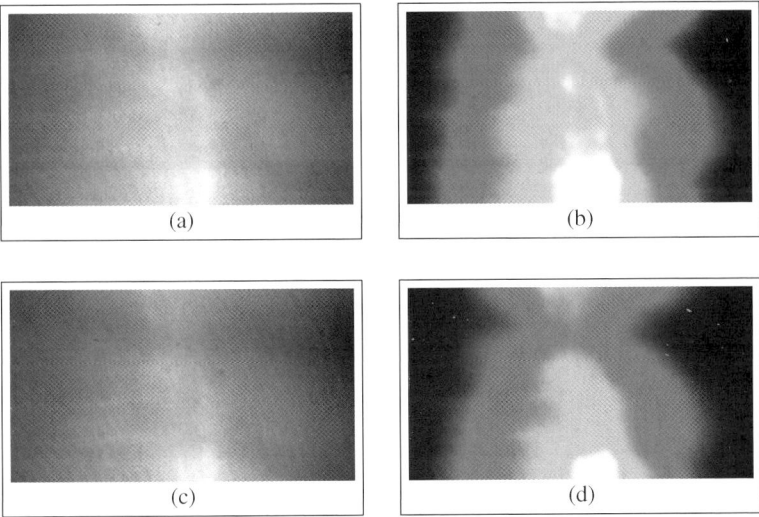

Fig. 8a–d. Progressing rheumatoid arthritis of a finger joint on the little finger of the patient's right hand: (**a**) Scattered-light image of an earlier date; (**b**) respective image segmentation result; (**c**) scattered-light image of a later date; (**d**) respective image segmentation result

Acknowledgements

We would like to express our thanks to the Federal Ministry of Education and Research (BMBF) for supporting our joint project (grant no. FKZ 03BENGB6 and 03GANGB5). In addition, our thanks are expressed to Dr. Scheel, of Georg-August-Universität Göttingen, Medizinische Klinik für Nephrologie und Rheumatologie, for the anonymized patient images provided to us.

References

1. Steinbrocker, O., Traeger, C.H., Batermann, R.C.: Therapeutic criteria in rheumatoid arthritis. *J. Amer. Med. Ass. 140*, 659–662 (1949).
2. Beuthan, J., Minet, O., Müller, G., Prapavat, V.: IR-Diaphanoscopy in Medicine. *SPIE Institute Series 11*, 263–282 (1993).
3. Prapavat, V., Runge, W., Mans, J., et al.: Development of a finger joint phantom for the optical simulation of early stages of rheumatoid arthritis. *Biomed. Tech. 42*, 319–326 (1997).
4. Tuchin, V.: *Tissue Optics: Light Scattering Methods and Instruments for Medical Diagnosis*. Bellingham: SPIE Press, 2000.
5. Minet, O., Beuthan, J.: Investigating laser-induced fluorescence of metabolic changes in Guinea pig livers during Nd:YAG laser irradiation. *Las. Phys. Lett. 2*, 39–42 (2005).
6. Minet, O., Dörschel, K., Müller, G.: Lasers in Biology and Medicine. In: Poprawe, R., Weber, H., Herziger, G. (eds.): *Laser Applications. Landolt–Börnstein Vol. VIII/1c*, 279–310. Heidelberg, New York: Springer, 2004.
7. Beuthan, J., Prapavat, V., Naber, R., Minet, O., Müller, G.: Diagnostic of Inflammatory Rheumatic Diseases with Photon Density Waves. *Proc. SPIE 2676*, 43–53 (1996).
8. Beuthan, J., Netz, U., Minet, O., Klose, A., Hielscher, A., Scheel, A., Henniger, J., Müller, G.: Light scattering study of rheumatoid arthritis. *Quantum Electronics 32*, 945–952 (2002).
9. Minet, O., Zabarylo, U., Beuthan, J.: Deconvolution of laser based images for monitoring rheumatoid arthritis. *Las. Phys. Lett. 2*, 556–563 (2005).
10. Scheel, A.K., Krause, A., Mesecke-von Rheinbaben, I., Metzger, G., Rost, H., Tresp, V., Mayer, P., Reuss-Borst, M., Müller, G.A.: Assessment of proximal finger joint inflammation in patients with rheumatoid arthritis using a novel laser-based imaging technique. *Arthritis Rheum. 46*, 1177–1184 (2002).
11. Gajewski, H., Griepentrog, J.A.: A descent method for the free energy of multicomponent systems. *Discrete Contin. Dynam. Systems 15*, 505–528 (2006).
12. Minet, O., Gajewski, H., Griepentrog, J.A., Beuthan, J.: The analysis of laser light scattering during rheumatoid arthritis by image segmentation. *Las. Phys. Lett. 4*, 604–610 (2007).
13. Ishimaru, A.: *Wave propagation and scattering in random media*. New York: Academic Press, 1978.
14. Henyey, L.G., Greenstein, J.L.: Diffuse radiation in the galaxy. *Astrophys. J. 93*, 70–83 (1942).
15. Wilson, B.C., Adam, G.: Monte Carlo model for the absorption and flux distributions of light in tissue. *Med. Phys. 10*, 824–830 (1983).

16. Henniger, J., Minet, O., Dang, H.T., Beuthan, J.: Monte Carlo Simulations in Complex Geometries: Modeling Laser Transport in Real Anatomy of Rheumatoid Arthritis. *Las. Phys. 13*, 796–803 (2003).
17. Roggan, A., Minet, O., Schröder, C., Müller, G.: Measurements of optical tissue properties using integrating sphere technique. *SPIE Institute Series 11*, 149–165 (1993).
18. Morel, J.-M., Solimini, S.: *Variational Methods in Image Segmentation*. Boston, Basel, Berlin: Birkhäuser, 1994.
19. Gajewski, H., Gärtner, K.: On a nonlocal model of image segmentation. *Z. Angew. Math. Phys. 56*, 572–591 (2005).
20. Broser, P.J., Schulte, R., Lang, S., Roth, A., Helmchen, F., Waters, J., Sakmann, B., Wittum, G.: Nonlinear anisotropic diffusion filtering of three-dimensional image data from two-photon microscopy. *J. Biomed. Optics 9*, 1253–1264 (2004).

Part V

Transport, Traffic, Energy

Dynamic Routing of Automated Guided Vehicles in Real-Time

Ewgenij Gawrilow[1], Ekkehard Köhler[2], Rolf H. Möhring[1], and Björn Stenzel[1]

[1] Technische Universität Berlin, Institut für Mathematik, MA 6–1,
  Straße des 17. Juni 136, 10623 Berlin, Germany
  {gawrilow,moehring,stenzel}@math.tu-berlin.de
[2] Brandenburgische Technische Universität Cottbus, Institut für Mathematik,
  Postfach 10 13 44, 03013 Cottbus, Germany
  ekoehler@math.tu-cottbus.de

Summary. Automated Guided Vehicles (AGVs) are state-of-the-art technology for optimizing large scale production systems and are used in a wide range of application areas. A standard task in this context is to find efficient routing schemes, i.e., algorithms that route these vehicles through the particular environment. The productivity of the AGVs is highly dependent on the used routing scheme.

In this work we study a particular routing algorithm for AGVs in an automated logistic system. For the evaluation of our algorithm we focus on Container Terminal Altenwerder (CTA) at Hamburg Harbor. However, our model is appropriate for an arbitrary graph. The key feature of this algorithm is that it avoids collisions, deadlocks and livelocks already at the time of route computation (conflict-free routing), whereas standard approaches deal with these problems only at the execution time of the routes. In addition, the algorithm considers physical properties of the AGVs and certain safety aspects implied by the particular application.

1 Introduction

Automation of large scale logistic systems is an important method for improving productivity. Often, in such automated logistic systems Automated Guided Vehicles (AGVs) are used for transportation tasks. Especially, so called free-ranging AGVs are more and more used since they add a high flexibility to the system. The control of these AGVs is the key to an efficient transportation system that aims at maximizing its throughput.

In this work we focus on the problem of routing AGVs. This means we study how to compute good routes on the one hand and how to avoid collisions on the other hand. Note that dispatching of AGVs, i.e., the assignment of transportation tasks to AGVs, is not part of the routing and therefore not considered in this paper.

Fig. 1. The HHLA Container Terminal Altenwerder (CTA). ©HHLA

Our application is the Container Terminal Altenwerder (see Fig. 1), which is operated by our industrial partner, the Hamburger Hafen und Logistik AG (HHLA).

We represent the AGV network by a particular grid-like graph that consists of roughly 10,000 arcs. and models the underlying street network of a traffic system consisting of a fleet of AGVs. The task of the AGVs is to transport containers between large container bridges for loading and unloading ships and a number of container storage areas. The AGVs navigate through the harbor area using a transponder system and the routes are sent to them from a central control unit. AGVs are symmetric, i.e., they can travel in both of the two driving directions equally well and can also change directions on a route.

Previous Work

First ideas for free-ranging AGV systems were introduced by Broadbent et al. [2]. Since then, several papers concerning this topic have been published [18]. In this paper we focus on routing approaches in the case where dispatching of AGVs is already made.

In so-called *offline* approaches all requests (transportation tasks) are known right from the beginning. Krishnamatury, Batta and Karwan [11] as well as Qui and Hsu [13] discuss the AGV routing problem in this case. While Krishnamatury, Batta and Karwan present a heuristic solution for general graphs (where this routing problem is $\mathcal{NP}$-hard [16]), Qui and Hsu consider a very simple graph and present a polynomial time algorithm.

In contrast, online approaches assume that requests appear sequentially. The *local* approach by Taghaboni-Dutta and Tanchoco [17] is such an online method. Here, a decentralized algorithm decides about the routes, based only on local information. In particular, it does not determine the whole path for an AGV, but iteratively computes sub-paths from checkpoint to checkpoint.

The *static* approach uses the full information about the already routed AGVs. Algorithms that are based on these approaches usually compute geographically shortest paths with optional additional penalty costs on congested arcs from the source to the destination of the current request [12]. This static formulation needs an additional collision avoidance system to make the routes collision-free, since time dependences are not represented in that model.

Our Contribution

In this work we study a dynamic online AGV routing model for an arbitrary graph. This approach is motivated by dynamic flow theory (see [6, 7, 10]) and several papers on the Shortest Path Problem with Time-Windows [3, 4, 5, 14]. The main advantage of our model and algorithm over the known online methods is that the time-dependent behavior of AGVs is fully modeled, such that both conflicts and deadlock situations can be prevented already at the time of route computation. The newly designed model is not only very accurate in the mapping of properties of the actual application but, as we show in our computational experiments, it is also well suited for being used in a real-world production system.

The paper is organized as follows. In Sect. 2 we describe how we model the AGV network. In Sect. 3 we introduce our algorithm and show that it runs in polynomial time. Section 4 explains how subtle technical characteristics of the particular application can be represented in our model to get it ready for being used in practice. The computational results are presented in Sect. 5.

2 The Model

We model the automated transportation system by a directed graph G representing the feasible lanes of the system. These lanes are given by certain transponder positions. In the application, this graph has about 10,000 arcs. Initially, we assume that every arc a has a fixed, constant transit time $\tau(a)$.

Transportation tasks are consecutively arriving over time and are modeled by a sequence $\sigma = r_1, \ldots, r_n$ of requests. Each request $r_j = (s_j, t_j, \theta_j)$ consists of a start node s_j, an end node t_j, and a desired starting time θ_j. The aim is to minimize the overall transit times, that is the sum of transit times over all requests. We approach this goal by iteratively computing a shortest path for each request, which is a natural method in this online setting.

The physical properties of the AGVs demand for a variety of special features of the model. Although each route of an AGV can be represented in the given graph, not every route in this graph can in fact be conducted by an AGV. The reason for this difficulty is the complicated turning behavior, which makes it necessary to start turning the wheel already long before the particular intersection is reached. As a consequence, an AGV needs a sufficiently long straight route segment between two consecutive curves. To cope with this rather complicated turning behavior we introduce in a preprocessing step a set of *artificial arcs* to the network, each representing a possible turn (see Fig. 2). In addition, at each node of the graph we introduce turning rules defining which out-going arcs of a node can be used from a particular in-going arc. As a result we get a much larger network (about 45,000 arcs) that captures all possible movements of an AGV — each feasible route in this network can be executed by an AGV.

Another complicating property is the size of the AGVs compared to the rather closely meshed network of lanes. If an AGV traverses or stands on an arc a, it affects a much larger portion of the network than only arc a, which is then blocked for other AGVs (see Fig. 3). In Sect. 2.2, we describe how we take this into account in our conflict-free (dynamic) approach.

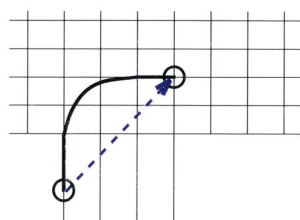

Fig. 2. The figure illustrates an artifical arc (*blue dotted arrow*) that models a curve. This is done for all permitted curves

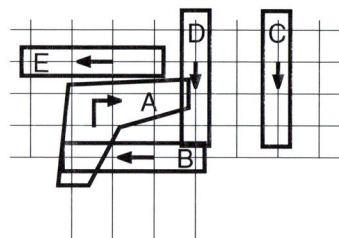

Fig. 3. The figure illustrates the polygons that are claimed by an AGV that moves in the indicated direction. Polygon A, B and D pairwise intersect each other while polygons C and E do not intersect any of the others, respectively

2.1 Static Routing and its Drawbacks

A standard approach for routing of AGVs in an online setting is the so-called static routing. In this case one only computes static routes in the network, ignoring their time dependent nature. More precisely, one computes a standard shortest path, e.g., using Dijkstra's algorithm, with respect to arc costs consisting of the transit times τ_a plus a load dependent penalty cost which is a function of the number of routes that are already using this arc (see [12]). The computed routes are, of course, not collision-free. Hence, one needs an additional conflict avoidance system that, at execution time of the routes, guarantees that there are no collisions. One way to do this is to iteratively allocate to an AGV the next part of its route (the "claim") and block it for all other AGVs ("claiming").

The advantage of the static approach is clear. It is easy to implement and allows very fast route computation. However, various drawbacks are caused by the collision avoidance at execution time. In particular, the claiming rules can cause deadlocks and have a deteriorating effect on the system performance.

Deadlocks (see Fig. 4) appear if a group of AGVs wish to claim an area which is already occupied by another AGV in this group. None of them is able to continue its route and thus the system is blocked. Algorithmic solutions for that problem are only suitable for a very small number of AGVs [9, 12].

In addition to deadlocks, a variety of other drawbacks like detours and high congestion occur in the static setting; again, since time-dependent behavior is not considered. This results in traveling times that can be far away from the shortest possible traveling time. Moreover, actual arrival times of the AGVs at their destinations are completely random and cannot be predicted at the time of route computation. This is a major drawback for other planning steps in the logistic chain that depend on the knowledge of these arrival times.

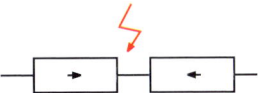

Fig. 4. Simplified deadlock situation. Both AGVs are trying to occupy the same arc of the network, thereby blocking each other

2.2 Dynamic Routing of AGVs

In order to avoid the problems of the simple model given in Sect. 2.1, we follow a completely different approach that computes shortest (w.r.t. traveling time) and conflict-free routes simultaneously.

There are two key ingredients which must be considered in our approach. On the one hand, one has to deal with the physical dimensions of the AGVs because they usually have to claim several arcs in the directed graph at the same time. On the other hand, the approach has to be time-dependent (dynamic).

Every arc can be seen as a set of time intervals, each representing a different AGV that is routed over this arc or, at least, blocks this arc during some time interval. Note that these intervals have to be mutually disjoint since an overlap would mean that the corresponding AGVs collide on this arc at the time of the overlap. In fact, in our algorithm, we will not maintain the set of intervals in which an arc is blocked, but the complementary set of free time-intervals (*time-windows*).

Maintaining these sets of intervals may be seen as a compact representation of the standard time-expanded graph, in which there is a copy of each vertex/arc for each point in time (with respect to some time discretization). In contrast, the set of time-windows of an arc a only models those times, in which there actually is no AGV on a. Similar compact representations of a time-expanded graph by time intervals have been studied before, see e.g. [3, 4, 5, 14] and Sect. 3.2.

For dealing with the physical dimensions of the AGVs we use polygons $P(a)$ for each arc a, which describe the blocked area when an AGV (the center of an AGV) is located on arc a (Fig. 3). Thus, it is prohibited to use two arcs at the same time if the corresponding polygons intersect. For each arc a, this leads to a set $confl(a)$ of so called *geographically dependent arcs* which must not be used at the same time. If an AGV travels along an arc a during the interval $[\theta_1, \theta_2]$, all geographically dependent arcs are blocked from θ_1 to θ_2. Note that in this approach there is no need to model traveling on nodes since each edges contains its end nodes.

After routing a request one has to readjust the time-windows according to the arc usage of the newly found route and their geographically dependent arcs. Note that this implies that one does not have to take care of the physical dimensions of the AGVs during route computation, since it is already fully represented by readjusting the time-windows on all affected arcs.

As mentioned before, the advantage of this approach is the fact that the problems of Sect. 2.1 are avoided because in a conflict-free approach there is no need for an additional collision avoidance since the routes are planned conflict-free in advance.

Additionally, as a welcome side effect, the completion time of a request is known immediately after route computation since the time-dependent behavior is fully modeled. This is a great advantage for a higher-level management system which plans the requests.

3 The Algorithm

The algorithm consists of two parts. The first part is a preprocessing step; during the second part all requests are routed iteratively in a real-time route

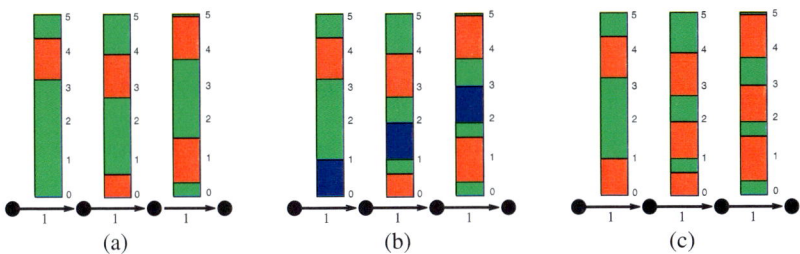

Fig. 5a–c. Illustration of the real-time computation on three consecutive arcs with transit time 1. **(a)** shows the situation before the new request arrives. There is a graph with some blockings (*red*) and some time-windows (*green*) on the time axis (*y axis*). The task is to compute a quickest path that respects the time-windows. This is illustrated in **(b)**. The chosen path is blocked afterwards (see **(c)**)

computation and for each computed route the time-windows of the affected arcs are adjusted.

The structure of the real-time computation (route computation and readjustment of time-windows) is illustrated in Fig. 5.

3.1 Preprocessing

The preprocessing step determines the conflict sets and the turning rules for each arc a. First, all polygons $P(a)$ (see Sect. 2.2) are compared pairwise. If the polygons $P(a)$ and $P(b)$ of arcs $a, b \in A(G)$ intersect, then a is added to $\mathit{confl}(b)$ and b is added to $\mathit{confl}(a)$. Second, one computes for each arc a a list $OUT(a)$ of arcs containing those arcs b that are permitted to be used after arc a on a feasible route respecting the physical properties of an AGV. This is done only once for a given layout (harbor).

3.2 Route Computation: Quickest Paths with Time-Windows

As pointed out in Sect. 2.2, the route computation can be done in an idealized model where the dimension of the AGV need no longer to be considered, since the conflict sets take care of this. Instead, one just has to compute a route for an infinitesimal mass point representing the center of the AGV.

This simplified problem is related to the *Shortest Path Problem with Time-Windows (SPPTW)* [3, 4, 5, 14] and can be formulated as follows: Given a graph G, a source node s, a destination node t, a start time θ, transit times $\tau(a)$, costs $c(a)$ and a set of time-windows $\mathcal{F}(a)$ on each arc a; compute a shortest path (w.r.t. arc costs $c(a)$) that respects the given time-windows.

Since AGVs are allowed to stop during their route, waiting is allowed on such a path. 'Respecting' the time-windows means that AGVs wait on an arc or traverse an arc a only during one of its "free" time-windows given by $\mathcal{F}(a)$.

The SPPTW and also our variant is $\mathcal{NP}$-hard. The hardness can be shown by reduction of the Constrained Shortest Path Problem (CSPP [1]).[3]

Our algorithm for this problem is a generalized arc-based label setting algorithm resembling Dijkstra's algorithm. A *label* $L = (a_L, c_L, I_L, pred_L)$ on an arc a_L consists of a cost value c_L, a predecessor $pred_L$ and a time interval I_L. Each label L represents a path from start node s to the tail of a_L, whereas c_L contains the cost value of the path up to the tail of a_L; the label interval $I_L = (A_L, B_L)$ represents an interval of possible arrival times at arc a_L (at the tail of a_L); $pred_L$ is the predecessor of a_L on that path.

We define an ordering for these labels. We say that a label L *dominates* a label L' if and only if

$$c_L \leq c_{L'} \quad \text{and} \quad I_{L'} \subseteq I_L.$$

The labels are stored in a priority queue H, e.g., a binary heap. The generalized arc-based Dijkstra algorithm works as follows.

- **Initialization**
  Create a label $L = (a, 0, (0, \infty), nil)$ for all out-going arcs a of s and add them to the priority queue H.
- **Loop**
  Take the label L with lowest cost value c_L from H. If there is no label left in the queue, output the information that there is no feasible path from s to t. If t is the tail of a_L, output the corresponding path.
  – **For** each time-window on arc a_L.
    • **Label Expansion**
      Try to expand the label interval along a_L through the time-window of the arc a_L (new label interval should be as large as possible, see Fig. 6), add the costs $c(a_L)$ to the cost value c_L and determine the new predecessor. If there is no possible expansion, consider the next time-window of arc a_L.
    • **Dominance Test**
      For each out-going arc a in $OUT(a_L)$, add the new label to the heap if it is not dominated by any other label on a. Delete the labels in the heap that are dominated by the new label.

Since the SPPTW is $\mathcal{NP}$-hard the algorithm cannot be executed in polynomial time, unless $\mathcal{P} = \mathcal{NP}$. However, the AGV routing problem differs from the SPPTW in a subtle point. In AGV routing, the cost of a path is the sum of the transit times of the arcs on the path plus waiting times on arcs which is the crucial property that makes the problem polynomial.

---

[3] The instance of the SPPTW is constructed by placing time-windows $[0, R]$ at each arc while R denotes the resource constraint in the CSPP instance.

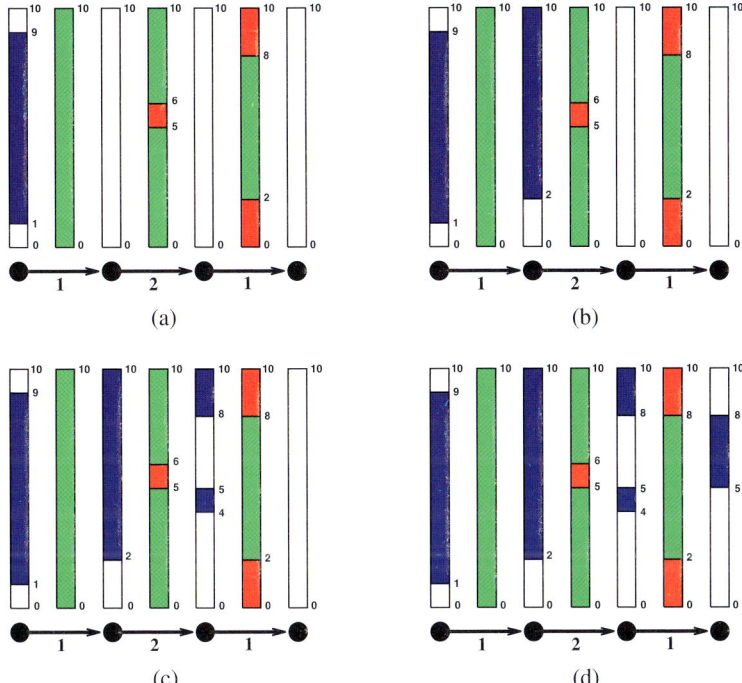

Fig. 6a–d. Label Expansion on three consecutive arcs. The label intervals are represented by *blue bars* and are placed above the nodes. The blockings are colored *red* (*arcs*). The *green* intervals between these blockings are the time-windows. The figures (**a**) to (**d**) show the successive expansion of the label intervals

Thus, the costs on an arc a (the transit time of a plus possible waiting time on a) depend no longer only on the arcs itself, but also on the routing history given by the label interval and the current time-window. We call the resulting problem the *Quickest Path Problem with Time-Windows (QPPTW)*.

For the QPPTW we can obtain a polynomial time algorithm, since here the costs correlate to the lower bounds of the label intervals.

Theorem 1. *The described generalized arc-based Dijkstra algorithm solves the QPPTW in polynomial time (in the number of time-windows).*

Proof. The algorithm computes all required paths since the expansion of the label intervals is maximal and no optimal path (label) will be dominated. Therefore, on termination the algorithm has computed an optimal path respecting the time-windows. That it terminates follows from the complexity analysis given below.

Consider the correlation between costs and traveling time (including waiting times): they differ only by an additive constant, namely the starting time.

Thus, for any two labels that are expanded w.r.t. the same time-window the cost value controls the dominance relation. Hence, one label dominates another if and only if it has a lower transit time.

Therefore, the number of possible labels on an arc a is bounded by the number of time-windows on all in-going arcs (in-going time-windows $\mathcal{F}^-(a)$). As a consequence, the number of iterations in each loop (the number of labels taken from the priority queue) is bounded from above by the product of the number of arcs and the maximum number of in-going time-windows over all arcs ($\sum_{a \in A} |\mathcal{F}^-(a)|$). In each iteration a label is expanded along at most the number of time-windows $|\mathcal{F}(a)|$ at a and each of the resulting labels is compared with at most $\sum_{b \in OUT(a)} |\mathcal{F}^-(b)|$ existing labels. If the priority queue is implemented as a heap, updating can be done in $O(\log(\sum_{a \in A} |\mathcal{F}^-(a)|))$. This leads to the following run time:

$$O\left(\left(\sum_{a \in A} |\mathcal{F}^-(a)|\right) \cdot \left(\max_{a \in A} |\mathcal{F}_a|\right) \cdot \left(\max_{a \in A} \left\{\sum_{b \in OUT(a)} |\mathcal{F}^-(b)|\right\}\right) \cdot \log\left(\sum_{a \in A} |\mathcal{F}^-(a)|\right)\right).$$

Hence, the algorithm terminates in polynomial time with an optimal path or the notification that there is no feasible path at all. □

For an additional acceleration of the algorithm we use *goal-oriented search* [8, 15].

After each route computation we verify for each arc a of the computed path, whether the time-windows on the arcs in *confl(a)* have to be readjusted.

4 Additional Practical Requirements

To make the algorithm practical, additional ingredients have to be taken into account.

4.1 Container Orientation

Since containers are not completely symmetric, it may be necessary to give an AGV an explicit target orientation. We model this orientation constraint by a flag, indicating whether the AGV is in the right driving direction to reach the target in the correct orientation without a turn.

To this end, we maintain labels at an arc for each of the two possible directions and define the domination rule accordingly. Using the observations of Theorem 1 we get the following result.

Theorem 2. *The described generalized arc-based Dijkstra algorithm solves the QPPTW in polynomial time (in the number of time-windows) even if the orientation of AGVs (containers) is taken into consideration.*

4.2 Safety Tubes and Re-Routing

In spite of the fact that the computed routes are conflict-free, additional safety is required in practice because the AGVs possibly deviate from the computed routes in time. Also technical problems may occur while traveling through the network. We have implemented two different safety tubes, a distance-dependent and a time-dependent one, and re-routing techniques to cope with this difficulty.

The distance-dependent tube blocks an area in front of the AGV. The length depends on the speed of the AGV and is at least the distance needed to come to a complete stop (braking distance). This allows the AGV to stop if something unexpected happens (for example an unexpected stop of another AGV) without causing a collision.

The time-dependent tube allows a little deviation from the computed time, i.e., the expected arrival time at a specific point. This is necessary because there will always be small differences between computed times in the model and the times when the AGVs reaches a point in reality.

In order to cope with more challenging perturbations as large deviations from the expected starting time, lower driving speeds than expected or vehicle breakdowns we also implemented re-routing strategies based on the described algorithm.

4.3 Non-Constant Transit Times

Instead of constant transit times as described in Sect. 2, we take the variable speed of the AGVs into account, i.e., we model the acceleration behavior and the different possible maximum speeds.

The maximum speed depends on the kind of movement (curve or straight section), the weather conditions, and the status of the AGV. The acceleration value depends on the current speed.

5 Computational Results

We now address two important questions with our approach.

- Is the approach better than the static one?
- Is the algorithm suitable for real-time computation?

Both questions can be answered in the affirmative. The comparison of both approaches shows that the conflict-free approach is superior to the static one (exact numbers at CTA have to be kept confidential). Additionally, the presented algorithm is able to provide fast answers. On average, the computation in all scenarios does not require more than a few hundredth of a second. And also the maximum values of less than half a second are small enough to ensure fast real-time computation in practice.

Acknowledgements

We are grateful to Andreas Parra and Kai-Uwe Riedemann from the Hamburger Hafen und Logistik AG (HHLA) and Boris Wulff from Container Terminal Altenwerder (CTA) for their valuable remarks and suggestions and for providing us with real world data.

References

1. Beasley, J. E., Christofides, N. (1989) An algorithm for the resource constrained shortest path problem. Networks 19, 379–394
2. Broadbent, A. J. et al. (1987) Free-ranging AGV and Scheduling System. In Automated Guided Vehicle Systems, 301–309
3. Desrosiers et al. (1986) Methods for routing with time windows. European Journal of Operations Research 23, 236–245
4. Desrosiers, M., Soumis, F. (1988) A generalized permanent labelling algorithm for the Shortest Path Problem with Time Windows, INFOR 26(3), 191–212
5. Desrosiers, J., Solomon, M. (1988) Time window constrained routing and scheduling problems, Transportation Science 22, 1–13
6. Ford, L. R., Fulkerson, D. R. (1959) Constructing maximal dynamic flows from static flows. Operations Research 6, 419–433
7. Ford, L. R., Fulkerson, D. R. (1962) Flows in Networks. Princeton University Press, Princeton, NJ
8. Hart, P., Nilsson, N., Raphael, B. (1968) A formal basis for the heuristic determination of minimum cost paths. In IEEE Transactions on Systems, Science and Cybernetics SCC-4, 100–107
9. Kim, K. H., Jeon, S. M., Ryu, K. R. (2006) Deadlock prevention for automated guided vehicles in automated container terminals. OR Spectrum 28 (4), 659–679
10. Köhler, E., Möhring, R. H., Skutella, M. (2002) Traffic networks and flows over time. In Jürg Kramer, Special Volume Dedicated to the DFG Research Center "Mathematics for Key Technologies Berlin", published by Berliner Mathematische Gesellschaft, 49–70, http://www.math.tu-berlin.de/coga/publications/techreports/2002/Report-752-2002.html
11. Krishnamurthy, N., Batta, R., Karwan, M. (1993) Developing conflict-free routes for automated guided vehicles. Operations Research 41, 1077–1090
12. Moorthy, K. M. R. L., Guan, W. H. (2000) Deadlock Prediction and Avoidance in an AGV System. SMA Thesis
13. Qui, J., Hsu, W.-J. (2000) Conflict-free AGV routing in a bi-directional path layout. In Proceedings of the 5th International Conference on Computer Integrated Manufacturing (ICCIM 2000), volume 1, 392–403
14. Sancho, N. G. F. (1994) Shortest path problems with time windows on nodes and arcs. Journal of mathematical analysis and applications 186, 643–648
15. Sedgewick, R., Vitter, J.S. (1986) Shortest paths in Euclidian graphs. Algorithmica 1, 31–48
16. Spenke, I. (2006) Complexity and Approximation of Static k-Splittable Flows and Dynamic Grid Flows. PhD Thesis Technische Universität Berlin

17. Taghaboni-Dutta, F., Tanchoco, J. M. A. (1995) Comparison of dynamic routeing techniques for automated guided vehicle system. International Journal of Production Research 33(10), 2653–2669
18. Vis, I. F. A. (2006) Survey of research in the design and control of automated guided vehicle systems. European Journal of Operational Research 170, 677–709

Optimization of Signalized Traffic Networks

Ekkehard Köhler[1], Rolf H. Möhring[2], Klaus Nökel[3], and Gregor Wünsch[2]*

[1] Technische Universität Cottbus, Institut für Mathematik, Postfach 10 13 44, 03013 Cottbus, Germany, `ekoehler@math.tu-cottbus.de`
[2] Technische Universität Berlin, Institut für Mathematik, MA 6–1, Straße des 17. Juni 136, 10623 Berlin, Germany
`{moehring,wuensch}@math.tu-berlin.de`
[3] PTV AG, Stumpfstraße 1, 76131 Karlsruhe, Germany
`klaus.noekel@ptv.de`

Summary. The coordination of signal controls in a traffic network has a significant impact on throughput and journey times of the vehicles in the network. We present a mixed-integer linear programming model optimizing the coordination, i.e., a model that minimizes waiting times of individual traffic by adjusting the offsets between signals. Compared to existing approaches, our model supports non-uniform cycle-lengths of signals and yields optimal solutions (w.r.t. the objective function of our model) in less computation time. We use microsimulation which shows that high quality solutions can be obtained with little computational effort.

1 Introduction

As traffic in urban street networks continues to increase, traffic engineers strive to manage capacity intelligently rather than merely add road space which may not be an option in many built-up areas. Signalized intersections are a critical element in such networks and care must be taken in the definition of the signal phasing. Not only do the cycle length and splits influence capacity, but so does signal coordination. In order to avoid (both objective and perceived) delays through frequent stops and to reduce queue waiting time and length, the signal timings at several intersections should be offset from each other in such a way, that platoons of cars can progress through the network with minimum impedance.

Optimization approaches that consider offsets between signals have been considered by Little [9], by Gartner et al. [2] and by Improta [7], among others. A genetic algorithm approach for network coordination has been developed

---
* Supported by the DFG Research Training Group GK-621 "Stochastic Modelling and Quantitative Analysis of Complex Systems in Engineering" (MAGSI). 03/2004–02/2007.

by Robertson [10] and successfully implemented in the still state-of-the-art software tool TRANSYT [10]. However, both, the absence of guarantees on the quality of solutions and the sometimes high running times are disadvantages of this tool.

In this paper we develop a mixed-integer linear program that minimizes the total delay occurring in a traffic network by adjusting optimal offsets and split modes. It is based on an approach by Gartner et al. [4], also [2] and [3] and produces high quality solutions with little computational effort.

In Sect. 2 we give a detailed description the model. In Sect. 3 we show via real-world instances that the model produces good solutions in very reasonable running times. Finally, in Sect. 3.1, we discuss different aspects of our optimization model in practice.

2 Our Approach

In our approach we consider an inner-city traffic network with fixed-time signal control at the intersections. Non-uniform cycle lengths are permitted in the network. Moreover, we assume the network to operate at near saturated condition. Hence, as already motivated by Wormleighton [13], it is justified to model the traffic macroscopically via platoons. With doing so we further assume the traffic volumes to be given link-wise. So, we set up a mathematical model that minimizes the sum of delays on all links in the network with offsets between the signals and split modes as decision variables. In Sects. 2.1 to 2.4 we discuss the meaning of the variables, the constraints that are formulated and how the evaluation of the waiting time on a link is estimated. However, it is not the scope of this paper to provide all details.

2.1 Preliminaries and the Offset Variables

We model the traffic network by a directed multi-graph $G = (V, A)$, where the vertices $v \in V$ represent the signalized intersections and the edges $a \in A$ stand for the links between the intersections. We will also use the terms node and link, respectively. The reason for allowing parallel edges is that we distinguish the traffic flow on a link w.r.t. the signal groups of the adjacent intersections that these vehicles come from and go to, respectively. However, in the following we omit the indices for different copies of a link to simplify notation.

Furthermore, we assume that each signal control in the network has a cycle length of either 60, 80, or 120 seconds, which is exogenously given. While not being overly restrictive, this constraint allows us to handle non-uniform cycle lengths.

As already mentioned above, there are two main characteristics of our model. First, it is assumed that vehicles move in platoons, see Figs. 2 and 3.

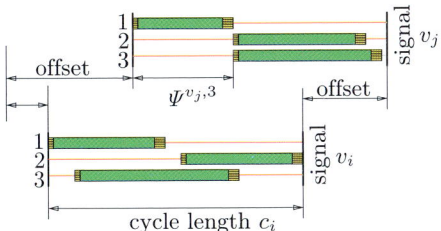

Fig. 1. An illustration of the so-called intra-node offset $\Psi^{v,i}$. It quantifies the shifting of the beginning of a green phase of a particular signal group of the signal relative to the beginning of the green of signal group 1 at that signal. In any mode $\Psi^{v,1} = 0$ for all $v \in V$

Within a platoon, non-uniform flow rates may exist. For expository reasons, this is not displayed in the figures. The model's second characteristic is that it does not use any time discretization. The offset at the signal is modelled as a continuous variable. Although in the application the offset has to be given for each intersection, it is defined here for each link. On link $e = (v_i, v_j)$ the arrival time of the platoon's head at intersection v_j is denoted by γ_{ij}. Clearly, $\gamma_{ij} \in [0, c_j]$, where c_j is the cycle length at intersection v_j. Whenever it is obvious from the context, we omit the indices of the parameters. Fig. 1 shows the intra-node offset Ψ, which is defined for each signal group at an intersection. In addition, this figure shows the two possible offset interpretations. In our approach, the offset ϕ_{ij} of link $e = (v_i, v_j)$ is defined as the distance in time between the start of the platoon at node v_i and the beginning of the green phase that defines the relevant interval for the arrival time of that platoon.

The connection between arrival time γ and offset ϕ and all other constraints are formalized in the next section.

2.2 The Constraints

Now we describe the constraints for our mixed-integer program formulation.

The first constraint specifies the dependency between the arrival time of the platoon and the associated offset. For each link $e = (v_i, v_j)$,

$$\tau_{ij} - \gamma_{ij} + r_{ij} = \phi_{ij}, \tag{1}$$

where τ_{ij} denotes the travel time on the link and r_{ij} is the length of the red phase at intersection v_j. Note that we use the notation r_{ij} instead of r_j in order to stress the assignment of a red phase to a link. However, the simple linear dependency of γ and ϕ enables us to use the arrival times γ instead of the offsets ϕ as decision variables for the model. This will be useful when we calculate the delay on a link.

Before we define the next group of constraints, we need to define the set of circuits of G. The set of circuits, denoted by $\mathcal{L}$, contains all cycles of the

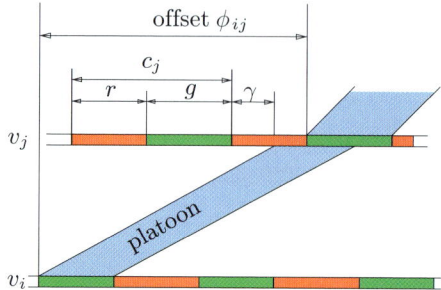

Fig. 2. The relation between parameters offset and arrival time

underlying undirected graph. Notice that a pair of parallel or anti-parallel edges form a circuit as well.

For a particular circuit $\ell \in \mathcal{L}$ that is traversed clockwise, $F(\ell)$ denotes the set of forward edges and $R(\ell)$ the set of reverse edges, respectively. Further, we define the circuit's cycle time c_ℓ as the greatest common divisor of the cycle lengths at the vertices, i.e., signals, of ℓ. So, each circuit $\ell \in \mathcal{L}$ is required to fulfill

$$\sum_{e \in F(\ell)} \phi_e - \sum_{e \in R(\ell)} \phi_e + \sum \Psi \;=\; n_\ell \cdot c_\ell, \qquad (2)$$

with $n_\ell \in \mathbb{Z}$. The periodicities expressed in (2) are physical constraints that are necessary to convert the link offsets into signal offsets.

Note that the intra-node offsets have to be included as well, since the normal offsets do not respect "changes" of signal groups at an intersection. This can be seen as follows. Consider a circuit that contains two links that share a node v. Further, assume that these edges have the same orientation with respect to the clockwise orientation of the circuit. If the edges do not share the same signal group at node v, the shifting between the two signal groups has to be taken into account.

At this point we already mention that providing good bounds for the integer variables in (2) is essential for a fast solving of the MIP. We will discuss this in more detail in Sect. 2.4.

In contrast to Gartner's approach [4], our model offers the possibility to choose between different red-green split modes. However, they need to be specified in advance. Let $\mathcal{R}_v$ be the set of signal groups at node v. Then, for a vertex $v \in V$ and a signal group $p \in \mathcal{R}_v$, the parameter $\overline{\Psi}_m^{v,p}$ denotes the intra-node offset at intersection v of signal group p in mode m, where $m = 1, \ldots, \alpha_v$. The parameter α_v stands for the number of predetermined modes at signal v. The binary variables $d_{v,m}$ then select a mode. Hence, we obtain the following two groups of constraints

$$\sum_{m=1}^{\alpha_v} d_{v,m} \cdot \overline{\Psi}_m^{v,p} \;=\; \Psi^{v,p} \qquad \forall v \in A, p \in \mathcal{R}_v, \qquad (3)$$

$$\sum_{m=1}^{\alpha_v} d_{v,m} = 1 \qquad \forall v \in A, \qquad (4)$$

where Equalities 4 ensure that exactly one mode is adjusted. In any of the predetermined modes the signal group 1 has intra-node offset 0.

2.3 The Objective

Our objective is to minimize the total network traffic delay, that is due to missing or bad coordination within the network. We still need to specify how this delay is actually determined from the arrival pattern of the platoons. Since non-uniform cycle lengths are allowed at an intersection we have to consider a time span equal to the least common multiple of the cycle lengths at the signals incident to the particular link.

Evaluating the arrival pattern means the following. Depending on the arrival time of a platoon – which itself depends on the offset of the link – vehicles may have to stop during the red phase and form a queue. Then, in the green phase they are released. Note that we require that waiting queues must be empty at the end of each circulation of the cycle length at the link's destination signal.

This mechanism is sketched in Fig. 3. So, the link's average waiting time z equals the size of these queues accumulated over time and divided by the number of vehicles. Unfortunately, this term is not linear in our decision variable γ, the arrival time of the platoon. We overcome this by piece-wise linearizing the delay function which we evaluate at characteristic intermediate points. We chose only a few intermediate points – between three and five – such that the linearization becomes convex. Unlike in MITROP [4], this can be done consistently and effectively, since we are considering a function of only one variable. So, for a link $e = (v_i, v_j)$, let g_k^e denote the line segments in a piece-wise linear approximation of the delay function, where $k = 1, \ldots, \beta_e$, and β_e stands for the number of line segments. Thus,

$$z_{ij} \geq g_k^e(\gamma_{ij}), \qquad (5)$$

together with the convexity of the linearization ensures a consistent setting

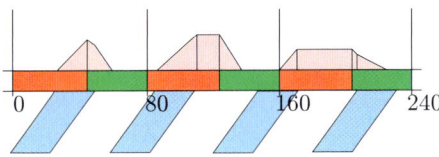

Fig. 3. The arrival pattern on a link $e = (v_i, v_j)$ with cycle lengths at the intersections of 60 and 80 seconds, respectively

for the delay variable z. Hence, the objective is formulated simply by

$$\sum_{(i,j) \in A} f_{ij} \cdot z_{ij}, \qquad (6)$$

with $f_{i,j}$ denoting the traffic volume on a link.

Although it is not explicitly stated in our MIP several other optimization criteria can be included. For example, the average number of stops can be incorporated in a very similar way.

2.4 The MIP

Now, we define the problem as a mixed-integer linear program with offsets and red-green split modes as decision variables. Again, for simplicity, we omit indices of different copies of link $e = (v_i, v_j)$ in the MIP formulation. However, the interaction between different traffic flows on parallel and anti-parallel links is considered in the linearization of the delay function.

$$\text{minimize} \sum_{(i,j) \in A} f_{ij} z_{ij}$$

$$\sum_{e \in F(\ell)} \phi_e - \sum_{e \in R(\ell)} \phi_e + \sum_{r=1}^{h_\ell} \Psi^{v_i,p} = n_\ell \cdot c_\ell \qquad \forall \ell \in \mathcal{C},$$

$$\sum_{m=1}^{\alpha_k} c_{k,m} \, \overline{\Psi}_m^{k,p} = \Psi^{k,p} \qquad \forall k \in \mathcal{K}, p \in \mathcal{R},$$

$$\sum_{m=1}^{\alpha_k} c_{k,m} = 1 \qquad \forall k \in \mathcal{K},$$

$$z_{ij} \geq g_r^e(\gamma_{ij}) \qquad \forall e, r,$$

$$\tau_{ij} - \gamma_{ij} + r_{ij} = \phi_{ij} \qquad \forall (i,j),$$

$$\underline{n_\ell} \leq n_\ell \leq \overline{n_\ell} \qquad \forall \ell,$$

$$n_\ell \in \mathbb{Z}, \quad \gamma_{ij} \in [0, T_{ij}], \quad c_{k,m} \in \{0,1\}.$$

Here v_i^ℓ, $i = 1, \ldots, h_\ell$ denote the vertices on circuit $\ell \in \mathcal{L}$. Note that the cycle equations, (2), do not have to be considered for all circuits of the underlying graph G. Rather, it suffices to require them to hold for all cycles of an integral cycle basis [8]. Hence, we have a polynomial number of constraints in the MIP.

For solving this mixed-integer linear program either standard tools like ILOG CPLEX [6] or academic software such as SCIP [1] can be used. However, any solver's performance strongly depends on the bounds ($\underline{n_\ell}$ and $\overline{n_\ell}$) that are provided for the integer variables n_ℓ. Although simple bounds for the n_ℓ are quite immediate – just take advantage of the bounds for the offsets – *optimizing* them over all integral cycle bases is often neglected. The advantages of strengthening bounds are well studied for the related problem of cyclic timetabling [8].

3 Application of the Model

We have tested our MIP model on several real world instances. Two prominent examples are inner-city area networks of Portland and Denver, as they have a regular structure and their fixed-timed signals correspond well to our model.

First, we applied the model to a section of the network of Portland with 16 fixed-time controlled signals and route volumes adapted to peak hour volumes. We observed results which are promising in two respects. First, by simulating the network and its signal control in VISSIM [12], we observed that the calculated offsets show a consistent character. For those paths through the network, where the optimizer had achieved compatible offsets, smooth progression was indeed reproduced within the simulation. Then we reconstructed the grid-like network in TRANSYT, ran its genetic algorithm to find good offsets and compared the results. Afterwards, we again used VISSIM to compare simulation results – one hour of simulation with an appropriate start-up phase – of both our model's offsets and the ones obtained by TRANSYT. The results are displayed in Table 1(a).

As a second real-world instance we chose the inner-city area network of Denver, see Fig. 4, with an original morning peak hour traffic. On this network with 146 fixed-time signals the MIP could be solved within 4 minutes (CPLEX) leaving an optimality gap of only 4%. The results in Table 1(b) show that this solution is comparable to the present coordination.

We are currently preparing more large real-world instances that allow us comparisons with TRANSYT without reconstructing the network.

Table 1. Comparison of delays from different coordinations.

(a) Portland (section), 16 signals

Coordination	Delay in s/veh	CPU time	Relative difference
best random	27.0	-	63%
present	16.6	-	0%
optimization	16.1	$< 1s$	−3%
TRANSYT	15.9 (22.2)	800s (10s)	−4% (34%)

(b) Denver, 146 signals

Coordination	Delay in s/veh	Relative difference
random	192.08	49.2%
best of 20 random	171.75	33.4%
present	128.70	0
optimization	132.85	3.2%

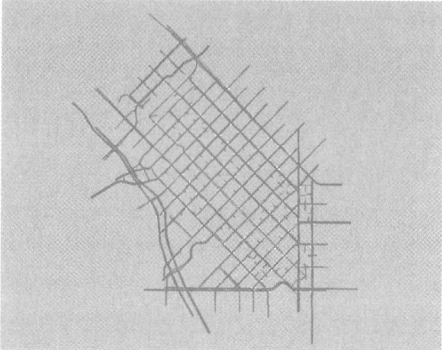

Fig. 4. A screenshot from the microsimulation VISSIM showing the Denver network

3.1 Different Aspects of the Optimization Algorithm in Practice

The optimization algorithm mainly has two different practical applications. First, the optimization of the network coordination can be used as a standalone optimizer. Here, the optimization is done once and interactions between signal-control and demand are neglected.

On the other hand, the model, i.e., the network coordination can be embedded within an optimization scheme that reflects the dependencies between signal-control and demand. We describe the latter of the two applications in more detail below.

Optimizers, such as TRANSYT [10] and SYNCHRO [11], accept as input a description of the supply, i.e. the street network (mainly links with travel times, permitted turns at intersections), and of the demand, i.e. traffic flows through the network. The demand can be either observed or derived from a demand model. Interestingly, demand models are themselves extending in the direction of traffic engineering, supporting models for signal control at intersections and offering analysis methods for them, e.g. by incorporating capacity analysis according to the HCM [5]. These models aim to provide a more realistic node impedance for route choice by incorporating capacity analysis into the assignment step. So far, most packages are limited to the optimization of cycle lengths c and green time fractions (g/c), as run time requirements have precluded network coordination. This is due to the nature of the solution methods used in practice (genetic algorithms or quasi-exhaustive search). A faster solution method would enable a closer integration between macroscopic demand modeling and network coordination. The network coordination could then be part of the assignment process, allowing to update route flows in response to changed offsets, and thus reach an equilibrium between demand and supply that includes all aspects of signal control.

This overall optimization process is illustrated by Fig. 5. Given OD flows (from previous model steps) are assigned to a network with signal control,

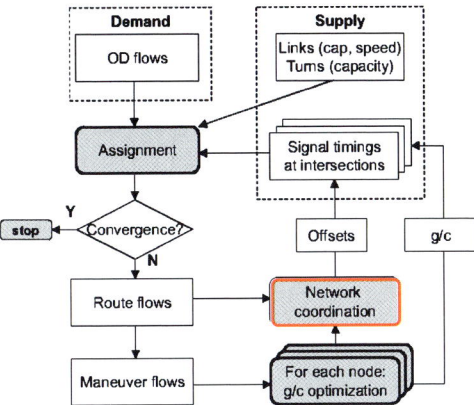

Fig. 5. The organization of the overall optimization process with components assignment, local optimization and network coordination. We focus on the latter one stressing the requirements in the context of the optimization scheme

using an assignment method which calculates node impedance from signal timings and offsets between different intersections. The assignment produces node flows (traffic volumes per maneuver at each intersection) which are input to the local optimization of g/c. Path-based assignment procedures also produce route flows on the sub-paths between successive signalized intersections, or more specifically, between successive signal groups or stages. A network coordination optimizer takes these sub-path flows and globally optimizes the offsets. Then, the set of offsets are fed back to the assignment. The loop is executed at least once and continues until route impedances (including loss times experienced at signal controls) and flows converge.

4 Conclusion

In this paper we focused on network coordination with minimizing delay as objective. Based on a macroscopic approach used in the software tool MITROP[4], we developed a mixed-integer linear program, thus enabling guarantees on the solution quality, which is a step forward compared with heuristic methods such as genetic algorithms.

Our mixed-integer linear program can handle non-uniform cycle lengths at intersections and a selection between different red-green split modes. Last, but not least, we suggested the application of new graph-theoretical insights, i.e. tightening bounds of variables via the use of good integral cycle bases to accelerate the running time of the MIP. Empirical studies showed promising results concerning both the time needed to find a good solution and the solution quality. All this encourages the use of our network coordination approach as a component in the overall optimization procedure sketched in Fig. 5.

References

1. Achterberg, Tobias (2004). SCIP – a framework to integrate constraint and mixed integer programming. Technical Report 04-19. Zuse Institute. Berlin.
2. Gartner, N.H., J.D.C. Little and H. Gabbay (1975). Optimization of traffic signal settings by mixed-integer linear programming, part i: The network coordination problem. *Transportation Science* **9**, 321–343.
3. Gartner, N.H., J.D.C. Little and H. Gabbay (1975). Optimization of traffic signal settings by mixed-integer linear programming, part ii: The network synchronization problem. *Transportation Science* **9**, 344–363.
4. Gartner, N.H., J.D.C. Little and H. Gabbay (1976). Mitrop: a computer program for simultaneous optimisation of offsets, splits and cycle time. *Traffic Engineering and Control* **17**, 355–359.
5. *Highway Capacity Manual* (2000). TRB, Washington, USA.
6. ILOG CPLEX 10.1. (2006) User's Manual.
7. Improta, G., A. Sforza (1982) Optimal Offsets for Traffic Signal Systems in Urban Networks. *Transportation Research Board 16B*, No. 2, 143–161.
8. Liebchen, Christian (2003). Finding short integral cycle bases for cyclic timetabling.. In: *ESA*. Vol. 2832 of *LNCS*. Springer. pp. 715–726.
9. Little, J. D. C. (1966) The Synchronization of Traffic Signals by Mixed-Integer Linear Programming. *Operations Research* 14, 568–594.
10. Robertson, D.I. (1969). Transyt, a traffic network study tool. Technical Report LR 253. Transport and Road Research Laboratory.
11. *SYNCHRO, User's Guide* (2000). Trafficware, Sugar Land, USA.
12. *VISSIM 4.10, User's Guide* (2005). ptv AG, Karlsruhe, Germany.
13. Wormleighton, R. (1965). Queues at a fixed time traffic signal with periodic random input. *CORS-Journal* **3**, 129–141.

Optimal Sorting of Rolling Stock at Hump Yards

Ronny S. Hansmann and Uwe T. Zimmermann

Institute of Mathematical Optimization, Pockelstraße 14, 38106 Braunschweig, Germany, {r.hansmann,u.zimmermann}@tu-bs.de

Summary. In this paper we provide a quite general description of a class of problems called *Sorting of **R**olling **S**tock **P**roblem(s)*. An **SRSP** consists in finding an optimal schedule for rearranging units of rolling stock (railcars, trams, trains, ...) at shunting yards, covering a broad range of specially structured applications. Here, we focus on versions of **SRSP** at particular shunting yards featuring a *hump*. We analyze the use of such a *hump yard* in our research project *Zeitkritische Ablaufbergoptimierung in Rangierbahnhöfen*[1] in cooperation with BASF, The Chemical Company, in Ludwigshafen. Among other results we present a remarkably efficient algorithm with linear running time for solving the practical **SRSP** at the BASF hump yard.

1 Practical Problem Description

In railroad shunting yards, see Fig. 1, incoming freight or passenger trains are split, parked and rearranged according to destinations or according to construction type of railcars. Uncertain arrival times, ad hoc changing orders of incoming railcars, the increasing number of rolling stock, sparse capacities and financial constraints complicate the process and offer large potential for optimization.

In the above context, we provide a description of a quite general class of problems called *Sorting of Rolling Stock Problems*, covering a broad range of special applications. In general, an **SRSP** consists of three processes: arrival, parking, and departure. At the beginning an ordered *input sequence* of units of rolling stock (railcars, trams, complete trains, ...) arrives at the shunting yard. Then the parking process starts and the units enter the tracks of the shunting yard. Here, incoming units have to be parked in such a way that at departure the parked units can leave the shunting yard in a structured *output sequence*. The difficulty of **SRSP** depends on the structural differences of the

---
[1] Funded by the German Federal Ministry of Education and Research (BMBF), grant no. 03ZINJBS

Fig. 1. One of the shunting yards at the site of our practical partner: BASF, The Chemical Company, Ludwigshafen

input sequence and the requested output sequence as well as on the structure and flexibility of the shunting yard.

As most practitioners will confirm the following rule of thumb holds in general: the less tracks used during the sorting procedure the lower the operational costs. Thus, a main goal of **SRSP** is to use as few tracks as possible. In other words, an optimal solution is a *schedule*, i.e., an assignment of tracks to units, using only a minimal number of tracks for the required sorting.

Structure of Output Sequence

As usual the incoming units are classified by a particular distinctive criterion, e.g., their destination or their construction type. As common in practice, we say that units satisfying the same criterion form a *group*.

We distinguish the following different structures of "labeled output sequences". At first, all positions of the output sequence are labeled (e.g. with integers). Secondly, all actually assigned units departing at positions with identical label have to be members of the same group and, vice versa, all members from a group have to be assigned to positions with identical labels. In particular, the number g of different groups is the same as the number of different labels. The labels of the positions of the output sequence form certain *patterns*. If the positions of the output sequence are labeled in a blockwise manner, i.e., the labeled output sequence contains no subsequence of positions labeled (u, v, u) for distinct labels $u \neq v$, we say that the labeled output sequence consists of *blocks*. In view of the number of different groups in the input, we call the labeled output sequence a g-**pattern** sequence or

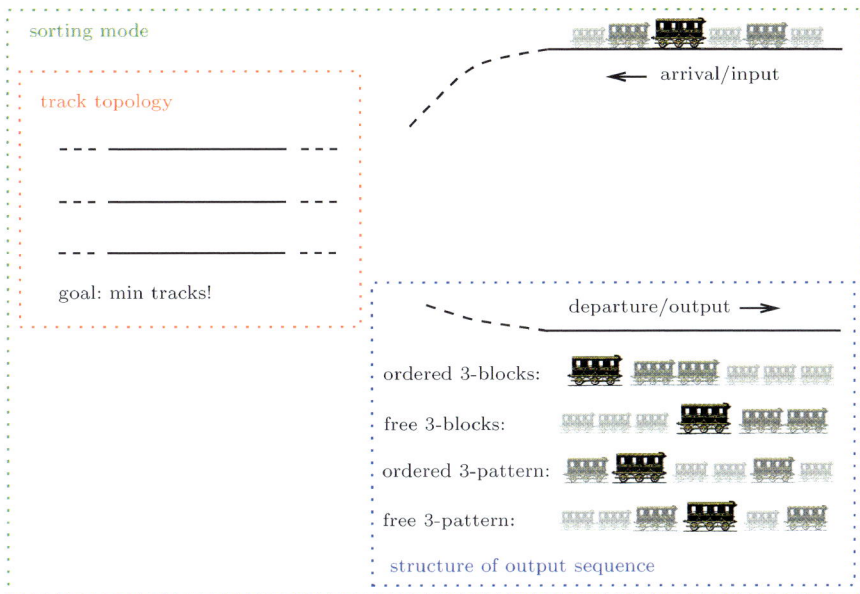

Fig. 2. Sorting of Rolling Stock at a shunting yard

Table 1. Parameters for **SRSP** ($b \geq 1$, $h \geq 0$, $s \geq 0$, $g \geq 1$)

track topology		sorting mode			structure of	
design	length	shunting	timing	splitting	output sequence	
<u>st</u>acks	<u>un</u><u>b</u>ounded	<u>no</u> <u>sh</u>unting	<u>se</u>quential	s-<u>sp</u>lit	<u>fr</u>ee	g-<u>bl</u>ocks
<u>qu</u>eues	b-<u>b</u>ounded	h-<u>h</u>ump-<u>sh</u>unting	<u>co</u>ncurrent	<u>sp</u>lit	<u>or</u>dered	g-<u>pa</u>ttern
<u>s</u>tacks/<u>q</u>ueues			time <u>w</u>indows	<u>c</u>hain-<u>sp</u>lit		
<u>si</u><u>d</u>o						
<u>d</u>i<u>s</u>o						
<u>d</u>i<u>d</u>o						

a g-**blocks** sequence. For **free** output sequences, there is no fixed assignment between groups and labels. Otherwise units departing at some position of the output sequence are members of a pre-defined group, see Fig. 2, and we speak of **ordered** output sequences.

Summarizing, we consider four types of structures of the output sequence: **free** g-**pattern** and **ordered** g-**pattern** and their special cases **free** g-**blocks** and **ordered** g-**blocks**. The departing order of units within a group is not fixed and offers potential for optimization.

Of course, the choice of tracks is affected by several other parameters of the shunting yard specifying either the track topology or the required sorting mode, see Table 1.

Track Topology

There are various track topologies depending on the design and the length of the tracks in the shunting yard.

Design. If the tracks may be accessed only from one side, that is, entrance and exit are on the same side of the tracks, such that the other end is a dead-lock, we speak of **stacks**. Note, that any two units parked on the same stack will change their order from arrival to departure if not additionally rearranged or shunted. Under the same assumption any two units preserve their order from arrival to departure when placed on a queue (**queues**), which is a one way track where the units arrive at one end and leave at the opposite side. In the case denoted as **stacks/queues** one may freely decide whether a track is used as queue or stack. In the above three cases the entrance as well as the exit are only on one (possibly differing) side of the track, which is known in literature as *siso* (single in single out). In the following further track designs units may arrive at or depart from both sides of the tracks: **sido** (single in double out), i.e., entrance is on one side, exit is on both sides; **diso** (double in single out), i.e., entrance is on both sides, exit is on one side; **dido** (double in double out), i.e., entrance and exit are on both sides.

Length. Of course, in real shunting yards, tracks are bounded in length. In the case b-**bounded**, at most b units may be placed on each track. Though **unbounded** track lengths are seemingly a rather theoretical issue, they may well be reasonable from a practical point of view. Firstly, in general, it is much harder and much more time-consuming to determine optimal schedules complying with the real track lengths. Secondly, solutions being optimal with respect to unbounded tracks, in practice seem to be easily transformable into practically good schedules on bounded tracks. For example, there are two well known approaches for handling "overfilled" tracks during actual operations. Immediately after a track gets filled to capacity, one may either empty it using an additional buffer, i.e., tracks beyond the shunting yard, or redirect units initially planned to go on the filled track to another track in the shunting yard.

Sorting Mode

The technical and organizational infrastructure of the yard leads to different constraints on the feasible movements of units. We mainly capture these differences in the distinctive feasible movements available in a sorting mode: a unit may (be) move(d) from the input(-track) to a track (*i-t-move*), from a track to another or the same track (*t-t-move* or *shunting-move*), from a track to the output(-track) (*t-o-move*) or directly from the input(-track) to the output(-track) (*i-o-move*). If the structures of the input sequence and the output sequence differ, at least some i-t-moves and t-o-moves are necessary.

Shunting. We consider special cases for which i-o-moves and/or different kinds of shunting-moves are permitted or not. For instance, in the **no shunting** case, no shunting-moves and no i-o-moves are permitted. Otherwise, depending on the actual infrastructure of the shunting yard, there are various constraints on the type of feasible shunting-moves. If the shunting yard features a so called *hump* for splitting trains, we speak of a *hump yard*. At the site of our project partner, the BASF AG in Ludwigshafen, the sorting and shunting operations are mainly performed using a large and expensive hump yard facility. Sorting or shunting over a hump is a common strategy to rearrange trains of units *without* own power units. At arrival such units are pushed over the hump before rolling down one by one into either appropriately chosen tracks (i-t-moves) or the output-track (i-o-moves). As a consequence, instead of many time-consuming pushing/pulling operations of units by locomotives on tracks we only need one pushing operation of the complete input sequence at arrival. In the same convenient manner one may use the hump for t-t-moves and t-o-moves. For example, at BASF t-t-moves and t-o-moves are performed in the following fashion. Per *humping step* all units placed on one track are pulled back over the hump. Then these units are again pushed over the hump either to the output-track (t-o-move) or to other tracks (t-t-move). For h-**hump-shunting** we allow at most h such humping steps. The only difference between **no shunting** and 0-**hump-shunting** is that i-o-moves are infeasible for **no shunting** but feasible for 0-**hump-shunting**. If shunting-moves are allowed, it is necessary to additionally describe the performance of the required shunting-moves within the schedules.

Timing. We distinguish cases in which i-t-moves and t-o-moves appear completely separated or mixed on the time line. For example, at night depots it is quite common that the first outgoing unit departs in the morning, long after the parking process is completely finished at night. In this case (first t-o-move after last i-t-move) we say that arrival and departure are **sequential**. Otherwise, if we allow that departure and arrival are **concurrent** (i-t-moves and t-o-moves are not consecutively), we may freely choose the departure time of any track-leaving unit. However, since we want to minimize the number of tracks used, a unit or group should obviously leave a track as soon as possible, since then the chance of blockades of departures of other units or groups is reduced. The input information for the **sequential** as well as the **concurrent SRSP** is the input sequence, i.e., one only knows in which order the units arrive. Contrary to these **sequence SRSP**, in the more general **time windows SRSP** more input information has to be taken into account. Here, the arrival time and departure time for each unit are exactly fixed in advance. We assume w.l.o.g. that units with identical departure time belong to the same group.

Splitting. Finally, the sorting mode is influenced by the way units may depart from tracks which is closely related to the type of the units. Suppose our problem consists in sorting units with own power units such as trams. Such units are able to leave the shunting yard without the help of other devices like locomotives. Thus, it is possible to split the units of one group arbitrarily over the tracks (**split**). However, this splitting might not be reasonable if we want to sort railcars without own power units into **blocks** since then one or more locomotives would have to collect the units of one group from several tracks. It would be much less time-consuming for the locomotive to pick all units of one group as a block from a single track. As a consequence, we also consider s-**split SRSP** where the units of one group may only be split up over at most $s+1$ tracks, $s \geq 0$. The particularly restrictive splitting condition **chain-split** is reasonable only for **sequential** and **no shunting**. In this case, units may be distributed over all tracks in such a way that collecting the units track by track, i.e., all units placed on a track depart completely before all units of the next track depart etc., leads to the required output sequence.

Similar to the notation of scheduling problems we propose an $\alpha|\beta|\gamma =: \mathcal{V}$ notation for the description of the many different *versions* $\mathcal{V}$ of **SRSP** which result from the specification of the various parameters. Here, α specifies the track topology, β the sorting mode, and γ the structure of the output sequence. A complete detailed list using the abbreviations defined in Table 1 reads as follows: $\alpha \in \{\{\mathbf{st}, \mathbf{qu}, \mathbf{sq}, \mathbf{sd}, \mathbf{ds}, \mathbf{dd}\} \times \{\mathbf{ub}, b\text{-}\mathbf{bd}\}\}$, $\beta \in \{\{\mathbf{nsh}, h\text{-}\mathbf{hsh}\} \times \{\mathbf{se}, \mathbf{co}, \mathbf{tw}\} \times \{s\text{-}\mathbf{sp}, \mathbf{sp}, \mathbf{csp}\}\}$, and $\gamma \in \{\{\mathbf{fr}, \mathbf{or}\} \times \{g\text{-}\mathbf{bl}, g\text{-}\mathbf{pa}\}\}$. Note, that the combination of **free** and **time windows** is not reasonable, since a priori known departure times of the units obviously result in a particular departure order as in the **ordered** case. In view of complexity results we will distinguish optimization and decision problems. By $\alpha|\beta|\gamma$, we denote the corresponding optimization problem, i.e., finding a correspondingly feasible schedule with minimal number of tracks. By $k - \alpha|\beta|\gamma$ we denote the corresponding decision problem, i.e., answering the question wether there is a correspondingly feasible schedule using at most k tracks.

In Sect. 2, we remind some notation for integer sequences and we give a short description of the corresponding **sequence** versions of **SRSP**. In Sect. 3, we briefly survey known results on **SRSP**. In Sect. 4, we present new results with direct practical relevance for our joint project with BASF. In particular, we discuss the versions **st,ub**|h-**hsh,se,sp**|**or**,g-**bl** and **st,ub**|h-**hsh,se,sp**|**fr**,g-**bl**. Finally, in Sect. 5, we summarize some computational and practical results.

2 Sequence Versions of SRSP

We denote an *integer sequence*, i.e., a *sequence* of integers, by $S = (s_{i_1}, \ldots, s_{i_n})$ where s is a surjective function mapping each integer $i \in \mathcal{I} = \{i_1, \ldots, i_n | i_1 < i_2 < \cdots < i_n\}$ to an integer $s_i \in \mathcal{G}$. In particular each integer $g \in \mathcal{G}$ occurs in S. A *permutation* $\Pi = (\pi_1, \ldots, \pi_n)$ of the integers $1, \ldots, n$ corresponds to an integer sequence with $\mathcal{G} = \{1, \ldots, n\}$. A *subsequence of* S is a sequence $S' = (s_{i_{j_1}}, s_{i_{j_2}}, \ldots, s_{i_{j_m}})$ such that $1 \leq j_k < j_l \leq n$ for each $k < l$.

We say that S *contains a subsequence* (u,v) (for example $(1,2)$), if there exists a subsequence (s_i, s_j) of S with $s_i = u$, $s_j = v$ (or $s_i = 1$, $s_j = 2$). A set of subsequences P_S of S is called *partition of S* if each element of S belongs to exactly one subsequence in P_S. With respect to a certain property of subsequences (for example monotonicity), we will call a subsequence of S feasible, if it has the property, or infeasible otherwise. Then, a *minimum (feasible) partition of S* is a partition of S containing a minimum number of feasible subsequences of S.

Two sequences $\bar{S} = (\bar{s}_{i_1}, \bar{s}_{i_2}, \ldots, \bar{s}_{i_n})$ and $\tilde{S} = (\tilde{s}_{j_1}, \tilde{s}_{j_2}, \ldots, \tilde{s}_{j_n})$ are said to be *equal* if $\bar{s}_{i_l} = \tilde{s}_{j_l}$ for all $l = 1, \ldots, n$. Otherwise, the two sequences are called *different*. The concatenation $S \oplus \bar{S} = (\tilde{s}_{l_1}, \ldots, \tilde{s}_{l_{n+\bar{n}}})$ of two sequences $S = (s_{i_1}, s_{i_2}, \ldots, s_{i_n})$ and $\bar{S} = (\bar{s}_{j_1}, \bar{s}_{j_2}, \ldots, \bar{s}_{j_{\bar{n}}})$ is a binary operation defined by $\tilde{s}_{l_k} := s_{i_k}$ for $k = 1, \ldots, n$ and $\tilde{s}_{l_k} := \bar{s}_{j_{n-k}}$ for $k = n+1, \ldots, n+\bar{n}$.

For **sequence** versions of **SRSP** an input sequence of n *units* (elements) is described by an integer sequence $S = (s_1, \ldots, s_n)$. Here, the i-th unit (unit at position i in S) belongs to the s_i-th *group* (integer). In particular in **ordered** versions the groups are numbered according to their departing order. For example $S = (2, 3, 1, 2, 1, 2, 3)$ implies that the third and the fifth incoming unit form the group containing the first outgoing unit. Otherwise, in **free** versions, the groups may be arbitrarily numbered and these numbers do not carry any information on the ordering of the outgoing units. The optimal value, i. e., the minimal number of tracks used in a feasible schedule, will be denoted by $z^*_\mathcal{V}(S)$ for the **sequence** version $\mathcal{V}$, or shortly by z^*.

The **sequence** versions correspond to minimum partition problems, i. e., find a partition of S into a minimal number of feasible subsequences $S_1, \ldots, S_{z^*}$ where the definition of feasibility depends on the version. Each subsequence S_k corresponds to the units placed on track k.

There are several versions of **SRSP** where we can characterize whether any two units (groups) may be placed on the same track or not. Then, the **SRSP** corresponds to a minimum coloring problem of a corresponding graph whose vertices are the units (groups). Two vertices in this graph are adjacent if and only if the two units (groups) may not be placed on the same track. In any feasible coloring, the vertices colored with color k correspond to the units (groups) placed on track k.

3 Overview of Results in Literature

In the following we will focus on publications containing results for versions listed in Table 1. Of course, there is a broad range of literature dealing with similar or related problems.

In [4], Di Stefano and Koči present results for some n-**blocks** versions. For the equivalent cases n-**pattern** and n-**blocks** the input sequence is a permutation and the splitting conditions coincide. It is shown that the equivalent versions **st,ub|nsh,se,sp|or,n-bl**, **qu,ub|nsh,se,sp|or,n-bl**, **qu,ub|nsh,tw,sp|**

or,n-bl, and **qu,ub|nsh,co,sp|or,n-bl** correspond to minimum coloring of permutation graphs and consequently are solvable in $\mathcal{O}(n \log n)$ time. It is mentioned that **st,ub|nsh,tw,sp|or,n-bl** is equivalent to minimum coloring of circle graphs and therefore $\mathcal{NP}$-hard. Furthermore, minimum coloring formulations in different hypergraphs for the versions **sd,ub|nsh,se,sp|or,n-bl**, **ds,ub|nsh,se,sp|or,n-bl**, and **dd,ub|nsh,se,sp|or,n-bl** are introduced. It is shown that version k–**sd,ub|nsh,se,sp|or,n-bl** is equivalent to deciding whether there exists a partition of the given permutation (input sequence) into at most k unimodal subsequences, which is shown to be $\mathcal{NP}$-complete in [5]. Since the versions **sd,ub|nsh,se,sp|or,n-bl** and **ds,ub|nsh,se,sp|or,n-bl** are equivalent (see [4]), they are both $\mathcal{NP}$-hard and both 3.42-approximable in polynomial time, see [6]. The version **sq,ub|nsh,se,sp|or,n-bl** is equivalent to minimum cocoloring of permutation graphs, which is proven to be $\mathcal{NP}$-hard in [13], and which is 1.71-approximable in polynomial time, see [7].

The version **st,ub|nsh,se,csp|or,g-bl** (**st,ub|nsh,se,csp|fr,g-bl**) is obviously equivalent to the problem of finding a minimum partition of the (input) sequence S into subsequences $S_1, \ldots, S_{z^*}$ such that the (output) sequence $S_1 \oplus S_2 \oplus \cdots \oplus S_{z^*}$ has the structure **ordered g-blocks** (**free g-blocks**). Dahlhaus et al. provide an algorithm in [3] which solves **st,ub|nsh,se,csp|or, g-bl** in $\mathcal{O}(n)$ time and in [2] the version **st,ub|nsh,se,csp|fr,g-bl** (called Train Marshalling Problem) is shown to be $\mathcal{NP}$-hard.

In [9], we derive results for further versions. Here, we summarize some of these results in the following table.

Version	Equivalence to Min Coloring of	Theoretical Complexity
st,ub|nsh,se,sp|or,g-bl	permutation graphs	$\mathcal{O}(n \log n)$
st,ub|nsh,se,0-sp|gr,g-bl	interval graphs	$\mathcal{O}(n \log n)$
st,ub|nsh,se,0-sp|or,g-bl	PI-graphs	$\mathcal{O}(n \log n)$
st,ub|nsh,tw,sp|or,g-bl	circle graphs	$\mathcal{NP}$-hard
st,ub|nsh,tw,0-sp|or,g-bl	circle-polygon graphs	$\mathcal{NP}$-hard
st,ub|nsh,co,0-sp|or,g-bl	circle-polygon graphs	$\mathcal{NP}$-hard
st,ub|nsh,co,0-sp|gr,g-bl	circle-polygon graphs	$\mathcal{NP}$-hard
st,ub|nsh,co,sp|or,g-bl	subclass of circle graphs	$\mathcal{NP}$-hard

Moreover, in [9] we introduce an Integer Programming model for minimum coloring of circle-polygon graphs based on a particular network flow model. Computational results show that the model can effectively be solved. In this way, optimal colorings of circle-polygon graphs and therefore optimal schedules for the above $\mathcal{NP}$-hard versions can be determined with surprisingly small computational effort.

Of course, if an **unbounded** version is $\mathcal{NP}$-hard then the respective **b-bounded** version is also $\mathcal{NP}$-hard. Using results for Mutual Exclusion Scheduling presented in [12] and [1], in [9] we show $\mathcal{NP}$-hardness of the versions **st,b-bd|nsh,se,sp|or,g-bl** and **st,b-bd|nsh,se,0-sp|gr,g-bl**. In [14] it is shown that **st,b-bd|nsh,se,sp|or,g-pa** is $\mathcal{NP}$-hard. A more comprehensive version of the proof is published in [15]. In the unpublished paper [11], Jacob proves $\mathcal{NP}$-hardness of the version **st,b-bd|h-hsh,se,sp|or,n-bl** which extends to the more general versions **st,b-bd|h-hsh,se,sp|or,g-bl** and **st,b-bd|h-hsh,se,sp|or,g-pa**.

Further results will be published in [8].

4 Versions of SRSP Applied at Hump Yards

Internal logistics at the site of our practical partner, BASF, The Chemical Company, in Ludwigshafen is mainly based on rail transport. The aim of our project was to provide our practical partner with effective schedules for rearranging trains. Though there are a few shunting yards at the site, nearly all rearrangements are performed using the main hump yard, see Fig. 3.

The tracks in the hump yard are **stacks**, railcars may arbitrarily **split** up over tracks, and arrival and departure of trains are **sequential**.

For each incoming railcar we know in which train and block within the train it has to leave. Each train serves several factory buildings and the railcars requested by one factory form a block. To assure a secure and efficient handling,

Fig. 3. Railcars being pushed over the hump (*left*) such that they roll on appropriately chosen tracks (*right*). Pictures of the hump yard at the site of BASF, The Chemical Company, Ludwigshafen

trains consist in suitably **ordered blocks** matching the served buildings; hence the railcars of the respective block can simply be decoupled at the rear of the train, preventing shunting in the streets. The trains leave the hump yard in a known order. Both ordering informations can easily be encoded into one integer input sequence. For example, the input information about (train numbers, block numbers) in $((2,1),(1,3),(1,2),(2,2),(1,1),(2,2),(1,3),(1,1))$, i. e., the first incoming railcar has to leave in the first block of the second outgoing train is encoded as $(4,3,2,5,1,5,3,1)$.

For h-**hump-shunting** the movement of railcars as already described in Sect. 1 proceeds as follows: per humping step all railcars placed on one track are pulled back over the hump and then each of these railcars is again pushed over the hump rolling either to the output-track (t-o-move) or to tracks in the shunting yard (t-t-moves).

Obviously, the practical problem corresponds to the version **st**,b-**bd**|h-**hsh, se,sp**|**or**,g-**bl**. However, dispatchers prefer the optimal schedules of the respective **unbounded** version, see Sect. 5 for some reasoning. In the following, we develop a computationally very fast, linear time algorithm for solving the version **st,ub**|h-**hsh,se,sp**|**or**,g-**bl**. We previously presented this algorithm at OR 2006, Karlsruhe, September 2006. In Jacob [11], seemingly the same algorithm is described in different terminology for the special case with permutations as input. Furthermore, we show that the version **st**,b-**bd**|h-**hsh,se,sp**|**fr**,g-**bl** is $\mathcal{NP}$-hard.

For bookkeeping purposes, we arbitrarily number the tracks and, then, we consider the *track execution order* $E = (e_1, \ldots, e_h)$. At humping step i, the e_i-th track is *executed*. For each railcar, we consider its *path* through the tracks, i. e., the integer sequence of the track numbers of visited tracks. In particular, the path of a railcar moving directly from the input to the output corresponds to the empty sequence (), denoted by $\emptyset$. Two easily derived observations (cf. the example shown in Fig. 4) are:

Observation 1. Railcars taking the same path correspond to a subsequence of the input sequence.

Observation 2. A path corresponds to a subsequence of the track execution order.

For fixed number of tracks and humping steps, the number of different paths, along which a railcar can move, depends on the choice of the track execution order. We call such a path realizable. For example for 2 tracks and 3 humping steps, we consider the two track execution orders $E_1 = (1,2,2)$ and $E_2 = (1,2,1)$. As shown in Fig. 4, E_2 admits seven different realizable paths; on the other hand, E_1 only admits the six paths $\emptyset$, (1), (2), (1,2), (2,2), (1,2,2). The following theorem contains an explicit formula for the maximum number of different realizable paths and shows that it is achieved for cyclic track execution.

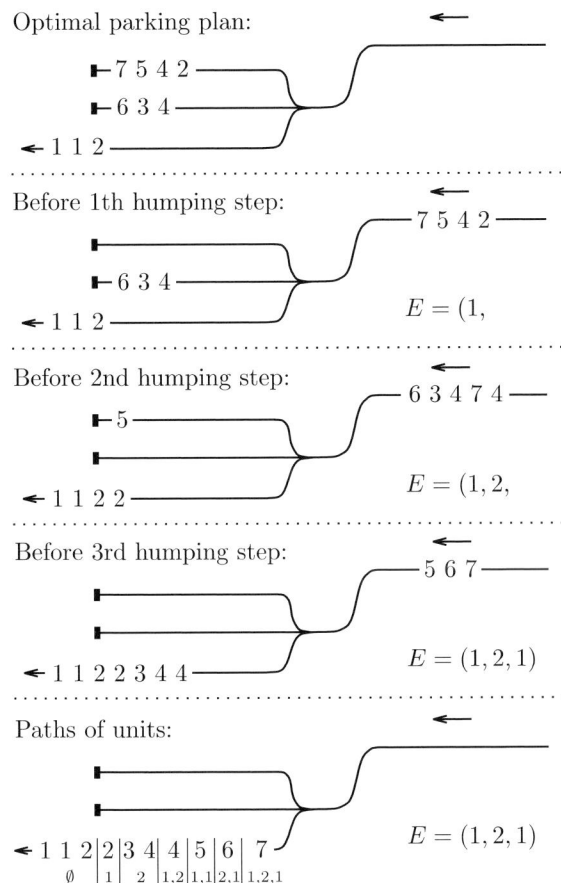

Fig. 4. Optimal schedule, shunting moves over the hump, and the paths of the railcars for the instance $(7, 6, 5, 4, 3, 4, 1, 2, 1, 2)$ for version **st,ub**$|h$**-hsh,se,sp**$|$**or,**g**-bl**

Theorem 1. *For $k \geq 1$ tracks in the hump yard and $h \geq 0$ humping steps the maximal number $f(k, h)$ of different realizable paths is achieved by performing the humping steps according to the **cyclic track execution order** $(1, 2, \ldots, k, 1, 2, \ldots)$ of length h. A recursion for $f(k, h)$ is $f(k, h+1) = 2 \cdot f(k, h) - f(k, h-k)$ for $h > k$ with starting values $f(k, h) = 2^h$ for $h \leq k$. An explicit formula reads*

$$f(k, h) = 2^h + \sum_{j=1}^{\lfloor \frac{h}{k+1} \rfloor} (-1)^j \cdot \binom{h - j \cdot k}{j} \cdot 2^{h - j(k+1)}.$$

Theorem 1 is an immediate consequence of Observation 2 and of the observations made in the proof of the following Corollary.

Corollary 1. *Among all integer sequences $S_{k,n}$ of length n containing integers from $\{1,\ldots,k\}$ the **cyclic sequence** $S_{k,n}^c = (1,2,\ldots,k,1,2,\ldots)$ contains the maximum number of different subsequences.*

Proof. We denote the number of different subsequences of an arbitrary sequence S by $\bar{f}(S)$, the number of different subsequences of S ending with integer t by $g(S,t)$, and the position of the rightmost integer t in S by $r(S,t)$.

It suffices to show $\bar{f}(S_{k,n}) \leq \bar{f}(S_{k,n}^c)$ for all subsequences $S_{k,n}$ with $n > k$ containing all integers $1,\ldots,k$.

For an inductive proof, let $S_{k,0} := \emptyset$, and let $S_{k,i} := (s_1,\ldots,s_i)$ for $i = 1,\ldots,n-1$. Thus, we have to show that $\bar{f}(S_{k,i}^c) \geq \bar{f}(S_{k,i})$ ① for some i with $k \leq i < n$ implies the corresponding inequality for $i+1$.

We observe the following properties of $S_{k,i}$:

② $\quad \bar{f}(S_{k,i+1}) = 2 \cdot \bar{f}(S_{k,i}) - g(S_{k,i}, s_{i+1})$,
③ $\quad \bar{f}(S_{k,i}) = 1 + \sum_{t=1}^{k} g(S_{k,i}, t)$,
④ $\quad g(S_{k,i}, t) = \bar{f}(S_{k,r(S_{k,i},t)-1})$, $\quad t = 1,\ldots,k$.

For easier comparison with $S_{k,i+1}^c$ we renumber the integers in $S_{k,i+1}$ and denote the result by $\tilde{S}_{k,i+1} := (\tilde{s}_1,\ldots,\tilde{s}_{i+1})$. Let $r(S_{k,i}, s_j)$ be the l-th greatest index in $\{r(S_{k,i}, t) \mid t = 1,\ldots,k\}$. Then, $\tilde{s}_j := s_{i+1-l}$ for $j = 1,\ldots,i+1$. For example, $\tilde{S}_{3,6} = (3,3,1,2,2,1)$ for $S_{3,6} = (2,2,3,1,1,3)$ due to $S_{3,5}^c = (1,2,3,1,2)$.

As a consequence, we get $r(\tilde{S}_{k,i}, s_{i+1}^c) \leq r(\tilde{S}_{k,i}, \tilde{s}_{i+1})$ which leads to $\bar{f}(\tilde{S}_{k,r(\tilde{S}_{k,i}, s_{i+1}^c)-1}) \leq \bar{f}(\tilde{S}_{k,r(\tilde{S}_{k,i}, \tilde{s}_{i+1})-1})$ and with ④ to

⑤ $\quad g(\tilde{S}_{k,i}, s_{i+1}^c) \leq g(\tilde{S}_{k,i}, \tilde{s}_{i+1})$.

Moreover, we observe

⑥ $\quad r(\tilde{S}_{k,i}, t) \leq r(S_{k,i}^c, t), \quad t = 1,\ldots,k$,
⑦ $\quad \bar{f}(\tilde{S}_{k,j}) = \bar{f}(S_{k,j}), \quad j = 1,\ldots,i+1$.

Finally, the following inequalities complete the proof.

$$\bar{f}(S_{k,i+1}) \stackrel{⑦}{=} \bar{f}(\tilde{S}_{k,i+1}) \stackrel{②}{=} 2 \cdot \bar{f}(\tilde{S}_{k,i}) - g(\tilde{S}_{k,i}, \tilde{s}_{i+1})$$
$$\stackrel{⑤}{\leq} 2 \cdot \bar{f}(\tilde{S}_{k,i}) - g(\tilde{S}_{k,i}, s_{i+1}^c)$$
$$\stackrel{③}{=} 2 \cdot \left(1 + \sum_{t=1}^{k} g(\tilde{S}_{k,i}, t)\right) - g(\tilde{S}_{k,i}, s_{i+1}^c)$$
$$\stackrel{④}{=} 2 + 2 \sum_{\substack{t=1 \\ t \neq s_{i+1}^c}}^{k} \bar{f}\left(\tilde{S}_{k,r(\tilde{S}_{k,i},t)-1}\right) + \bar{f}\left(\tilde{S}_{k,r(\tilde{S}_{k,i}, s_{i+1}^c)-1}\right)$$
$$\stackrel{⑥}{\leq} 2 + 2 \sum_{\substack{t=1 \\ t \neq s_{i+1}^c}}^{k} \bar{f}\left(\tilde{S}_{k,r(S_{k,i}^c,t)-1}\right) + \bar{f}\left(\tilde{S}_{k,r(S_{k,i}^c, s_{i+1}^c)-1}\right)$$

$$\stackrel{\text{\textcircled{7}}}{=} 2 + 2 \sum_{\substack{t=1 \\ t \neq s_{i+1}^c}}^{k} \bar{f}\left(S_{k,r(S_{k,i}^c,t)-1}^c\right) + \bar{f}\left(S_{k,r(S_{k,i}^c,s_{i+1}^c)-1}^c\right)$$

$$\stackrel{\text{\textcircled{1}}}{\leq} 2 + 2 \sum_{\substack{t=1 \\ t \neq s_{i+1}^c}}^{k} \bar{f}\left(S_{k,r(S_{k,i}^c,t)-1}^c\right) + \bar{f}\left(S_{k,r(S_{k,i}^c,s_{i+1}^c)-1}^c\right)$$

$$\stackrel{\text{\textcircled{4}}}{=} 2 \cdot \left(1 + \sum_{t=1}^{k} g(S_{k,i}^c, t)\right) - g(S_{k,i}^c, s_{i+1}^c)$$

$$\stackrel{\text{\textcircled{3}}}{=} 2 \cdot \bar{f}(S_{k,i}^c) - g(S_{k,i}^c, s_{i+1}^c) \stackrel{\text{\textcircled{2}}}{=} \bar{f}(S_{k,i+1}^c) \qquad \square$$

In [10], Hirschberg and Régnier prove that among all integer sequences $S_{k,n}$ of length n containing integers from $\{1,\ldots,k\}$ the cyclic sequence $S_{k,n}^c$ contains the maximum number $\bar{f}_t(S_{k,n}^c)$ of different subsequences with cardinality $n-t$ for $t = 0,\ldots,n$. Obviously this implies the above Corollary 1. They also provide recursions for computing each of these maximum numbers in $\mathcal{O}(t + (k-1)n^2)$ time. In our direct proof with respect to subsequences of arbitrary cardinality, we obtain a recursion for the total number $\bar{f}(S_{k,n}^c)$ of all different subsequences of $S_{k,n}^c$ and we derive an explicit formula for this maximum number, cf. Theorem 1. Applying our recursion, we can compute $f(k,n) = \bar{f}(S_{k,n}^c)$ significantly faster in $\mathcal{O}(n)$ time.

Observations 1 and 2 together with Theorem 1 imply the validity of the following Algorithm 1 for solving the versions **st,ub**|h-**hsh,se,sp**|**or**,g-**bl** (**st,ub**|h-**hsh,se,sp**|**fr**,g-**bl**).

Algorithm 1:

begin

    Step 1: Find a minimum partition of the input sequence S into subsequences $S_1,\ldots,S_{z^*}$ such that $S_1 \oplus S_2 \oplus \cdots \oplus S_{z^*}$ has the structure **ordered** g-**blocks** (**free** g-**blocks**)

    Step 2: For given number h of humping steps determine the minimal number k^* of tracks with at least z^* different realizable paths, i.e., with $f(k^*,h) \geq z^*$

    Step 3: For all $i = 1,\ldots,z^*$ assign a suitable path to subsequence S_i (railcars moving along path i)

end

For fixed number h of humping steps, Algorithm 1 computes an optimal schedule for **st,ub**|h-**hsh,se,sp**|**or**,g-**bl** in $\mathcal{O}(n)$ time: Step 1 can be solved in $\mathcal{O}(n)$, see [3], in Step 2 the minimal number k^* of tracks can be computed recursively, see Theorem 1, in $\mathcal{O}(\log n \cdot h)$, and in Step 3 the assignment of paths to subsequences again can be done in $\mathcal{O}(n)$. For a fixed number k of tracks, a schedule with minimal number of humping steps can be computed in linear time applying Algorithm 1 with a suitably modified Step 2.

Theorem 2. *The version **st,ub**|*h*-**hsh,se,sp**|**fr**,*g*-**bl** is $\mathcal{NP}$-hard.*

Proof. If 1-**st,ub**|($h-1$)-**hsh,se,sp**|**fr**,*g*-**bl** is decidable in polynomial time, then, due to previous results and due to the fact that $f(1, h-1) = h$, we can decide in polynomial time whether the input sequence S can be partitioned into subsequences $S_1, \ldots, S_h$ such that $S_1 \oplus S_2 \oplus \cdots \oplus S_h$ has the structure **free** *g*-**blocks**. However, the latter decision problem is known to be $\mathcal{NP}$-complete, see [2]. Thus, 1-**st,ub**|*h*-**hsh,se,sp**|**fr**,*g*-**bl** is $\mathcal{NP}$-complete. □

5 Practical Results

The task of our project at BASF, i. e., to provide an optimal schedule realizing the required rearrangements of trains at the hump yard, corresponds to the version **st,*b*-bd**|*h*-**hsh,se,sp**|**or**,*g*-**bl** and is therefore a hard optimization problem ($\mathcal{NP}$-hard, see [11]). However, for practical data from real instances per day with up to 600 incoming railcars we can compute optimal schedules via solving our Integer Programming model with the commercial software CPLEX 10.0 within half an hour, see [8] for detailed computational results. Although these solutions comply with the modeled fixed track lengths, they are mostly not directly applicable since in practice the number of railcars that can be placed on a track depends on several "soft" parameters. For example, railcars in fact vary in length and railcars rolling down to tracks are slowed down by automatic brakes which may lead to gaps between the railcars on the track. Thus, again, it seems to be reasonable to compute optimal solutions of the respective **unbounded** version **st,ub**|*h*-**hsh,se,sp**|**or**,*g*-**bl** and leave it to the dispatcher to handle "full" tracks adequately, see also paragraph on *Track Topology* in Sect. 1. Another practical advantage of the unbounded version is that the running time of Algorithm 1 for practical data is less than one second allowing quick reactions on real time changes in the predicted input sequence.

Although we implemented and demonstrated a prototype, the approach will not be put into daily action before the dispatcher can use it directly within the Monitoring and Control System VICOS, developed by SIEMENS and installed at BASF. Following the proposal of our practical partners at BASF, we contacted SIEMENS Braunschweig and we hope that the method may be added as tool within VICOS in due course.

Acknowledgements

We are very grateful for the many motivating and supporting practical discussions as well as the real world data made available by Holger Schmiers, Matthias Ostmann, and Rainer Scholz (BASF, Service Center Railway). We would also like to express our sincere thanks to Dr. Anna Schreieck and Prof. Dr. Josef Kallrath (BASF, Scientific Computing) for their steady support and encouragement from the very first conceptual ideas until the not yet reached successful end of the project.

References

1. H.L. Bodlaender and K. Jansen. Restrictions of graph partition problems. Part I. *Theoretical Computer Science*, 148(1):93–109, 1995.
2. E. Dahlhaus, P. Horak, M. Miller, and J.F. Ryan. The train marshalling problem. *Discrete Applied Mathematics*, 103(1–3):41–54, 2000.
3. E. Dahlhaus, F. Manne, M. Miller, and J.F. Ryan. Algorithms for combinatorial problems related to train marshalling. In *Proceedings of AWOCA 2000*, pages 7–16, Hunter Valley, 2000.
4. G. Di Stefano and M.L. Koči. A graph theoretical approach to the shunting problem. *Electronic Notes in Theoretical Computer Science*, 92:16–33, 2004.
5. G. Di Stefano, St. Krause, M.E. Lübbecke, and U.T. Zimmermann. On minimum k-modal partitions of permutations. In J.R. Correa, A. Hevia, and M. Kiwi, editors, *Latin American Theoretical Informatics (LATIN2006)*, volume 3887 of *Lecture Notes in Computer Science*, pages 374–385. Springer-Verlag, Berlin, 2006.
6. G. Di Stefano and U.T. Zimmermann. Short note on complexity and approximability of unimodal partitions of permutations. Technical report, Inst. Math. Opt., Braunschweig University of Technology, 2005.
7. F.V. Fomin, D. Kratsch, and J.-Ch. Novelle. Approximating minimum cocolourings. *Information Processing Letters*, 84(5):285–290, 2002.
8. R.S. Hansmann. *Optimal sorting of rolling stock (in preparation)*. PhD thesis, Inst. Math. Opt., Braunschweig Technical University Carolo-Wilhelmina.
9. R.S. Hansmann and U.T. Zimmermann. The sorting of rolling stock problem (in preparation).
10. D.S. Hirschberg and M. Régnier. Tight bounds on the number of string subsequences. *Journal of Discrete Algorithms*, 1(1):123–132, 2000.
11. R. Jacob. On shunting over a hump. unpublished. 2007.
12. K. Jansen. The mutual exclusion scheduling problem for permutation and comparability graphs. *Information and Computation*, 180(2):71–81, 2003.
13. K. Wagner. Monotonic coverings of finite sets. *Elektronische Informationsverarbeitung und Kybernetik*, 20(12):633–639, 1984.
14. T. Winter. *Online and real-time dispatching problems*. PhD thesis, Inst. Math. Opt., Technical University Carolo-Wilhelmina, 2000.
15. T. Winter and U.T. Zimmermann. Real-time dispatch of trams in storage yards. *Annals of Operations Research*, 96:287–315, 2000.

Stochastic Models and Algorithms for the Optimal Operation of a Dispersed Generation System Under Uncertainty

Edmund Handschin[1], Frederike Neise[2], Hendrik Neumann[1], and Rüdiger Schultz[2]

[1] University of Dortmund, Institute of Energy Systems and Energy Economics, Emil-Figge-Str. 70, 44227 Dortmund, Germany
{edmund.handschin,hendrik.neumann}@edu.udo
[2] University of Duisburg-Essen, Department of Mathematics, Forsthausweg 2, 47048 Duisburg, Germany, {schultz,neise}@math.uni-duisburg.de

Summary. Due to the impending renewal of generation capacities and present decisions concerning energy policy, dispersed generation systems become more and more important. The optimal operation of such a system and corresponding trading activities are substantially influenced by uncertainty and require powerful optimization techniques. We present expectation-based as well as risk-averse stochastic mixed-integer linear optimization models using risk measures and dominance constraints. Two case studies show the benefit of stochastic optimization in power generation and the superiority of tailored solution methods over standard solvers.

1 Introduction

Technical, economical and also political developments have lead to substantial changes in the field of the power industry in recent years. Politically motivated decisions such as the nuclear power phase-out, the support of renewable energies as well as the amendment of the energy industry law (ENWG) and the internationally declared Kyoto protocol will significantly influence the realignment of the future energy supply. Technical aspects as the obsolescence of German and also European power plants and the intention of saving CO_2 without use of nuclear energy pose big challenges for the future energy supply.

Owing to the planned and partly started new building of conventional power plants one can assume that the predominant part of our electrical demand will be supplied by these conventional power plants. Due to the permanent further technical development the dispersed generation (DG) close to the consumer will admittedly play an important role in the future and therefore meet a significant part of the generation capacity to be substituted. Especially the combined heat and power production (CHP) becomes very important because of high overall efficiencies of these units.

The profitability of a single DG-unit is exclusively determined by the amount of electricity and heat it produces. Contrary to a single DG-unit the coordinated operation of several units offers an additional potential for maximizing their profitability. Such an interconnection of DG-units partly combined with storage devices is called a Virtual Power Plant (VPP). This VPP is integrated into the existing power economical structures and can be considered as a participant at the electricity market. Hence this VPP offers a broad potential of optimization to its operator.

The paper is organized as follows: In Sect. 2 we describe the current developments of the electric power industry which offer many opportunities for an economical operation of VPP's. Further, Sect. 3 deals with an appropriate mixed-integer linear modeling of all technical and economical constraints of a VPP. In Sect. 4 we discuss uncertainties in the forecasted input data and how they can be handled in the spirit of two-stage stochastic optimization. Section 5 contains mathematical details of the different applied stochastic optimization models and deterministic equivalents which are derived if we assume discrete probability distributions of the uncertain input data. In Sect. 6 algorithms are developed which exploit the deterministic equivalents' special structure. We finally present two case studies showing the benefits of stochastic optimization for the operation of VPP's and the superiority of tailored algorithms over standard solvers.

2 Current Developments of the Electric Power Industry

Investigating the current aging structure of the German generation system it becomes clear that a big need for action exists. More than 40 % of the thermal power plants will be elder than 35 years in the year 2010. Many of these power plants will reach the end of their technical life time within the next few years. Besides this technical aspect the politically motivated decision of the nuclear power phase-out causes an additional deficit of installed generation capacity of 20 GW. The increasing demand of replacing generation capacity offers a big opportunity for a supply structure by use of dispersed generation.

The present electrical energy supply system is based on a central structure with conventional power plants feeding into the high and extra high voltage levels. This present situation causes a vertical power flow from the transmission networks via the distribution networks to the consumer. A significant rise in DG of electricity and heat is prognosticated for the future which reduces this vertical power flow [5].

DG is characterized by the following properties:

- Location of the DG-unit close to the consumer (at least in the same distribution network and/or local heat network)

- Nominal capacity of the DG-unit smaller than 5 MW (Integration into distribution networks only)
- Frequently coupled generation of electricity and heat

By use of CHP the primary energy is simultaneously converted into mechanical or electrical energy and useful heat. Energy conversion in CHP-units has clearly higher efficiency factors than the comparable separated production of electricity and heat. That? why a better use of the loaded primary energy and lower CO_2 emissions can be achieved.

A basic advantage of DG is the reduction of electrical network losses caused by a reduced power flow over the network. In certain situations with low local load an inversion of the load flow from the DG-unit into the network is even possible [2]. Compared to conventional power plants DG-units require low investment costs. Therefore the associated entrepreneurial risk of the investment is relatively low. Further advantages of DG are the opportunity to delay investments in the networks and the possibility to deliver ancillary services. The deficit power in case of an outage of a DG-unit is relatively small because of the low capacity of the DG-unit. These advantages come along with technical problems concerning the connection to the network [17] and disturbances during network operation [33].

2.1 Structure of the Electric Energy Supply System

Whereas a big part of the activities in the field of DG deals with the technical aspects of a area-wide integration of DG-units, important tasks concerning the coordinated operation under consideration of economical aspects are only insufficiently investigated. That is the reason why in this project an optimization model is developed which allows the short and very short term optimization of a VPP with respect to typical trade relations and contracts in the context of a liberalized electricity market. The uncertainties in this time range are of big importance for the development of the model and the selection of a suitable optimization method.

2.2 Virtual Power Plant

Current publications show that only a few projects have been realized in the field of Virtual Power Plants [1]. One of the first projects named VPP dealt with the network integration and the required communication technology of 31 fuel cells in the range of some kW in the year 2001. Besides the field test of the fuel cell, it was the aim of this project to prove the central management and control of all facilities by use of modern technology [21, 20]. A coordinated operation of the fuel cells in terms of an economical optimization was not aspired and would not have been sensible because of the long distance between the DG-units. Another project analyzed and developed technical concepts to enable a safe and reliable energy supply with a high penetration of renewable energies [6]. These investigations concentrate on the integration of

DG-units into the public low voltage networks. The aim of the optimization was to minimize the operating costs from the network operator perspective [10]. A project initiated in 2002 considered the operation of DG-units in connection with gas, heat and electricity networks. The DG-units in this project comprise two boilers, a district heating power station operated with biomass and a battery station as well as two stochastic infeeds from wind and photovoltaic. The objective of this project was the calculation of optimal operation schedules of the DG-units and controllable loads to reach the supply of the electrical and thermal loads in terms of households, industry and trade at minimal costs. A survey of the units considered in these projects shows that the number of DG-units used for an optimal operation is restricted to only a low quantity [4].

The aforementioned projects had a research character exclusively and were used to try out the technical feasibility. In contrast to that in the two subsequently described projects the operator aspires an increase of economic efficiency during practical operation. A VPP operated by a municipal utility comprises 20 DG-units with 5.1 MW total electrical and 39.2 MW total thermal capacity. The installed CHP-units include nine gas motors and one micro turbine. These CHP-units are installed in five local heat networks. Each of them is equipped with at least one thermal storage device [19]. The day-ahead forecasts of the electrical and thermal load as well as of the gas demand provide the basis for the calculation of optimal schedules of the DG-units. By use of the thermal storages the CHP-units are used to avoid expensive peaks in the electrical import.

A project called Virtual Control Power Plant has been realized in September 2003. It comprises the pooling of different power plants and big customers spread over the whole country with the objective to provide minute reserve. The bundled capacity has increased to about 1400 MW within the last four years. A bundled capacity of at least 30 MW allows the participation in the market of ancillary services.

A VPP is defined as an interconnection of DG-units and storage devices whose operation is optimized by a superior entity. Figure 1 illustrates this definition.

Besides always existing non influenceable electrical and thermal loads also controllable loads can be part of the VPP. The VPP is integrated into the existing technical and economical structures. Therefore an interaction with energy markets, e.g. in terms of energy contracts, trade at the spot market and the provision of ancillary services can be used to achieve an optimal operation of the VPP. The required processes are controlled by an Energy Management System (EMS). This EMS contains the forecast of all required data as the electrical and thermal demand of the customers belonging to the VPP and the stochastic infeed from wind turbines and photovoltaic as well as the spot market prices. The core of the EMS is the optimization module. Within this module the short term optimization of the VPP is performed based on models of the DG-units, storages and other available instruments.

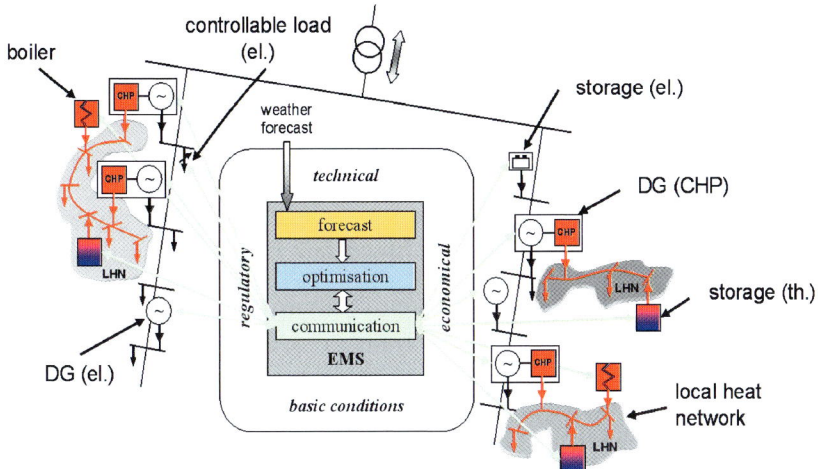

Fig. 1. Illustration of the definition of a VPP

Furthermore the communication technology which is needed to submit the optimal schedules to the different units and to control their operation is an important part of the EMS [31].

VPP as a Market Participant

Among the operation of his DG-units the operator of the VPP is also able to trade at different markets to maximize his profit resp. minimize his costs. He is at least a buyer at the market of electrical energy because the total demand within the VPP is generally not exclusively supplied by the DG-units (see Definition VPP in Sect. 2.2). This demand can be met by bilateral contracts at the OTC-market and the spot market as well as the intraday trading. In addition to that the operator is also able to sell ancillary services to the transmission system operator (TSO). All in all the following trading platforms are available:

- OTC-market (bilateral contracts)
- Stock exchange (spot market, intraday trading)
- Market of ancillary services (regulation and reserve power)

DG-Units and Storage Devices as Components of a VPP

Principally DG-units with different technologies are available on the market. The following listing gives a rough survey about the most important technologies. Among gas or oil driven motors also gas turbines are used as CHP-units. These technologies are already available whereas fuel cells are in general not ready for the market yet. One exception is a phosphor acid fuel cell (PAFC) which is commercially available with an electrical capacity of 200 kW and a thermal capacity of 220 kW. To meet the thermal peak demand gas or oil fired peak load boilers are used. Additionally hot water storages are applied as thermal storage devices. The much more difficult storage of electricity is normally realized by use of lead-acid batteries.

3 Dispersed Generation Model

An integral component for an economical optimization of the operation of technical equipment for energy supply comprises the mathematical modeling. The modeling has to cover all specific technical properties of the units and all hereby associated basic conditions. As mentioned before the economical boundary conditions are of great importance for the operation of a VPP and therefore have to be considered also within the model.

3.1 Modeling of the Technical Basic Conditions

The modeling of the technical basic conditions comprises the following aspects:

- Technical limits and gradients of the power output of DG-units and storage devices
- Efficiency characteristics of fuel cells with different technology, gas turbines and gas motors
- Technical specifications concerning operation times, down times and switching frequencies of DG-units
- Start-up and warm-up behavior of the different DG-units
- Diverse effects in conjunction with storage losses of electrical and thermal storage devices (self discharge losses, charge and discharge losses)
- Controllable electric loads

As an example of modeling technical aspects within a mixed-integer linear programming model (MILP) an efficiency characteristic model of a polymer electrolyte membrane (PEM) fuel cell is subsequently described. Most DG-units have a non-constant efficiency behavior which depends on the current power output of the unit. These characteristics can be described with efficiency characteristic curves. The developed model features by exclusively linear dependencies and is therefore very well applicable for the implementation within a linear program. If the achieved accuracy of the model is not sufficient in individual cases an extended model based on a segmented efficiency characteristic curve can be applied [25].

The principal approach for modeling typical efficiency characteristic curves becomes clear by equation (1), where P_i^t denotes the power ouptut and G_i^t the fuel input of unit i. The variable s_i^t distinguishes between an on- and off-status of unit i. The model can be adapted to the real efficiency characteristic curve derived from measurements by adjusting the parameters $k_{i,\eta 1}$ and $k_{i,\eta 2}$.

$$P_i^t = \eta_{i,\max} \cdot \left(G_i^t - \left(k_{i,\eta 2} \cdot s_i^t \cdot P_{i,\max} - k_{i,\eta 1} \cdot P_i^t \right) \right) \quad (1)$$

This approach can be applied to DG-units with exclusively thermal or electrical output as well as for CHP-units. To model the entire efficiency characteristic of a CHP-unit the modeling of the electrical and the total efficiency characteristic is sufficient. The resulting expression of the electrical efficiency characteristic in (2) can be achieved by transforming equation (1).

$$\eta_{i,el}(P_{i,el}^t) = \frac{P_{i,el}^t}{G_i^t} = \frac{\eta_{i,el\,\max}}{1 - k_{i,el\eta 1} \cdot \eta_{i,el\,\max} + k_{i,el\eta 2} \cdot \eta_{i,el\,\max} \cdot s_i^t \cdot \frac{P_{i,el\,\max}}{P_{i,el}^t}} \quad (2)$$

By use of equation (2) the power from fuel of a DG-unit i can be expressed with the following linear equality constraint:

$$G_i^t = P_{i,el}^t \cdot \left(\frac{1}{\eta_{i,el\,\max}} - k_{i,el\eta 1} \right) + k_{i,el\eta 2} \cdot s_i^t \cdot P_{i,el\,\max}$$

The thermal power output $\dot{Q}_i^t$ of a CHP-unit arises according to equation (3) as the difference of the total power output and the electric power output.

$$P_{i,total}^t = P_{i,el}^t + \dot{Q}_i^t \quad (3)$$

The equality constraint for a CHP-unit can be formulated by means of the expression for the entire efficiency (analogous to (2)) and equation (3) in the following way:

$$\dot{Q}_i^t = G_i^t \cdot \eta_{i,total}(P_{i,total}^t) - P_{i,el}^t$$

$$= \frac{\eta_{i,total\,\max} \left(G_i^t - k_{i,total\eta 2} \cdot s_i^t \cdot P_{i,total\,\max} \right)}{1 - k_{i,total\eta 1} \cdot \eta_{i,total\,\max}} - P_{i,el}^t$$

This procedure allows the modeling of output-variable CHP-coefficients. In case of constant electrical, thermal or entire efficiency factors this can be considered as a special case with $k_{i,el\eta 1} = k_{i,el\eta 2} = 0$, $k_{i,th\eta 1} = k_{i,th\eta 2} = 0$ or $k_{i,ges\eta 1} = k_{i,ges\eta 2} = 0$, respectively, within the presented approach.

Figure 2 shows the real and the modeled efficiency characteristic curve of a PEM fuel cell with a nominal electrical power $P_{\max}$ of 212 kW and a nominal thermal power $\dot{Q}_{\max}$ of 237 kW. The fuel cell has a minimum electrical output of 10 kW. The parameters $k_{i,el\eta 1}$ and $k_{i,el\eta 2}$ have been calculated to 0.565 and 0.502 by means of the least square errors method. The dashed line in Fig. 2 shows the error of the linear model compared to the values taken from the manufacturer's data. Due to a maximum error of $-1.6\,\%$ and an average absolute error of $0.6\,\%$ the model offers a very good accuracy.

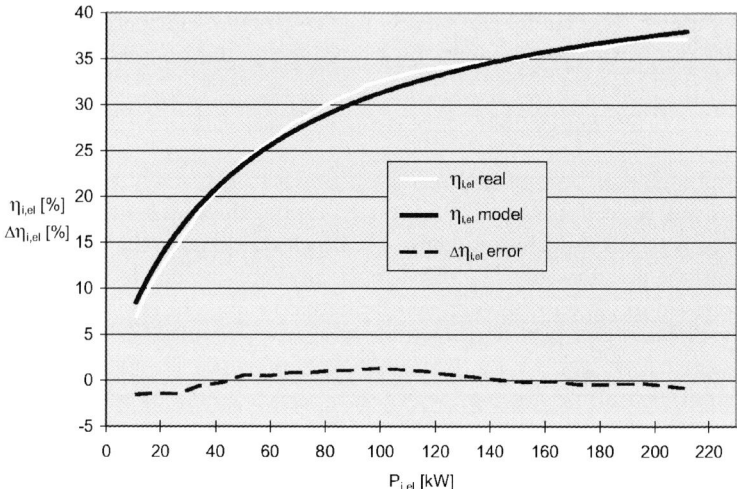

Fig. 2. Real and modeled electrical efficiency characteristic curve of the PEM fuel cell

3.2 Modeling of the Economical Basic Conditions

As economical basic conditions the following aspects have to be considered:

- Network usage fees of the gas and electric network (price per kWh and price per kW)
- Network usage fees avoided by infeed of DG-units
- Legally fixed feed-in tariffs according to the Renewable Energy Law (EEG) and CHP Law (KWK-G)
- Bilateral contracts for electricity and gas
- Trading at the German spot market (EEX)
- Costs of positive and negative balance energy

4 Uncertain Input Data

For optimizing the operation of a VPP the forecast of all input data is needed which influence the operation of the VPP during the time range of the optimization. The following analysis of the input data and especially their forecast errors is the basis for the selection of a capable optimization method.

4.1 Analysis of the Forecasted Input Data

The definition of a VPP explained in Sect. 2.2 directly reveals the data which are required in form of their forecasts. Some decisions which significantly influence the operation of a VPP, like the bids at the spot market or the adjustment of flexible energy contracts, have to be taken day-ahead. Therefore the forecasts have to be duly available. The optimization time range has to span at least a period up to the end of the next working day. Hence the results of the following forecasts are relevant [14]:

- Forecast of the electrical load
- Forecast of the thermal load in each local heat network
- Forecast of the stochastic infeed from Wind energy and Photovoltaic
- Forecast of the spot market prices

Subsequently the analysis of the electrical load forecast is exemplarily explained. Forecasted and instantaneous values provided by a municipal utility over a time range of two years form the basis for this investigation. The forecasts for the following working day are calculated each morning at 7 a.m. on the basis of the actual weather forecast. In general they comprise a time range of 41 hours (short term forecast). The distribution of the standard deviation σ of the relative forecast error over all following days was calculated and is depicted in Fig. 3. The dashed marked best fit straight line shows a nearly constant distribution over the day with an average standard deviation of approximately 3.1 %. Furthermore the very short term forecasts have been calculated by means of the Box–Jenkins-method at 0 and 6 a.m.. These results are also shown in Fig. 3. The analysis of Fig. 3 reveals that the forecast quality can be increased over the first five to six hours compared to the short term forecast by use of the simple Box-Jenkins-method.

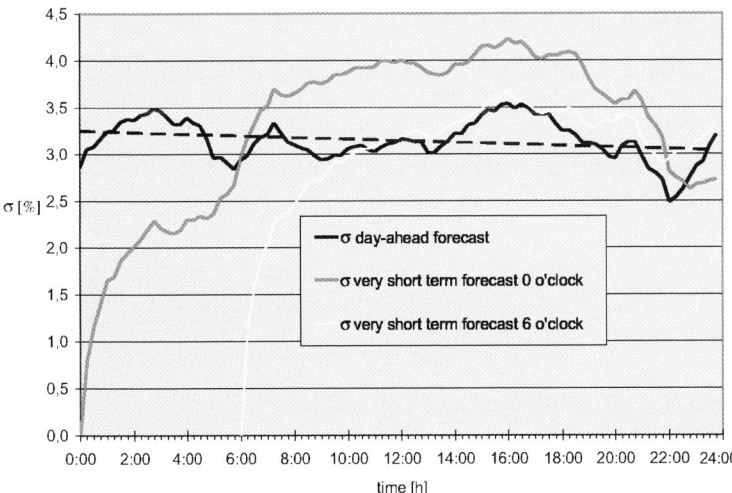

Fig. 3. Standard deviations of the forecast error of the short term forecast and very short term forecast

4.2 Handling of the Uncertain Input Data

Although well developed forecast tools are available nowadays the performed analysis of the forecast data shows that forecast errors of different quantities appear which can not be ignored. In addition to that some decisions have to be taken day-ahead in a liberalized energy market. In a VPP these decisions comprise e.g. the bids for the spot market and the adjustment of flexible energy contracts. Principally the influence of uncertainties in this system can be described like this: Some decisions have to be taken at a time at which the consequences of these decisions as well as their assessment can not be exactly determined because of uncertainties at the time of realization. This phenomenon can easily be described for trading at the spot market. To get a physical delivery of energy at day d + 1 the operator has to send his bids up to 12 a.m. of day d latest to the European Energy Exchange (EEX). At that time neither the spot market prices for the energy nor the electrical and thermal load and the stochastic infeed of day d + 1 are known precisely. Due to the different accuracies of the short and very short term forecast and the temporally limited validity of the spot market prices, which are only known up to the end of the actual day, a two-stage structure arises. This information structure can be handled mathematically with a two-stage stochastic programming model. It distinguishes between first-stage decisions that have to be made without anticipation of the future and second-stage decisions that can observe the realization of uncertainties. For the first-stage the input data is assumed to be certainly known, while the data concerning the second-stage is modeled by scenarios.

Input Data of the First Stage

The analysis of the forecast errors has shown that performing a very short term forecast of the electrical and thermal load and the stochastic infeed from wind leads to a highly reduced forecast error within the following five to nine hours. Therefore these input data are assumed to be known when the first-stage decisions are made. The time range of certain spot market prices depends on the starting time of the optimization. In general it does not match the time range where the loads and the stochastic infeed are known with certainty. Thus, the spot market prices up to the end of the actual day are supposed to be deterministic.

Scenario Generation for the Second Stage

The probability density functions of the forecast errors form the basis of the development of data scenarios. In the following a normal distribution of the forecast errors is assumed. We aim at the transformation of the continuous probability density function of each forecast error in each time interval t into a predefined number of scenarios with minimal loss of information. As an ex-

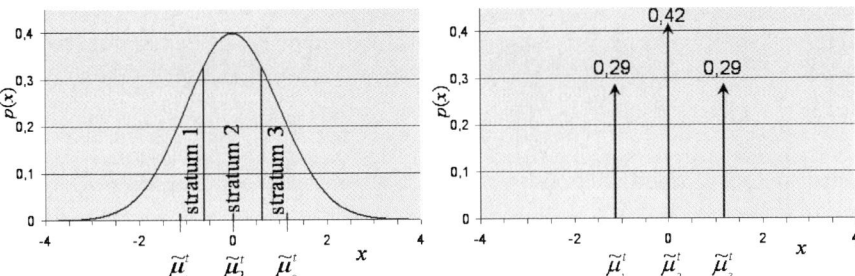

Fig. 4. Continuous (*left*) and discretized normal probability density function (*right*)

ample the discretization of a normal distribution into three strata is demonstrated in Fig. 4.

The conditional expected value $\tilde{\mu}_j^t$ and its occurrence probability $\pi(\tilde{\mu}_j^t)$ of stratum j are used as the substitute value of each scenario. The optimal strata limits are calculated by means of statistical methods in dependence on the predefined number of scenarios [25]. By calculating the optimal strata limits and the resulting conditional expected values $\tilde{\mu}_j^t$ it is ensured that all appearing forecast errors minimally diverge from their substitute values $\tilde{\mu}_j^t$. The number of scenarios in practice has to be determined individually in each single case. This selection will always be a compromise between accuracy of modelling and computational manageability. With this knowledge the scenarios of the second-stage can be generated under consideration of the different analysed forecast errors. According to the afore explained method the electrical load forecast modelled with three scenarios is shown in Fig. 5 for an optimization starting at 9 a.m.. Here the development of three scenarios by superposition of the day-ahead forecast and the conditional expected values $\tilde{\mu}_j^t$ is illustrated at 1 p.m. of the second day.

5 Stochastic Optimization

In this section, we introduce the mathematical optimization models which are able to reflect the above described real-world situation. We start out from the random optimization problem

$$\min \{ c^\top x + q^\top y \,:\, Tx + Wy = z(\omega),\, x \in X,\, y \in Y \}, \qquad (4)$$

together with the information constraint that x has to be selected without anticipation of the realization of the random data $z(\omega)$. This yields a two-stage scheme of decision and observation: The first-stage decision x is made before the observation of $z(\omega)$ which is followed by the second-stage decision y, thus depending on x and z. Since $X \subseteq \mathbb{R}^m$ and $Y \subseteq \mathbb{R}^{\bar{m}+m'}$ are polyhedra, possibly involving integer requirements to x and y, (4) is a mixed-integer linear program.

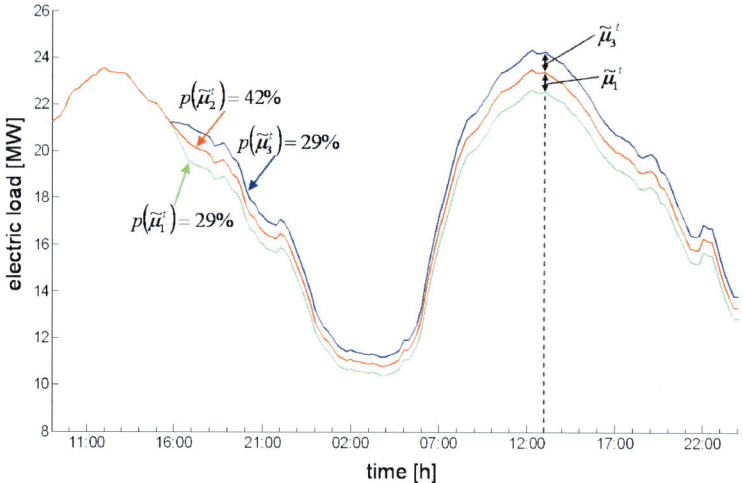

Fig. 5. Exemplary illustration of modelling the first and second-stage of the electrical load with three scenarios

The two-stage dynamics of (4) becomes explicit by a reformulation:

$$\min_x \left\{ c^\top x + \min_y \left\{ q^\top y \;:\; Wy = z(\omega) - Tx,\; y \in Y \right\} x \in X \right\}$$
$$= \min_x \left\{ c^\top x + \Phi(x,z) \;:\; x \in X \right\}. \tag{5}$$

This formulation is known as the above mentioned two-stage stochastic optimization problem. Obviously, each first-stage decision x induces a random variable and the selection of a decision x corresponds to the selection of a member of the family

$$\left(f(x,z) := c^\top x + \Phi(x,z) \right)_{x \in X} \tag{6}$$

of random variables.

Also the input data c, q, T, W could be assumed to be influenced by uncertainty. In view of the optimal management of dispersed generation, the uncertain spot market prices and fuel costs are reflected by c and q, respectively, the forecasted infeed from renewable resources is included in W and probabilistic load profiles go into z. For simplicity, we restrict ourselves to a random right hand side.

It remains to define the criterion for the selection of the random variable. A risk neutral approach is the application of the expectation $\mathbb{E}$. This leads to the expectation-based stochastic optimization problem

$$\min \left\{ \mathbb{E}\bigl(f(x,z)\bigr) \;:\; x \in X \right\}, \tag{7}$$

which is very common in stochastic programming, see [27, 28].

If one aims at risk averse decisions, mean-risk models as well as dominance constrained problem formulations come to the fore, which shall be described in detail below.

5.1 Mean-Risk Models

If the ranking of the random variables in (6) is done via a weighted sum of the expectation $\mathbb{E}$ and some risk measure $\mathcal{R}$, we arrive at the basic mean-risk model

$$\min \left\{ \mathbb{E}\big(f(x,z)\big) + \rho \cdot \mathcal{R}\big(f(x,z)\big) \: : \: x \in X \right\} \tag{8}$$

with a fixed parameter $\rho > 0$. The specification of the risk measure $\mathcal{R}$ should be done according to the decision maker's preferences, but also with respect to the mathematical structure it induces to (8) and the algorithmic possibilities for solving (8). As the initial random problem (5) is mixed-integer linear, we should choose specifications that maintain the mixed-integer linear characteristics in the case of finite, discrete distributions of $z(\omega)$. Moreover, for dimensionality reasons, our algorithms will be able to exploit a special block-structure of these mixed-integer linear programs which arises for many specifications of $\mathcal{R}$, but not for all. The following risk measures, for instance, are in this sense mathematically and structurally sound:

Expected Excess:
$$\mathbb{Q}_{\mathcal{D}^\eta}(f(x,z)) := \mathbb{E}\Big(\max\big\{f(x,z) - \eta, 0\big\} \Big),$$

which computes the excess of the random variable $f(x,z)$ over a prefixed threshold $\eta \in \mathbb{R}$.

Excess Probability:
$$\mathbb{Q}_{\mathbb{P}^\eta}(f(x,z)) := \mathbb{P}\Big(\{\omega \: : \: f(x,z) > \eta\} \Big),$$

which is the probability that the random variable $f(x,z)$ exceeds a given threshold $\eta \in \mathbb{R}$.

Conditional Value-at-Risk:
$$\mathbb{Q}_{\alpha\text{CVaR}}(f(x,z)) := \min_{\eta \in \mathbb{R}} g(\eta, f(x,z)),$$

where
$$g(\eta, f(x,z)) := \eta + \frac{1}{1-\alpha} \mathbb{E}\Big(\max\big\{f(x,z) - \eta, 0\big\} \Big),$$

which reflects the expectation of the $(1-\alpha) \cdot 100\,\%$ worst outcomes of the random variable $f(x,z)$ for a fixed probability level $\alpha \in (0,1)$.

Counterexamples for sound specifications of the risk measure are for example the semideviation or the variance. For more details concerning the mathematical structure of different mean-risk models and algorithms associated with (8) including different risk measures $\mathcal{R}$ see [23, 29, 30], for instance.

5.2 Deterministic Equivalents for Mean-Risk Models

For computational reasons, we assume the random input data $z(\omega)$ to be discretely distributed with finitely many scenarios z_l and probabilities π_l, $l = 1,\ldots,L$. The risk neutral model (7) can be equivalently restated as

$$\min\left\{ c^\top x + \sum_{l=1}^{L} \pi_l q^\top y_l : \begin{array}{l} Tx + Wy_l = z_l \ \forall l \\ x \in X,\ y_l \in Y \ \ \forall l \end{array} \right\}, \tag{9}$$

which again is mixed-integer linear. Furthermore, the constraint matrix of (9) shows the block-structure depicted in Fig. 6.

The constraint matrix has an L-shaped form, which results from the fact, that there are no constraints including second-stage variables y_{l_1} and y_{l_2} belonging to different scenarios l_1, l_2. Only the implicit coupling due to the nonanticipativity (NA) of x prevents (9) from decomposition into L scenario-specific subproblems. Similar results can be obtained for the above introduced risk measures:

For instance, let $\mathcal{R} = \mathbb{Q}_{\mathcal{D}^\eta}$ (*Expected Excess*), then (8) is equivalent to

$$\min\left\{ c^\top x + \sum_{l=1}^{L} \pi_l q^\top y_l + \rho \cdot \sum_{l=1}^{L} \pi_l v_l : \begin{array}{l} Tx + Wy_l = z_l \ \forall l \\ c^\top x + q^\top y_l \leq v_l \ \forall l \\ y_l \in Y,\ v_l \in \mathbb{R}_+ \ \forall l \\ x \in X \end{array} \right\}. \tag{10}$$

For all three risk measures deterministic equivalents can be derived that maintain the structure depicted in Fig. 6. This is important in view of the algorithmic methods developed in Sect. 6.

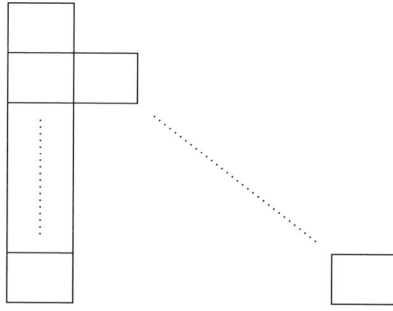

Fig. 6. Block-structure of the constraint matrix of (9)

In general, the deterministic equivalents of all mentioned mean-risk models could be solved by standard mixed-integer linear programming solvers (MILP). But, since their dimension increases substantially with an increasing number of scenarios L, the solution by a decomposition of the equivalents seems reasonable. The idea behind this decomposition is to exploit the above described structure of the constraint matrix.

5.3 Dominance Constrained Model

A more recent approach to tackle (5) is the application of stochastic dominance constraints. In contrast to mean-risk modeling, this approach aims at the identification of acceptable members among the family of random variables (6). Therefore, a reference benchmark $a(\omega)$ is introduced, which mathematically is again a random variable. It represents a random profile of the total costs that is just acceptable in context of the application at hand and can be derived from former optimization or experience of the decision maker, for example. This approach is a more appropriate reflection of the stochastic nature inherent in (4) and offers more flexibility to the decision maker to define what is an acceptable solution to him. Mathematically the acceptability of a random variable can be expressed by different notions of stochastic dominance, for details see for instance [11, 15, 24].

Let μ and ν denote the Borel probability measures induced by $z(\omega)$ and $a(\omega)$, respectively. Then $f(x, z)$, with a fixed $x \in X$, is said to dominate the reference benchmark $a(\omega)$ to first-order, denoted by $f(x, z) \succeq_{(1)} a$, if

$$\mu(\{z : f(x,z) \leq \eta\}) \geq \nu(\{a : a \leq \eta\}) \quad \forall \eta \in \mathbb{R}.$$

This means, that $f(x, z)$ takes smaller values with a higher probability than a. Note that the first-order dominance relation is suitable to reflect the behavior of a rational decision maker. The underlying theory of utility functions is established in [32].

Applying the first-order dominance to the two-stage random optimization problem (5) leads to the following stochastic optimization problem with first-order dominance constraints induced by mixed-integer linear recourse:

$$\min \{ g^\top x \ : \ f(x,z) \succeq_{(1)} a, \ x \in X \} \tag{11}$$

where $g^\top x$ is supposed a new linear objective function that can be chosen appropriate to the decision makers prerequisites. For example, in the context of optimal operation of an energy generation system it could count the start-ups of the single units and hence aim at minimizing abrasion.

Dominance constraints for stochastic optimization have already been established in [7, 8, 9]. There, more general random variables are involved, whereas in our case, $f(x, z)$ results from the specific two-stage stochastic programming context.

5.4 Deterministic Equivalents for the First-Order Dominance Constrained Model

As done in case of mean-risk models, we assume the random variables z, a to be discretely distributed with finitely many realizations z_l with probabilities π_l, $l = 1, \ldots, L$ and a_k with probabilities p_k, $k = 1, \ldots, K$. The first-order dominance constraint then simplifies:

$$f(x,z) \succeq_{(1)} a \Leftrightarrow \mu(\{z : f(x,z) \leq a_k\}) \geq \nu(\{a : a \leq a_k\}) \quad \forall k. \quad (12)$$

For a proof see [12, 13, 26].

In view of (12) and assuming the discretization of the distribution of z, we obtain a deterministic equivalent of (11):

There exists a constant $\mathsf{M} > 0$ such that the dominance constrained program (11) can be equivalently restated as

$$\min \left\{ g^\top x : \begin{array}{ll} c^\top x + q^\top y_{lk} - a_k \leq \mathsf{M}\theta_{lk} & \forall l\, \forall k \\ \sum_{l=1}^{L} \pi_l \theta_{lk} \leq \bar{a}_k & \forall k \\ Tx + Wy_{lk} = z_l & \forall l\, \forall k \\ x \in X,\ y_{lk} \in Y,\ \theta_{lk} \in \{0,1\}\ \forall l\, \forall k \end{array} \right\}, \quad (13)$$

where $\bar{a}_k := 1 - \nu(\{a : a \leq a_k\})$, $k = 1, \ldots, K$.

The constraint matrix of this equivalent shows a similar desirable block-structure, as the deterministic equivalents for mean-risk models. The structure of the constraint matrix is displayed in Fig. 7.

As for the mean-risk models introduced above, we have the implicit coupling of different scenarios by the nonanticipativity (NA) of x. Additionally,

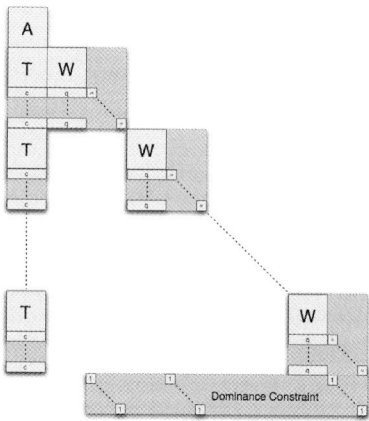

Fig. 7. Block-structure of the constraint matrix of (13)

there are K constraints, represented by the block called "Dominance Constraint" in Fig. 7, which couple second-stage variables belonging to different scenarios:

$$\sum_{l=1}^{L} \pi_l \theta_{lk} \leq \bar{a}_k \quad \forall k \tag{14}$$

This structure gives rise to a relaxation of coupling constraints and the establishment of lower bounds by some scenario-wise decomposition of the arising problems. This is described in detail both, for the mean-risk models and the first-order dominance constrained model, in the following section. As computational results presented in Sect. 7 will show, such a decomposition is superior to the application of standard MILP solvers to (13).

6 Algorithmic Issues

To reasonably reduce the presentation of algorithmic issues, we restrict ourselves to the mean-risk model with the Expected Excess as risk measure (10) and the first-order dominance constrained model (13). The results can then be transferred to the other stochastic models introduced above.

The overall idea to solve the above established deterministic equivalents is to calculate lower and upper bounds iteratively and to embed this procedure into a branch-and-bound scheme in the spirit of global optimization. The lower bounds are derived by different relaxations of coupling constraints, which is presented in the subsequent section. The next but one subsection deals with the selection of upper bounds and a final subsection describes the applied branch-and-bound method.

6.1 Lower Bounds

A first step towards lower bounds is to introduce copies $x_1, x_2, \ldots, x_L$ of x and to add the explicit NA-constraint $x_1 = x_2 = \ldots = x_L$, which can be restated by $\sum_{l=1}^{L} H_l x_l = 0$ with suitable matrices H_l, $l = 1, \ldots, L$. For the

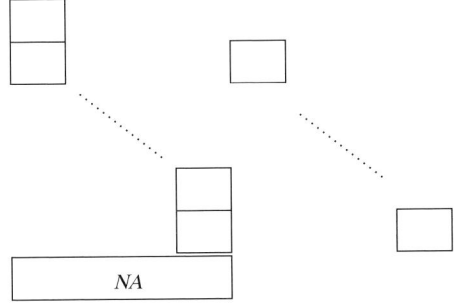

Fig. 8. Block-structure of the constraint matrix of (10) with explicit NA

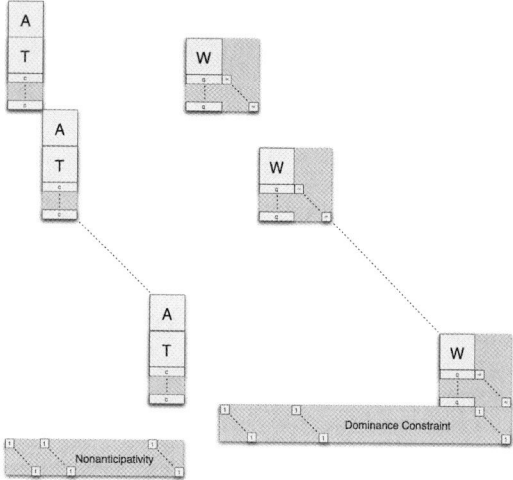

Fig. 9. Block-structure of the constraint matrix of (13) with explicit NA

deterministic equivalents (10) and (13) this yields the constraint matrices displayed in Figs. 8 and 9, respectively.

Obviously, (10) readily decomposes into scenario-specific subproblems, if the NA-constraint is relaxed, whereas for a decomposition of (13) the scenario-coupling dominance constraints have to be relaxed, too.

Lower Bounds for Mean-Risk Models

The Lagrangean relaxation of the NA-constraint in (10) yields the Lagrangean function

$$L(x,y,v,\lambda) := \sum_{l=1}^{L} \pi_l \left(c^\top x_l + q^\top y_l + v_l + \lambda^\top H_l x_l \right)$$

$$= \sum_{l=1}^{L} L_l(x_l, y_l, v_l, \lambda)$$

We obtain the Lagrangean dual

$$\max \left\{ D(\lambda) \ : \ \lambda \in \mathbb{R}^{L \cdot (m-1)} \right\}, \tag{15}$$

with

$$D(\lambda) := \min \left\{ L(x,y,v,\lambda) : \begin{array}{ll} T x_l + W y_l = z_l & \forall l \\ c^\top x_l + q^\top y_l \leq v_l & \forall l \\ x_l \in X, \ y_l \in Y, \ v_l \in \mathbb{R}_+ & \forall l \end{array} \right\}$$

$$= \sum_{l=1}^{L} \min \left\{ L_l(x_l, y_l, v_l, \lambda) : Tx_l + Wy_l = z_l \right. \\ \left. c^\top x_l + q^\top y_l \leq v_l \right. \\ \left. x_l \in X,\ y_l \in Y,\ v_l \in \mathbb{R}_+ \right\}. \quad (16)$$

The Lagrangean dual (15) is a nonsmooth concave maximization problem which is tackled with bundle subgradient methods from nondifferentiable optimization, see [18]. In each step of such a method, we have to compute the objective function of (15) for a fixed λ and one subgradient. At this point, the above presented block-structure of (10) plays an important role: Together with the Lagrangean relaxation of the NA-constraint it allows the computation of the objective value $D(\lambda)$ by a decomposition into small subproblems. This is reflected by (16) which shows that $D(\lambda)$ can be restated as the sum of L independent subproblems corresponding to the different scenarios.

Lower Bounds for the First-Order Dominance Constrained Problem

The complexity of the Lagrangean relaxation mainly results from the number of relaxed constraints. Since usually the number of reference profiles K in (13) is much smaller than the number of scenarios L, we apply the Lagrangean relaxation only to the linking dominance constraints and ignore the NA-constraint. Moreover, the NA-constraint can be recovered much easier than the dominance constraints.

We obtain the following Lagrangean function

$$L(x, \theta, \lambda) := \sum_{l=1}^{L} \pi_l \cdot g^\top x_l + \sum_{k=1}^{K} \lambda_k \left(\sum_{l=1}^{L} \pi_l \theta_{lk} - \bar{a}_k \right)$$

$$= \sum_{l=1}^{L} L_l(x_l, \theta_l, \lambda).$$

The Lagrangean dual then reads

$$\max \left\{ D(\lambda) : \lambda \in \mathbb{R}_+^K \right\}, \quad (17)$$

with

$$D(\lambda) := \min \left\{ L(x, \theta, \lambda) : Tx_l + Wy_{lk} = z_l \quad \forall l, \forall k \right. \\ \left. c^\top x_l + q^\top y_{lk} - a_k \leq M\theta_{lk} \quad \forall l, \forall k \right. \\ \left. x_l \in X,\ y_{lk} \in Y,\ \theta_{lk} \in \{0,1\} \quad \forall l, \forall k \right\}$$

$$= \sum_{l=1}^{L} \min \left\{ L_l(x_l, \theta_l, \lambda) : \begin{array}{ll} Tx_l + Wy_{lk} = z_l & \forall k \\ c^\top x_l + q^\top y_{lk} - a_k \leq M\theta_{lk} & \forall k \\ x_l \in X,\ y_l \in Y,\ \theta_{lk} \in \{0,1\} & \forall k \end{array} \right\}. \quad (18)$$

Again, the Lagrangean dual decomposes and can be computed by the solution of scenario-specific subproblems.

6.2 Upper Bounds

Both described lower bounding procedures provide $\bar{x}_1, \ldots, \bar{x}_L$ which can be understood as proposals for a feasible first-stage solution $\bar{x}$ of (10) and (13), respectively. There are different possibilities to derive $\bar{x}$ which means to recover the relaxed nonanticipativity. For example, we can compute the mean of the $\bar{x}_1, \ldots, \bar{x}_L$ and round the result to the next integer if required, we can choose the proposal $\bar{x}_{l'}$ belonging to the scenario with the highest probability $\pi_{l'}$ or we choose the proposal $x_{l''}$ if scenario l'' incurs the highest costs. Having selected $\bar{x}$ we check its feasibility.

In case of the dominance constrained model, this is included into a procedure that tries to find $\bar{\theta}_{lk}$ which – together with the $\bar{x}$ – fulfill the relaxed dominance constraints. In a first step, we solve for all $l = 1, \ldots, L$ the problem

$$\min \left\{ \sum_{k=1}^{K} \theta_{lk} : \begin{array}{ll} c^\top \bar{x} + q^\top y_{lk} - a_k \leq M\theta_{lk} & \forall k \\ T\bar{x} + Wy_{lk} = z_l & \forall k \\ y_{lk} \in Y,\ \theta_{lk} \in \{0,1\} & \forall k \end{array} \right\}$$

which aims at pushing as many θ_{lk} as possible to zero.

The second step is to check for all $k = 1, \ldots, K$, if the found $\bar{\theta}_{lk}$ fulfill

$$\sum_{l=1}^{L} \pi_l \bar{\theta}_{lk} \leq \bar{a}_k.$$

If they do, $(\bar{x}, \bar{\theta}_{lk})$ is a feasible point for (13).

6.3 A Branch-and-Bound Algorithm

The iterative calculation of lower and upper bounds is embedded into a branch-and-bound scheme which establishes partitions of the feasible set X with increasing granularity. This is done via additional linear inequalities to maintain the mixed-integer linear description of the constraint set. In the procedure, elements of the partition of X, which correspond to nodes of the arising branching tree, can be pruned because of infeasibility, inferiority, or optimality. In each node the gap between the currently best solution value and the currently best lower bound is computed and serves as a stopping criterion. This results in the following algorithm for the dominance constrained problem, which can simply be transferred to mean-risk models.

Let $\mathbf{P}$ denote a list of problems and $\varphi_{LB}(P)$ a lower bound for the optimal value of $P \in \mathbf{P}$. Furthermore, $\bar{\varphi}$ denotes the currently best upper bound to the optimal value of (13) and $X(P)$ is the element in the partition of X belonging to P.

Branch-and-Bound Algorithm:

STEP 1 (INITIALIZATION):
Let $\mathbf{P} := \{(13))\}$ and $\bar{\varphi} := +\infty$.

STEP 2 (TERMINATION):
If $\mathbf{P} = \emptyset$ then the feasible point $\bar{x}$ that yielded $\bar{\varphi} = g^\top \bar{x}$ is optimal.

STEP 3 (BOUNDING):
Select and delete a problem P from $\mathbf{P}$. Compute a lower bound $\varphi_{LB}(P)$ solving the Lagrangean dual (17) and find a feasible point $\bar{x}$ of P with the above described upper bounding procedure.

STEP 4 (PRUNING):
If $\varphi_{LB}(P) = +\infty$ (infeasibility of a subproblem in (18)) or $\varphi_{LB}(P) > \bar{\varphi}$ (inferiority of P), then go to Step 2.
If $\varphi_{LB}(P) = g^\top \bar{x}$ (optimality of P), then check whether $g^\top \bar{x} < \bar{\varphi}$. If yes, then $\bar{\varphi} := g^\top \bar{x}$. Go to Step 2.
If $g^\top \bar{x} < \bar{\varphi}$, then $\bar{\varphi} := g^\top \bar{x}$.

STEP 5 (BRANCHING):
Create two new subproblems by partitioning the set $X(P)$. Add these subproblems to $\mathbf{P}$ and go to Step 2.

7 Computational Results – Case Study 1

To analyze the optimization results a VPP consisting of two local heat networks is taken as a basis (see Fig. 10). The thermal demand is supplied by four CHP-units and two peak load boilers as well as two hot water storages. To exhaust excessive heat in case of an emergency two cooling devices are installed. The electrical demand is covered by the four CHP-units, a battery storage and a flexible energy contract (10 % of fixed schedule adaptable day-ahead) as well as by energy bought from the spot market. The complete block diagram is depicted in Fig. 10.

To determine the optimal adaption of the flexible energy contract and the optimal bids at the spot market for the following day the results of an

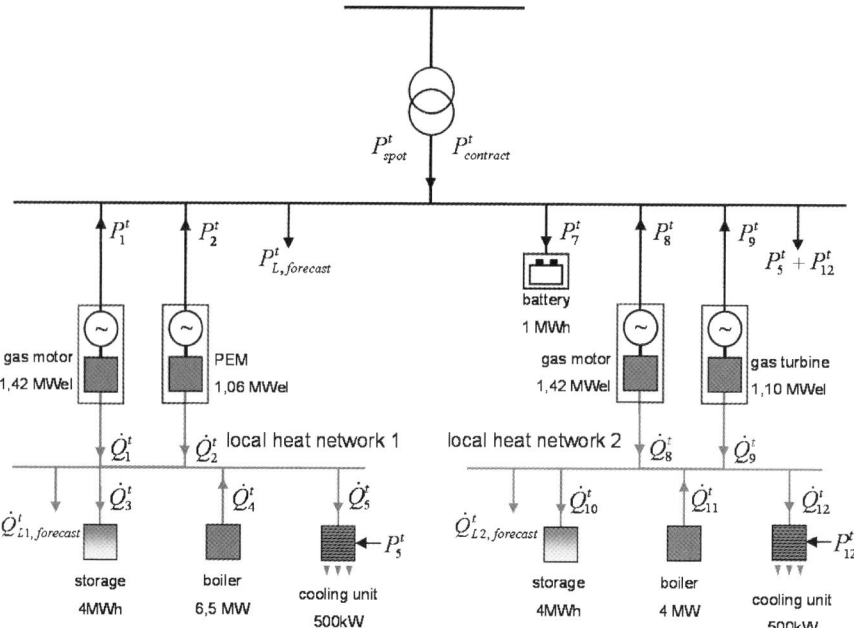

Fig. 10. VPP Configuration

optimization started at 9 a.m. are analyzed. The electrical demand (see also Fig. 11), the thermal demand and the spot market prices are modeled by four scenarios each. Thus 64 scenarios are considered in total.

Figure 11 shows a selection of optimal first-stage decisions under consideration of the uncertain electrical and thermal demand as well as the uncertain spot market prices. Among the optimal adaption of the flexible contract the optimal spot market bids are displayed for a period of two consecutive working days in winter. The power delivered by the contract is during the whole time at its lowest limit excepting the period between 9 a.m. and 3 p.m.. Due to the high spot market price to be expected during this period only very little energy is bought at the spot market instead it is taken out of the contract. Between 11 and 12 a.m. the high spot market price is used to gain profit by trading at the spot market (sale of energy).

7.1 Benefit of the Stochastic Optimization

When solving optimization problems with stochastic input data in practice the uncertainties are often substituted by their expected value, i.e. the forecast [3]. Hence, in the following the benefit is identified which can be achieved performing the stochastic optimization for the above configuration. Therefore, the objective function value of the stochastic optimization (Value of the recourse problem, VRP) is compared to the value of the deterministic objec-

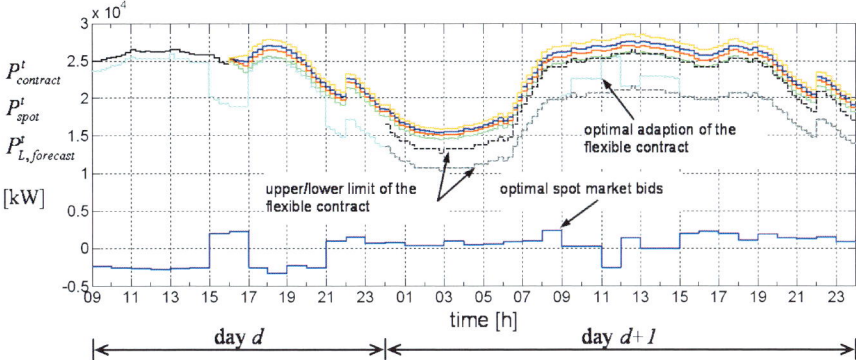

Fig. 11. Optimization results for trading at the spot market and the day-ahead adaption of the flexible contract

tive function when all uncertainties are substituted by their expected values (Expected result of the expected value solution, EEV). The difference of both values (Value of stochastic solution, VSS) can be distinguished as a benchmark of the stochastic optimization [3].

Table 1 shows an excerpt of the results achieved for different constellations in which the volatility of the spot market prices, the price of balance energy and the forecast accuracy have been varied. The so called gap quantifies the maximum decrease of the objective function value if longer computation time would be admitted.

The results indicate that in each of the constellations investigated the stochastic optimization leads to a better result (lower costs) than the conventional deterministic alternative. It becomes apparent that the overall operating costs over a time period of 39 hours can be reduced by 1–2 % applying the stochastic optimization. Especially the increasing benefit of the stochastic optimization in context with the decreasing quality of the electrical load forecast becomes apparent (see instances **2** and **5** in Table 1). In addition to that the price of balance energy also has a big influence on the benefit of the stochastic optimization. In this case an increase of 5 ct/kWh to 30 ct/kWh leads to a VSS which is 295 € higher (see instances **2** and **4** in Table 1).

8 Computational Results – Case Study 2

We consider the optimal operation of a dispersed generation system with respect to the supply of uncertain heat and power demand. The system, which is run by a German utility, consists of five engine-based cogeneration (CG) stations which produce thermal and electrical energy simultaneously. The CG stations altogether include nine gas motors, eight boilers and one gas turbine. They are each equipped with a thermal storage and a cool-

Table 1. Results for different parameters and input data

Instance No.	Constellation	VRP (€)	VSS (€)	VSS (%)	Gap (%)
1	winter, low volatility	45939	409	0.88	3.20
2	winter, high volatility	45259	538	1.17	3.21
3	winter, high volatility $c_{bal\,Energy,+} = 20$ ct/kWh	45025	397	0.87	2.85
4	winter, high volatility $c_{bal\,Energy,+} = 30$ ct/kWh	45284	833	1.81	3.25
5	winter, high volatility $\sigma_{el,\,error}$ +1%	45680	1041	2.23	3.54
6	winter, high volatility $\sigma_{th,\,error}$ +1%	45282	513	1.12	3.26
7	winter high volatility $\sigma_{Spot,\,error}$ +1%	45164	602	1.32	3.03

Table 2. Increasing dimensions of the expectation-based model

Scenarios	Boolean variables	Continuous variables	Constraints
1	8.959	8.453	22.196
5	38.719	36.613	96.084
10	75.919	71.813	188.444
50	373.519	353.413	927.324

ing device, such that excessive heat can be stored or exhausted. The system is completed by twelve wind turbines and one hydroelectric power plant. Whereas the thermal energy is distributed locally around each CG station, the electrical energy is fed into the distribution network to supply the total demand.

As shown in Sect. 3 the operation of such a system with all its technical characteristics can be described as a mixed-integer linear model. Assuming a time horizon of 24 hours, divided into 96 quarter-hourly time intervals, and that all input data is deterministic, we obtain a problem formulation with about 9000 boolean and 8500 continuous variables and 22000 constraints. Though already large-scale, such a deterministic problem can be handled with standard MILP solvers like Ilog-Cplex ([22]). Computational tests for different deterministic load profiles showed that we are able to find operational schedules inducing minimal costs with an optimality gap below 0.1% for all test instances in less than 20 seconds on a Linux-PC with a 3GHz processor and 2GB ram. For details see [16].

The situation changes, if we include uncertain load profiles and use a mean-risk formulation of the optimization problem. Dimensions for an expectation based model, which can be compared to the dimensions of the mentioned mean-risk formulations, are displayed in Table 2.

Table 3. Computation times for the expectation-based model

Number of	Ilog-Cplex			Decomposition		
scenarios	Objective	Gap (%)	Time (sec.)	Objective	Gap (%)	Time (sec.)
5	7.150.945	0,0072	11	7.150.903	0,0067	8
10	5.987.642	0,0099	253	5.987.578	0,0087	52
20	5.837.584	0,0093	1.769	5.837.559	0,0088	172
30	5.928.500	0,0091	9.486	5.928.542	0,0093	364
40	5.833.156	0,0107	26.286	5.833.191	0,0093	487
50	5.772.335	0,0129	22.979	5.772.268	0,0094	611

Table 4. Results for the pure risk model with Conditional Value-at-Risk

Number of	Ilog-Cplex			Decomposition		
scenarios	$\mathcal{R}$	Gap (%)	Time (sec.)	$\mathcal{R}$	Gap (%)	Time (sec.)
5	9.256.610	0,0007	23	9.256.642	0,0010	37
10	9.060.910	0,0033	67	9.060.737	0,0013	79
20	8.522.190	0,0040	504	8.522.057	0,0023	436
30	8.996.950	0,0049	5.395	8.996.822	0,0033	594
40	8.795.120	0,0049	7.366	8.795.064	0,0039	1.038
50	8.557.813	0,0050	9.685	8.557.755	0,0039	1.286

For the solution of the expectation-based problem the usage of the dual decomposition algorithm, described in Sect. 6, already is preferable over the solution of the corresponding deterministic equivalent by Cplex. Table 3 compares computational results for expectation-based instances with 5 up to 50 scenarios with a stopping criterion of a gap smaller than 0.1 % gained from Cplex and the decomposition method, respectively. The computation times of Cplex are up to 35 times of the computation times of the decomposition.

Similar results can be observed for the models including risk measures. Table 4 displays computations for a pure risk model with the α-Conditional Value-at-Risk and with an optimality gap falling below 0.005 % as stopping criterion. Obviously, Cplex is superior only in cases with 5 or 10 scenarios and the decomposition is preferable if more scenarios are included.

As an example for computations including the expected costs and a risk measure, Table 5 presents results for the mean-risk model with the Excess Probability. Here, we apply a timelimit of 5 up to 25 minutes and can therefore use the reached gap as a criterion to compare the two solution methods. While for the instance with 5 scenarios Cplex beats the decomposition method, in all instances with more scenarios the decomposition reaches better solutions. In cases with 20 scenarios or more, Cplex even fails to find one feasible solution.

Moreover, computations for a first-order dominance constrained problem were made. As objective function $g^\top x$ we considered the sum of all start-ups of the generation units in the first-stage which means, that the program aims

Table 5. Results for the mean-risk model with Excess Probability

No. scen.	Time (sec.)	ρ	Ilog-Cplex $\mathbb{E}$	$\mathcal{R}$	Gap (%)	Decomposition $\mathbb{E}$	$\mathcal{R}$	Gap (%)
5	300	0	7.150.850	–	0,0057	7.150.866	–	0,0062
		0,0001	7.150.900	0,400	0,0064	7.150.913	0,400	0,0066
		10.000	7.150.880	0,400	0,0342	7.150.930	0,400	0,0348
10	450	0	5.987.670	–	0,0104	5.987.526	–	0,0078
		0,0001	5.987.630	0,370	0,0098	5.987.554	0,370	0,0084
		10.000	5.987.630	0,370	0,0315	5.987.556	0,370	0,0302
20	600	0	infeasible		infinity	5.837.500	–	0,0078
		0,0001	infeasible		infinity	5.837.560	0,205	0,0089
		10.000	infeasible		infinity	5.837.556	0,205	0,0183
50	1500	0	infeasible		infinity	5.772.226	–	0,0087
		0,0001	infeasible		infinity	5.772.271	0,205	0,0096
		10.000	infeasible		infinity	5.772.287	0,205	0,0151

Table 6. Dimensions of the equivalent of the dominance constrained problem

Number of	10 scenarios	20 scenarios	30 scenarios	50 scenarios
Boolean variables	299159	596799	894439	1489719
continuous variables	283013	564613	846213	1409413
constraints	742648	1481568	2220488	3698328

at the minimization of abrasion of the units. The benchmark profile is derived from the solution of the expectation-based problem. We pick some of the solution's single scenario costs, cluster all other scenarios around the chosen scenarios and assign as benchmark probabilities the sum of the scenario probabilities. The dimensions of the arising deterministic equivalents for $K = 4$ benchmark profiles and $L = 10 - 50$ scenarios are displayed in Table 6.

Tables 7 and 8 picture the solution process of five problem instances with different benchmark profiles and 10 or 30 data scenarios, respectively. The status of the solution – represented by the currently best upper and lower bounds – is given for those points in time, when one of the solvers reaches a feasible solution, when a solver proves optimality of a currently best solution, or when a solver runs out of memory. The timelimit for all computations is eight hours.

In case of 10 scenarios the decomposition finds the first feasible solution faster than Cplex and for instances 1, 4 and 5 it also proves optimality before Cplex does. In contrast, for instances 2 and 3 Cplex is able to prove optimality faster than the decomposition. Hence, for 10 scenarios the decomposition is preferable if we aim at feasibility, but there is no obvious superiority.

For instances including 30 data scenarios the superiority of the decomposition gets evident. In these cases the decomposition is still able to solve all instances up to optimality, while Cplex reaches the limits of available memory before it finds a first feasible solution, which is marked by 'mem.' in Table 8.

Table 7. Results for instances with 10 data scenarios and 4 benchmark scenarios

Number of scenarios	Instance	Benchmarks Probability	Benchmark Value	Time (sec.)	Ilog-Cplex Upper Bound	Ilog-Cplex Lower Bound	ddsip.vSD Upper Bound	ddsip.vSD Lower Bound
10	1	0.12	2895000	430.43	–	29	29	15
		0.21	4851000	899.16	–	29	29	29
		0.52	7789000	15325.75	29	29	29	29
		0.15	10728000					
	2	0.12	2900000	192.48	–	27	28	15
		0.21	4860000	418.90	28	28	28	15
		0.52	7800000	802.94	28	28	28	28
		0.15	10740000					
	3	0.12	3000000	144.63	–	21	21	12
		0.21	5000000	428.61	21	21	21	18
		0.52	8000000	678.79	21	21	21	21
		0.15	11000000					
	4	0.12	3500000	164.34	–	11	13	10
		0.21	5500000	818.26	–	12	13	13
		0.52	8500000	28800.00	13	12	13	13
		0.15	11500000					
	5	0.12	4000000	171.52	–	7	8	8
		0.21	6000000	3304.02	8	8	8	8
		0.52	9000000					
		0.15	12000000					

Table 8. Results for instances with 30 data scenarios and 4 benchmark scenarios

Number of scenarios	Instance	Benchmarks Probability	Benchmark Value	Time (sec.)	Ilog-Cplex Upper Bound	Ilog-Cplex Lower Bound	ddsip.vSD Upper Bound	ddsip.vSD Lower Bound
30	1	0.085	2895000	473.27	–	28	29	12
		0.14	4851000	1658.02	–	29	29	29
		0.635	7789000	3255.99	–	29 mem.	29	29
		0.14	10728000					
	2	0.085	2900000	1001.53	–	26	28	18
		0.14	4860000	2694.93	–	27	28	28
		0.635	7800000	3372.24	–	27 mem.	28	28
		0.14	10740000					
	3	0.085	3000000	469.93	–	17	23	10
		0.14	5000000	3681.15	–	18 mem.	21	20
		0.635	8000000	28800.00	–	–	21	20
		0.14	11000000					
	4	0.085	3500000	618.21	–	10	14	8
		0.14	5500000	3095.02	–	11 mem.	14	10
		0.635	8500000	28800.00	–	–	14	13
		0.14	11500000					
	5	0.085	4000000	672.73	–	7	8	8
		0.14	6000000	8504.88	–	8 mem.	8	8
		0.635	9000000					
		0.14	12000000					

If one tries to solve instances with 50 scenarios, the situation for Cplex gets even worse, because then the available memory is not sufficient to build up the (lp-)model file which is needed as input for Cplex. In contrast, for the decomposition only a (lp-)model file including one single-scenario is needed.

All in all, the decomposition is obviously preferable over standard solvers when it comes to computations with a high number of data scenarios and benchmark profiles.

Acknowledgement

This research has been funded by the German Federal Ministry of Education and Research under grants 03HANIVG and 03SCNIVG.

References

1. U. Arndt, S. von Roon, and U. Wagner. Virtuelle Kraftwerke: Theorie oder Realität? *BWK, das Energie-Fachmagazin*, 58(6), Juni 2006.
2. R. Becker, E. Handschin, E. Hauptmeier, and F. Uphaus. Heat-controlled combined cycle units in distribution networks. *CIRED 2003*, Barcelona, 2003.
3. J. R. Birge and F. Louveaux. *Introduction to Stochastic Programming*. Springer, New York, 1997.
4. B. Buchholz, N. Hatziargyriou, I. Furones, and U. Schlücking. Lessons Learned: European Pilot Installations for Distributed Generation – An Overview by the IRED Cluster. *Cigré Session*, Paris, August 2006.
5. European Commission. EU-15 Energy and Transport – Outlook to 2030, part II, October 2006.
6. T. Degner, J. Schmid, and P. Strauss, editors. *Final Public Report, Distributed Generation with high Penetration of Renewable Energy Sources (DISPOWER)*, Juni 2006.
7. D. Dentcheva, R. Henrion, and A. Ruszczyński. Stability and sensitivity of optimization problems with first order stochastic dominance constraints. *SIAM Journal on Optimization*, 18(1):322–337, 2007.
8. D. Dentcheva and A. Ruszczyński. Optimization with stochastic dominance constraints. *SIAM Journal on Optimization*, 14(2):548–566, 2003.
9. D. Dentcheva and A. Ruszczyński. Portfolio optimization with stochastic dominance constraints. *Journal of Banking and Finance*, 30:433–451, 2006.
10. T. Erge et. al. Reports on Improved Power Management in Low Voltage Grids by the Application of the PoMS System, December 2005.
11. P. C. Fishburn. *Utility Theory for Decision Making*. Wiley, New York, 1970.
12. R. Gollmer, U. Gotzes, and R. Schultz. Second-order stochastic dominance constraints induced by mixed-integer linear recourse. *Preprint Series, Department of Mathematics, University of Duisburg-Essen*, 644–2007, 2007.
13. R. Gollmer, F. Neise, and R. Schultz. Stochastic programs with first-order dominance constraints induced by mixed-integer linear recourse. *Preprint Series, Department of Mathematics, University of Duisburg-Essen*, 641–2006, 2006.
14. G. Gross and F. Galiana. Short-Term Load Forecasting. In *Proceedings of the IEEE*, volume 75, December 1987.
15. J. Hadar and W. R. Russell. Rules for ordering uncertain prospects. *The American Economic Review*, 59:25–34, 1969.
16. E. Handschin, F. Neise, H. Neumann, and R. Schultz. Optimal operation of dispersed generation under uncertainty using mathematical programming. *International Journal of Electrical Power and Energy Systems*, 28:618–626, 2006.
17. E. Hauptmeier. *KWK-Erzeugungsanlagen in zukünftigen Verteilungsnetzen – Potenzial und Analysen –*. Dissertation, University of Dortmund, 2007.
18. C. Helmberg and K. C. Kiwiel. A spectral bundle method with bounds. *Mathematical Programming*, 93:173–194, 2002.

19. E. Hennig. Betriebserfahrungen mit dem virtuellen Kraftwerk Unna. *BWK*, 7/8:28–30, 2006.
20. http://ec.europa.eu/research/energy/pdf/efchp_fuelcell16.pdf. Europe's first virtual fuel cell power plant, October 2006.
21. http://www.cogen.org/projects/vfcpp.htm. The virtual fuel cell power plant, October 2006.
22. ILOG. CPLEX Callable Library 9.1.3, 2005.
23. A. Märkert and R. Schultz. On deviation measures in stochastic integer programming. *Operations Research Letters*, 55:441–449, 2005.
24. A. Müller and D. Stoyan. *Comparison Methods for Stochastic Models and Risks*. John Wiley and Sons, Chichester, UK, 2002.
25. H. Neumann. *Zweistufige stochastische Betriebsoptimierung eines Virtuellen Kraftwerks*. Dissertation, University of Dortmund, July 2007.
26. N. Noyan, G. Rudolf, and A. Ruszczyński. Relaxations of linear programming problems with first order stochastic dominance constraints. *Operations Research Letters*, 34:653–659, 2006.
27. A. Ruszczyński and A. Shapiro, editors. *Stochastic Programming*, volume 10 of *Handbooks in Operations Research and Management Science*. Elsevier, Amsterdam, 2003.
28. R. Schultz. Stochastic programming with integer variables. *Mathematical Programming*, 97:285–309, 2003.
29. R. Schultz and S. Tiedemann. Risk aversion via excess probabilities in stochastic programs with mixed-integer recourse. *SIAM Journal on Optimization*, 14(1):115–138, 2003.
30. R. Schultz and S. Tiedemann. Conditional value-at-risk in stochastic programs with mixed-integer recourse. *Mathematical Programming*, 105(2/3):365–386, 2006.
31. F. Uphaus. *Objektorientiertes Betriebsführungssystem zur Koordinierung dezentraler Energieumwandlungsanlagen*. Dissertation, University of Dortmund, May 2006.
32. J. von Neumann and O. Morgenstern. *Theory of Games and Economic Behaviour*. Princeton University Press, Princeton, 1947.
33. Th. Wiesner. Technische Aspekte einer großflächigen Integration dezentraler Energieversorgungsanlagen in elektrische Verteilungsnetze. *Fortschritt Berichte VDI*, Reihe 21, 313, 2001.

Parallel Adaptive Simulation of PEM Fuel Cells

Robert Klöfkorn[1*], Dietmar Kröner[1], Mario Ohlberger[2]

[1] Department for Applied Mathematics, University of Freiburg,
Hermann-Herder-Str. 10, 79104 Freiburg i. Br., Germany
{robertk,dietmar}@mathematik.uni-freiburg.de,
http://www.mathematik.uni-freiburg.de/IAM/
[2] Institute for Numerical and Applied Mathematics, University of Münster,
Einsteinstr. 62, 48149 Münster, Germany
mario.ohlberger@uni-muenster.de,
http://wwwmath1.uni-muenster.de/u/ohlberger/

1 Introduction

Polymer electrolyte membrane (PEM) fuel cells are currently being developed for production of electricity in stationary and portable applications. They benefit from pollution free operation and a potential for high energy conversion efficiency. As PEM fuel cells are currently operated within low temperature and pressure ranges, water management is one of the critical issues in performance optimization.
In this paper we present numerical simulations for liquid water and gas flow in the cathodic gas diffusion layer of a PEM fuel cell. Hereby, we focus on resolved three dimensional simulations of the two phase flow regime using modern numerical techniques like higher order discontinuous Galerkin discretizations, local grid adaptivity, and parallelization with dynamic load-balancing. A detailed model for the simulation of PEM fuel cells, including two water transport modes in the membrane was given in [16]. Here, we restrict to the transport mechanisms within the cathodic gas diffusion layer and extract a suitable sub-model, including two-phase flow and species transport in the gas phase. Details of this simplified model problem are presented in Sect. 2. In Sect. 3 we comment on the discretization schemes that were used for the simulation including remarks on adaptation and parallelization. In Sect. 4 numerical results for an instationary parallel adaptive simulation in three space dimensions are presented.

---

[*] Robert Klöfkorn was supported by the German Ministry of Education and Research under contract 03KRNCFR.

2 The Reduced Model Problem

The three-dimensional, coupled, adaptive, and parallel simulation software is tested on a reduced model problem which describes the fluid flow within the cathodic gas diffusion layer (GDL). The reduced model considers the following physical processes:

- Two-phase flow with phase transition in a porous medium.
- Transport of the gas species O_2 and H_2O with reaction.

2.1 Two-Phase Flow with Phase Transition

A PEM fuel cell is operating at relatively low temperatures (around 60–80 degrees Celsius). Therefore, at the cathode side of the fuel cell liquid water is produced. As a consequence, the flow in the gas diffusion layers (GDL) is modeled via two-phase flow in porous media taking into account the phases, liquid water and gas mixture, consisting of the species oxygen, hydrogen, water vapor, and some rest mostly consisting of nitrogen. In the following the index g denotes the gaseous phase whereas the index w denotes the liquid water phase.

From the balance of the volume saturations s_w, s_g of the two phases we get the two-phase flow in porous media (see [14, 11]):

$$\partial_t(\Phi \rho_i s_i) + \nabla \cdot (\rho_i \mathbf{v}_i) = q_i, \tag{1}$$

$$\mathbf{v}_i = -K \frac{k_{ri}}{\mu_i}(\nabla p_i - \rho_i \mathbf{g}), \quad i = w, g. \tag{2}$$

Here ρ_i denotes the density, s_i the saturation, $\mathbf{v}_i$ the Darcy velocity, p_i the pressure of the phase $i = w, g$ respectively, and $\mathbf{g}$ the gravity vector. Furthermore, K denotes the absolute permeability tensor, μ_i the viscosity of phase i, k_{ri} the relative permeability of phase i, and Φ the porosity of the porous medium. Additionally, q_i denotes the source term modeling the phase transition of the phase i which is defined as follows (see [14, 11]):

$$q_w := r_{\text{phase}}, \qquad q_g := -r_{\text{phase}}, \tag{3}$$

$$r_{\text{phase}} := \begin{cases} k_c \frac{M_g \Phi s_g c^{H_2O}}{RT}(p_{g,H_2O} - p_w^{\text{sat}}), & \text{if } p_{g,H_2O} \geq p_w^{\text{sat}} \\ k_v \Phi s_w \rho_w (p_{g,H_2O} - p_w^{\text{sat}}), & \text{else} \end{cases}. \tag{4}$$

Thereby, p_{g,H_2O} denotes the partial pressure of the water vapor. For the description of the other physical parameters see Table 1.

The following constitutive conditions close the two-phase flow system

$$s_w + s_g = 1, \qquad p_g - p_w = p_c(s_w). \tag{5}$$

Here $p_c(s_w)$ denotes the capillary pressure. Whereas in the liquid phase there exists only the species H_2O, in the gaseous phase we have the species $k =$

O_2, H_2O, and R. Here, R denotes all other existing species (mostly nitrogen). As a consequence, in the gaseous phase ($i = g$) the transport of species has to be taken into account separately. Following [11, 14] we get from the mass balances of the species in the gaseous phase the equations

$$\partial_t \left(\Phi \rho_g s_g c^k\right) + \nabla \cdot \left(\rho_g \mathbf{v}_g c^k\right) - \nabla \cdot \left(\rho_g D_{eff}\left(\Phi, s_g\right) \nabla c^k\right) = q_g^k. \tag{6}$$

Here c^k denotes the mass concentration of the k-th species, D_{eff} the effective diffusion coefficient, and q_g^k the source term of the k-th species in the gaseous phase. As all the species together form the hole gaseous phase, we get the following constitutive condition

$$c^{O_2} + c^{H_2O} + c^R = 1. \tag{7}$$

Therefore, the transport equation for c^R will be dropped as the concentration can be calculated using the constitutive condition from equation (7). The source terms $q_g^{H_2O}$ and $q_g^{O_2}$ are modeled as follows

$$q_g^{H_2O} := q_g, \qquad q_g^{O_2} := -r_{reac}\, c^{H_2O}. \tag{8}$$

The equations (1) to (8) describe the two-phase flow with species transport including phase transition and reactions in the gas diffusion layer (GDL) of a fuel cell. See Sect. 4 for a detailed description of the domain for the PDEs as well as the description of initial and boundary conditions.

2.2 Physical Parameters

The physical parameters in the two-phase flow equations and the transport equations are chosen from Table 1.

2.3 Global Pressure Formulation and Resulting Equations

First, the two-phase flow system in (1) and (2) is reformulated with $s := s_w$ and $p := p_g - \pi_w(s)$ as independent variables. Therefore, we introduce the following notation

global velocity: $\mathbf{u} := \mathbf{v}_w + \mathbf{v}_g$,
phase mobility: $\lambda_i := k_{ri}/\mu_i$,
total mobility: $\lambda := \lambda_w + \lambda_g$,
fractional flow: $f_i := \lambda_i/\lambda$,
phase velocity water: $\mathbf{v}_w := f_w \mathbf{u} + \lambda_g f_w K \nabla p_c$,
phase velocity gas: $\mathbf{v}_g := f_g \mathbf{u} - \lambda_g f_w K \nabla p_c$,
global pressure: $p := p_g - \pi_w(s)$, $\pi_w(s) := \int_0^s f_w(z) p_c'(z)\, dz + p_c(0)$,

Furthermore, the densities ρ_w and ρ_g are assumed to be constant and the influence of the gravity is neglected. The equation for the global pressure is

Table 1. Physical parameters for the model problem

parameter	symbol	value	unit
porosity	Φ	0.7	
abs. permeability	K	$5 \cdot 10^{-11}$	[m^2]
rel. permeability of gas	$k_{rg}(s)$	$\sqrt{1-s}\,(1-(1-(1-s)^{1/m})^m)^2$	
rel. permeability of water	$k_{rw}(s)$	$\sqrt{s}\,(1-(1-s^{1/m})^m)^2$	
Van Genuchten coefficient	m	0.95	
capillary pressure	p_c	$p_c(s) = 5300(1-s)^{\frac{-1}{2.3}}$	[Pa]
viscosity water	μ_w	$1.002\ 10^{-3}$	[Pa s]
viscosity gas	μ_g	$1.720\ 10^{-5}$	[Pa s]
density water	ρ_w	998.2	[kg/m^3]
density air	ρ_g	$\rho_g(p_g, T) = M_g p_g / RT$	[kg/m^3]
temperature	T	343.15	[K]
molar mass gas	M_g	$M_g(\mathbf{c}) = 1/(c^{H_2O}/M_{H_2O} + c^{O_2}/M_{air})$	[kg/mol]
molar mass water	M_{H_2O}	0.018	[kg/mol]
molar mass of dry air	M_{air}	0.02897	[kg/mol]
diffusions coefficient of water vapor in air	$D_g^{H_2O}$	$0.345 \cdot 10^{-4}$	[m^2/s]
effective diffusion coefficient	D_{eff}	$D_{eff}(\Phi, s_g) = (\Phi s_g)^3 D_g^{H_2O}$	[m^2/s]
gas constant	R	8.3144	[J/mol K]
condensation rate	k_c	$1\ 10^6$	[1/s]
vaporization rate	k_v	$1\ 10^{-2}$	[1/Pa s]
saturation vapor pressure	$p_w^{\text{sat}}(T)$	$a\exp(\frac{b}{T} + c - dT + eT^2 + f\,\ln(T))$	[Pa]
coefficient for p_w^{sat}	a	1.00519	
coefficient for p_w^{sat}	b	-6094.4642	
coefficient for p_w^{sat}	c	21.1249952	
coefficient for p_w^{sat}	d	$2.724552\ 10^{-2}$	
coefficient for p_w^{sat}	e	$1.6853396\ 10^{-5}$	
coefficient for p_w^{sat}	f	2.4575506	

obtained by summing up the equation (1) for $i = w, g$, applying the constitutive conditions (5), and inserting the above notations. Finally, we obtain the

pressure equation

$$-\nabla \cdot (K\lambda(s_w)\nabla p) = 0, \qquad (9)$$

and the

velocity equation

$$\mathbf{u} = -K\lambda(s_w)\nabla p. \qquad (10)$$

Assuming that the saturation s_w is given, equation (9) can be used to calculate the global pressure, and finally equation (10) to compute the global velocity.

Assuming the global pressure, the global velocity, and the species concentrations are given, then, since the density is assumed to be constant, from equation (1) including the definition of $\mathbf{v}_w$ we obtain for s_w the

saturation equation

$$\partial_t(\Phi s_w) + \nabla \cdot (f_w(s_w)(\mathbf{u}(s_w) + \lambda_g(s_w)K\nabla p_c(s_w))) = q_w. \quad (11)$$

With the pressure p, the velocity of the gaseous phase $\mathbf{v}_g$, and the saturation of the gaseous phase s_g the transport of species can be described by inserting these three values into equation (6). For $k = O_2, H_2O$ we obtain the

transport equations

$$\partial_t \left(\Phi s_g c^k\right) + \nabla \cdot \left(\mathbf{v}_g c^k\right) - \nabla \cdot \left(D_{eff}\left(\Phi, s_g\right) \nabla c^k\right) = q_g^k. \quad (12)$$

Now the considered model problem consists of the equations (9)–(12). Suitable boundary and initial conditions will be presented in the description of the simulated test problem in Sect. 4.2. Throughout the rest of this paper numerical simulations using this model problem will be presented.

3 Discretization of the Model Problem and Implementation

The discretization of the model problem (9)–(12) uses Discontinuous Galerkin methods. With these methods on one hand, higher order discretizations can be achieved without increasing the stencil of the methods which is an appealing feature when one wants to do parallel computations. Furthermore, there arise no special difficulties when dealing with non-conform grids. Non-conform grids on the other hand have the very nice feature, that the refinement zone stays local unlike for example when conformal bisection refinement is applied. This is again very useful for parallel computations. On the other hand higher order Discontinuous Galerkin methods have a higher number of unknowns than for example continuous Finite Element methods.

The implementation of the discretizations uses the software package DUNE [2, 7, 6]. DUNE has a modular structure and in the following the DUNE modules DUNE-Common, DUNE-Grid, DUNE-Istl, and DUNE-Fem have been used. A detailed description of the modules can be found on the Dune homepage [2]. DUNE-Common provides basic classes. DUNE-Grid defines the abstract grid interface and provides its implementations for several different grids[*]. The following numerical simulations use ALUCubeGrid which is the grid interface implementation of ALUGrid (see [1]) using hexahedral elements. DUNE-Istl (see [5]) is an Iterative Solver Template Library.

---

[*] For a complete list we refer to [2].

It provides classes for matrix – vector handling and solvers. For the solution of the pressure equations the BCRSMatrix and the BiCG-Stab solver from DUNE-Istl have been used. DUNE-Fem provides several implementations of discrete functions spaces such as Lagrange spaces or Discontinuous Galerkin spaces. The Discontinuous Galerkin space, i.e. the base functions and a mapping from local number of degrees of freedom to global number which is needed to store the data in vectors, have been used for the discretization of the model problem. Furthermore, DUNE-Fem provides mechanisms for projections of data and re-arrangement of memory during adaption. For solving the considered time depended problems the ODE Solvers implemented in DUNE-Fem or in the software package ParDG (see [13]) are available.

3.1 Discretization of the Pressure Equation

The equation (9) is discretized by using the Discontinuous Galerkin method. In [4] a variety of Discontinuous Galerkin methods for elliptic problems of the form $-\triangle u = f$ are presented and analyzed. For the following numerical simulation the Oden–Baumann method with polynomial degree 2 has been chosen. This method has been applied to two-phase flow in porous media for example in [8] and led to good results. The resulting linear system is stored using the block wise compressed row storage matrices (BCRSMatrix) from DUNE-Istl [5]. For preconditioning a block diagonal preconditioner is applied and the system is solved using the BiCG-Stab solver, both implemented in DUNE-Istl [5].

3.2 Discretization of the Velocity Equation

The equation (10) is discretized by using the Local Discontinuous Galerkin method [10]. Consider φ_j with $j = 1,...,N$ the vectorial basis functions of the discrete functions space consisting of piecewise polynomial functions of degree $q \in \{1,2,3\}$. Multiplying (10) with φ_j, integrating equation (10) over the domain Ω, integration by parts and taking into account that the integral over Ω can be split into integrals over all grid cells, on a single cell T we get

$$\lambda(s)^{-1} \int_T \mathbf{u} \cdot \varphi_j = \int_T -Kp\nabla \cdot \varphi_j + \int_{\partial T} K\widehat{p}\mathbf{n} \cdot \varphi_j, \quad \forall\, j = 1,...,N. \quad (13)$$

Here we used that the saturation s_w is constant on a single cell and $\widehat{p}$ is a numerical flux as described in [4, Table 3.1]. Here $\mathbf{n}$ denotes the outward normal with respect to ∂T. In our computation we have chosen the LDG flux $\widehat{p} := \{p\} - \beta \cdot [\![p]\!]$, where $\{.\}$ denotes the mean value of p on ∂T and $[\![p]\!]$ the jump in normal direction across ∂T which is a vector parallel to the normal (see [4]). Thereby, β was chosen as $\beta := |\partial T_k|^{10}\mathbf{n}$ where

∂T_k is a face segment of ∂T. In these definitions we follow the notation in [4]. As the $\beta \cdot [\![p]\!]$ in the LDG flux works as a penalty term for discontinuities, the velocity field has almost continuous normal components across element intersections. This is a very appealing feature for higher order discretizations of the saturation and transport equation. In the following numerical examples the polynomial degree for the discretization of the velocity equation was chosen to be 2. The implementation is based on the general framework for discretizing evolution equations presented in [9]. The implementation of this framework is part of the DUNE-Fem module (see [3]).

3.3 Discretization of the Saturation Equation

The saturation equation (11) is also discretized by using the Local Discontinuous Galerkin approach described in [10]. Here, as conservative numerical flux the Engquist-Osher flux is taken. Although in DUNE-Fem higher order LDG discretizations are implemented up to order 3, this equation is discretized by using piecewise constant base functions, i.e. polynomial degree 0. The reason is, as the non-linearity of the flux function is self-compressive, also the first order method produces a satisfactory result. For the time discretization an explicit or implicit Runge-Kutta solver of order $q + 1$, is applied, where q is the polynomial degree of the DG base functions. As for the velocity equation the discretization of the saturation equation is also implemented using the general framework for discretizing evolution equations presented in [9]. The used implicit ODE solvers are part of the DUNE module DUNE-Fem.

3.4 Discretization of the Transport Equation

The transport equation is discretized in the same way as the saturation equation. The only difference is that in this case a simple linear upwind flux can be chosen as the numerical flux. Due to the fact that we have a linear flux function the polynomial degree of the base functions can be chosen larger than 0. Although higher order LDG discretizations for this type of equation are stable, one can get oscillations which are even higher if the discontinuities of the velocity field are strong. To overcome this problem here the factor β is chosen sufficiently large which then works as a penalty term for jumps of the velocity **u** in normal direction across cell boundaries. In the following numerical example the polynomial order of the DG space for the transport equations is also 0. For the time discretization an explicit or implicit Runge-Kutta solver of order $q + 1$ is applied, where q is the polynomial degree of the DG base functions. Again the discretization uses the general framework for discretizing evolution equations presented as in [9]. Also the same implicit ODE solvers which are implemented in DUNE-Fem are used.

3.5 Operator Splitting

Since the equations (9) to (12) are non-linearly coupled, we apply an operator splitting to decouple the equations and solve each equation separately. This is done as follows. Assume that the unknowns s_w and c^k, $k = O_2, H_2O$ are given. Then one time step is solved as follows

1. for given s_w the pressure p can be calculated using equation (9),
2. for given saturation s_w and given pressure p, the velocity $\mathbf{u}$ can be calculated using equation (10),
3. for given saturation s_w, given pressure p, given velocity $\mathbf{u}$, and given concentration $\mathbf{c}$ the new saturation can be calculated using equation (11),
4. for given saturation s_w, given pressure p, given velocity $\mathbf{u}$, and given concentration $\mathbf{c}$ the new concentration of species can be calculated using equation (12).

3.6 Parallelization

The parallelization follows the concept of single program multiple data, meaning that one and the same program is executed on multiple processors. The computational domain is distributed via domain decomposition. Here the graph partitioner METIS has been applied to calculate these distributions. Due to the distribution of the data to multiple processes communication between processes sharing data is necessary during computation of the unknowns. Here the DG methods have the nice property to be local methods, which means that during calculation only neighbor elements have to be available. For parallelization this is a very appealing feature. Furthermore, due to the discontinuity of the methods only element data have to be communicated to neighboring cells located on other processes. All communications during the solution process therefore are interior – ghost communications.

Pressure Equation:
One communication for each iteration step of the BiCG-stab solver is necessary. Furthermore, each evaluation of a scalar product within the solver needs a global sum operation.

Velocity Equation:
One communication after calculation of the velocity is applied.

Saturation and Transport Equation:
One communication during each iteration step of the ODE solver is needed.

3.7 Adaptivity and Load Balancing

During the computation of the unknowns p, $\mathbf{u}$, s_w, and $\mathbf{c}$ an error indicator is evaluated to monitor the local errors introduced by a too coarse grid.

Error Indicators:
For the pressure and velocity together this indicator consists of the jump of the DG velocity in normal direction across cell boundaries in the normal direction. For the saturation equation and the transport equations we use the error indicators described in [15]. In [15] an a-posteriori error estimator is developed for advection-diffusion problems with source terms discretized by an implicit finite volume scheme. Although, the theoretical proof only holds for this type of discretization for weakly coupled systems, the described local error indicators work very well also for similar discretization schemes.

Marking Strategy:
Given a global adaptation tolerance the local cell tolerance is obtained by an equi-distribution strategy. Cells where the local indicator violates the local tolerance are marked for refinement. Cells, where the local indicator is 100 times smaller then the local tolerance are marked for coarsening.

Adaptation and Load Balancing:
After marking of elements is done a grid adaptation is performed. Elements that are marked for refinement are refined. Elements that were marked for coarsening are coarsened if possible. After each adaptation step the load of each processor is checked. For each macro grid cell the load consists mainly of the number of leaf cells that have its source in the macro cell. If the given imbalance tolerance is violated, a re-balancing, i.e. re-construction of a new well balanced distribution of the cells to the processors, is performed. This process is described in detail in [12].

4 Numerical Results

In this section simulation results for a model problem are presented. The model problem consists of water and gas transport with phase transition and reaction within the cathodic gas diffusion layer of a PEM fuel cell.

4.1 Geometry of the Model Problem

As the computational domain we consider $\Omega := \left[0, 2 \ 10^{-4}\right] \times \left[0, 6 \ 10^{-4}\right] \times \left[0, 2 \ 10^{-4}\right]$ m^3 in three space dimensions.

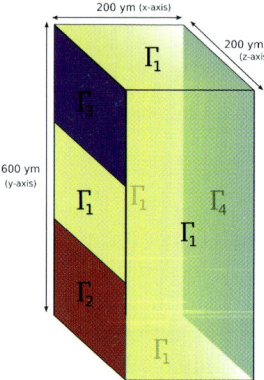

Fig. 1. Sketch of the computational domain

The boundaries are defined as follows

$$\Gamma_1 := \partial\Omega \setminus (\Gamma_2 \cup \Gamma_3 \cup \Gamma_4),$$
$$\Gamma_2 := \{0\} \times [0, 2 \cdot 10^{-4}] \times [0, 2 \cdot 10^{-4}] \text{ m}^2,$$
$$\Gamma_3 := \{0\} \times [4 \cdot 10^{-4}, 6 \cdot 10^{-4}] \times [0, 2 \cdot 10^{-4}] \text{ m}^2,$$
$$\Gamma_4 := \{2 \cdot 10^{-4}\} \times [4 \cdot 10^{-4}, 6 \cdot 10^{-4}] \times [0, 2 \cdot 10^{-4}] \text{ m}^2.$$

The domain Ω with boundaries $\Gamma_1, ..., \Gamma_4$ represents the GDL of a PEM fuel cell. Figure 1 shows a sketch of the computational domain.

4.2 Boundary Conditions and Initial Values

On the boundaries $\Gamma_1, ..., \Gamma_4$ the following boundary conditions were defined

Γ_1: (No-flow boundary – bipolar plate)

    Two-phase flow: no-flux boundary condition.
    Transport equation: no-flow boundary condition.

Γ_2: (Inflow boundary – channel)

    Two-phase flow: $s_w = 0.0$, $p = p_{cha}^1$.
    Transport equation: $c^{H_2O} = 0.2$, $c^{O_2} = 0.8$.

Γ_3: (Outflow boundary – channel)

    Two-phase flow: $s_w = 0.0$, $p = p_{cha}^2$.
    Transport equation: Outflow boundary condition.

Γ_4: (Catalyst layer)

    Two-phase flow: $s_w = 1.0$, no-flow boundary condition for p.
    Transport equation no-flow boundary condition.

The following initial values have been chosen

$$s_w(x,0) = s_0(x) = 0.1 \quad \text{in } \Omega.$$
$$c(x,0) = c_0(x) = (c_0^{H_2O}, c_0^{O_2})^T \quad \text{in } \Omega \quad \text{with } c_0^{H_2O} = 0.2,\ c_0^{O_2} = 0.8.$$

4.3 Technical Data of the Simulation

The following numerical simulations where performed on the $XC4000$ Linux Cluster of the Scientific Supercomputing Center of the University Karlsruhe with 32 processors. Table 2 shows the discretization method and polynomial order for equations (9)–(12). In addition, the resulting number of degrees of freedom (DOF) per element is shown.

To solve the resulting linear system from equation (9), a BiCG-Stab solver and a block diagonal pre-conditioner were applied. To solve the saturation equation (11) and the transport equation (12), an implicit ODE solver of order 1 has been used. The macro grid contains 98.304 hexahedrons, 32 in x and z-direction and 96 in y-direction. The edge length of one macro hexahedron is $6.25\mu m$. The resulting dynamically adapted mesh contains 543.448 hexahedrons. Thus, 43 DOFs per element lead to an overall number of 23.368.364 degrees of freedom per time step. To store the resulting matrix of the linear system from equation (9) a 10×10 block matrix has to be stored per element. This leads to an additional memory requirement for 54.344.800 DOFs.

The simulation took approximately 2 days and 4613 time steps were calculated. To compute one time step approximately 25 seconds were needed. Part of the time is spend in setting up the matrix and in the linear solver which took about 8.5 seconds each time step. The solution of the saturation equation took approximately 4.5 seconds. The transport equations needed 11 seconds and the adaptation step took 0.2 seconds. The missing difference was spend in other parts of the program or in waiting for other processes. The load balancing process costs between 10 and 20 seconds but is applied very rarely.

Table 2. Discretization method, polynomial order and degrees of freedom per element for the model problem.

Equation	Discr. Method	pol. deg. q	DOFs
pressure equation (9)	Oden-Baumann	2	10
velocity equation (10)	LDG	2	30
saturation equation (11)	LDG	0	1
transport equation (12)	LDG	0	2

4.4 Simulation Results

Figures 2, 3 show snapshots of the solution of the model problem taken at time step 4613 which corresponds to simulation time $T = 0.00022$. In addition, Figs. 4 and 5 show the time evolution of the saturation distribution from $T = 0.00011$ to $T = 0.00022$.

In Fig. 2, on the left hand side the pressure distribution is shown. As expected due the choice of the boundary values there is a continuous pressure drop from Γ_2 to Γ_3. The refinement level of the adapted mesh is shown in the

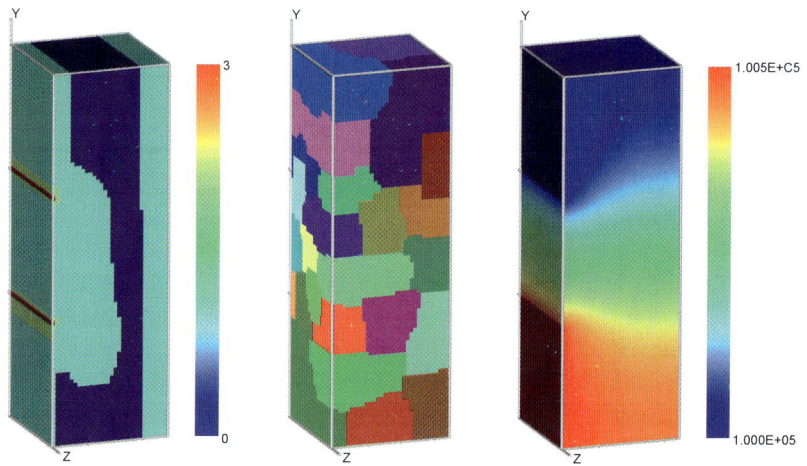

Fig. 2. Level of refinement (*left*), partitioning of the grid (*middle*), and pressure distribution (*right*) at computational time $T = 0.00022$

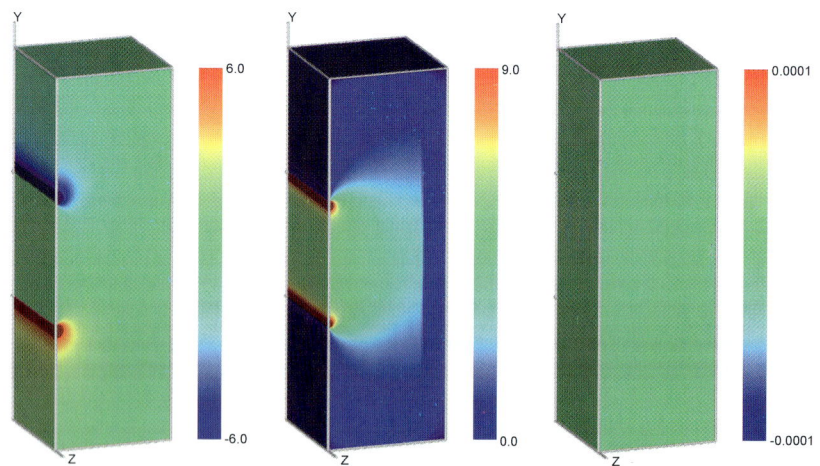

Fig. 3. Global velocity **u**. Components $\mathbf{u}_x$ (*left*), $\mathbf{u}_y$ (*middle*), and $\mathbf{u}_z$ (*right*) at computational time $T = 0.00022$

middle. One can see that the higher levels (red color) are located in the area where Γ_2 and Γ_3 are connected to Γ_1 on the left hand side of the geometry. Finally, on the right hand side of Fig. 2 the partitioning of the grid for the 32 processors is shown.

In Fig. 3 the components of the global velocity **u** are shown. One can see that in the areas where the velocity has strong variations, the grid is refined (see Fig. 2, middle). We state that the error indicator applied to monitor the velocity errors works well. In the middle of Fig. 3 also the influence of the water saturation to the y-component $\mathbf{u}_y$ of the velocity field near the membrane can be seen. The velocity is reduced in this area due to the presence of both phases, liquid water and gas. On the right side of Fig. 3 the z-component of the velocity is shown which is as expected close to zero.

The saturation s_w at time $T = 0.00011$ is illustrated on the left side of Fig. 4. The initial value was 0.1. One can see that phase transition has taken place in the area near the gas channels, as only dry air is entering the cell at Γ_2. Water is entering from the side of the membrane (Γ_4). The water is slowly moving towards the gas channels. Also for the saturation and transport equation the applied error indicator is able to monitor the zones of higher activity (see Fig. 2, left side). On the middle part of Fig. 4 the concentration of water c^{H_2O} in the gaseous phase is shown. In the area where saturation is non-zero, phase transition takes place. The mass concentration of water in the gas phase is increased due to vaporization. This, on the other hand decreases the concentration of oxygen, as both concentrations should sum up to 1. The middle and right picture also demonstrate that the concentrations sum up nicely to 1 as required.

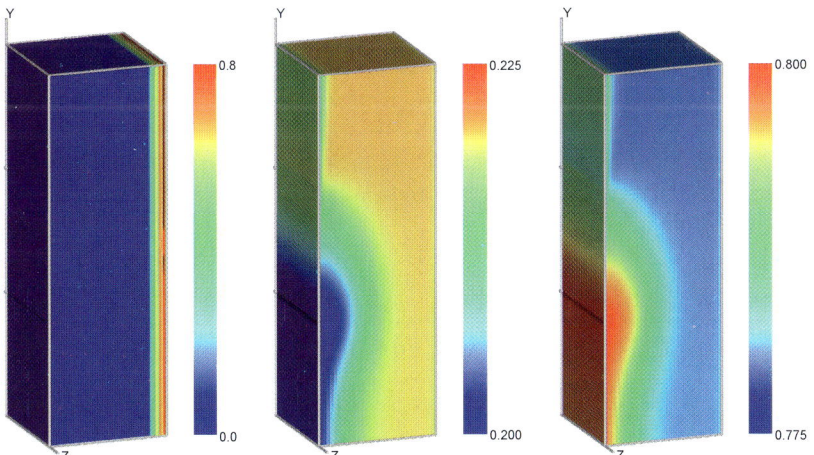

Fig. 4. Saturation s_w (*left*), concentration of water c^{H_2O} (*middle*), and concentration of oxygen c^{O_2} (*right*) at computational time $T = 0.00011$

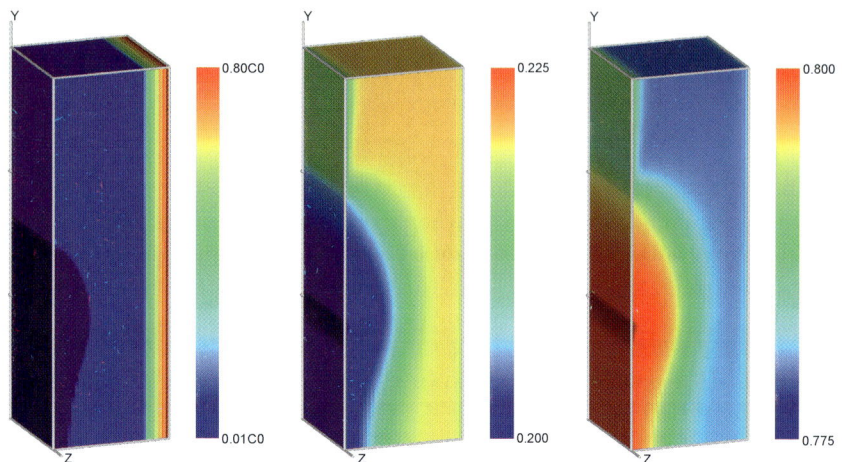

Fig. 5. Saturation s_w (*left*), concentration of water c^{H_2O} (*middle*), and concentration of oxygen c^{O_2} (*right*) at computational time $T = 0.00022$

Figure 5 shows the same variables as Fig. 4 but at a later time, i.e. $T = 0.00022$. One can see that more of the water has been vaporized but also that more and more dry air is covering the left side of the cathodic gas diffusion layer. From Γ_4 more and more water is entering the gas diffusion layer and moving towards the gas channel.

5 Conclusion and Outlook

We could show that our developed software is able to handle complex time-dependent models in three space dimension. Simulations done in 3d usually lead to a large number of unknowns. To cope with such large systems, the simulation tool can be used in parallel. In order to increase the efficiency while keeping the accuracy of the simulation, local grid adaptivity and dynamic load balancing is included.

The implementation of the detailed fuel cell model described in [16] will be subject of future work. Furthermore, the use of higher order methods will be extended to the transport equations.

References

1. *ALUGrid.*
   http://www.mathematik.uni-freiburg.de/IAM/Research/alugrid/.
2. *DUNE.* http://www.dune-project.org.

3. *DUNE-Fem*.
   http://www.mathematik.uni-freiburg.de/IAM/Research/projectskr/dune/feminfo.html.
4. D.N. Arnold, F. Brezzi, B. Cockburn, and L.D. Marini. Unified analysis of discontinuous galerkin methods for elliptic problems. *SIAM J. Num. Anal*, 39:1749–1779, 2002.
5. P. Bastian and M. Blatt. The Iterative Template Solver Library. In *Proc. of the Workshop on State-of-the-Art in Scientific and Parallel Computing, PARA '06*. Springer, 2006.
6. P. Bastian, M. Blatt, A. Dedner, C. Engwer, R. Klöfkorn, R. Kornhuber, M. Ohlberger, and O. Sander. A Generic Grid Interface for Parallel and Adaptive Scientific Computing. Part II: Implementation and Tests in DUNE. Preprint 404, DFG Research Center MATHEON, 2007.
7. P. Bastian, M. Blatt, A. Dedner, C. Engwer, R. Klöfkorn, M. Ohlberger, and O. Sander. A Generic Grid Interface for Parallel and Adaptive Scientific Computing. Part I: Abstract Framework. Preprint 403, DFG Research Center MATHEON, 2007.
8. P. Bastian and B. Riviere. Discontinuous galerkin methods for two-phase flow in porous media. Technical Report 28, IWR (SFB 359), Universität Heidelberg, 2004.
9. A. Burri, A. Dedner, D. Diehl, R. Klöfkorn, and M. Ohlberger. A general object oriented framework for discretizing nonlinear evolution equations. In *Proc. of The 1st Kazakh-German Advanced Research Workshop on Computational Science and High Performance Computing*, 2005.
10. B. Cockburn and C.-W. Shu. The Local Discontinuous Galerkin Method for Time-Dependent Convection-Diffusion Systems. *SIAM J. Numer. Anal.*, 35(6):2440–2463, 1998.
11. M.Y. Corapcioglu and A. Baehr. A compositional multiphase model for groundwater contamination by petroleum products: 1. theoretical considerations. *Water Resource Research*, 1987.
12. A. Dedner, C. Rohde, B. Schupp, and M. Wesenberg. A parallel, load balanced mhd code on locally adapted, unstructured grids in 3d. *Computing and Visualization in Science*, 7:79–96, 2004.
13. D. Diehl. Higher Order Schemes for Simulation of Compressible Liquid – Vapor Flows with Phase Change. Dissertation, Institute of Mathematics, University Freiburg, 2007.
14. R. Helmig. *Multiphase Flow and Transport Processes in the Subsurface: A contribution to the modeling of hydrosystems*. Springer, Berlin, Heidelberg, 1997.
15. M. Ohlberger and C. Rohde. Adaptive finite volume approximations for weakly coupled convection dominated parabolic systems. *IMA J. Numer. Anal.*, 22(2):253–280, 2002.
16. K. Steinkamp, J. Schumacher, F. Goldsmith, M. Ohlberger, and C. Ziegler. A non-isothermal pem fuel cell model including two water transport mechanisms in the membrane. Preprint 4, Mathematisches Institut, Universität Freiburg, 2007.

Part VI

Risk Management in Finance and Insurance

Advanced Credit Portfolio Modeling and CDO Pricing

Ernst Eberlein[1], Rüdiger Frey[2], and Ernst August von Hammerstein[1]

[1] Department of Mathematical Stochastics, University of Freiburg, Eckerstraße 1, 79104 Freiburg, Germany
{eberlein,hammerstein}@stochastik.uni-freiburg.de
[2] Department of Mathematics, University of Leipzig, 04081 Leipzig, Germany
ruediger.frey@math.uni-leipzig.de

1 Introduction

Credit risk represents by far the biggest risk in the activities of a traditional bank. In particular, during recession periods financial institutions loose enormous amounts as a consequence of bad loans and default events. Traditionally the risk arising from a loan contract could not be transferred and remained in the books of the lending institution until maturity. This has changed completely since the introduction of credit derivatives such as credit default swaps (CDSs) and collaterized debt obligations (CDOs) roughly fifteen years ago. The volume in trading these products at the exchanges and directly between individual parties (OTC) has increased enormously. This success is due to the fact that credit derivatives allow the transfer of credit risk to a larger community of investors. The risk profile of a bank can now be shaped according to specified limits, and concentrations of risk caused by geographic and industry sector factors can be reduced.

However, credit derivatives are complex products, and a sound risk-management methodology based on appropriate quantitative models is needed to judge and control the risks involved in a portfolio of such instruments. Quantitative approaches are particularly important in order to understand the risks involved in portfolio products such as CDOs. Here we need mathematical models which allow to derive the statistical distribution of portfolio losses. This distribution is influenced by the default probabilities of the individual instruments in the portfolio, and, more importantly, by the *joint behaviour* of the components of the portfolio. Therefore the probabilistic dependence structure of default events has to be modeled appropriately.

In this paper we use two different approaches for modeling dependence. To begin with, we extend the factor model approach of Vasiček [32, 33] by using more sophisticated distributions for the factors. Due to their greater flexibility these distributions have been successfully used in several areas of finance (see e.g. [9, 10, 11]). As shown in the present paper, this approach leads to a substantial improvement of performance in the pricing of synthetic CDO tranches. Moreover, in the last section we introduce a dynamic Markov chain model for the default state of a credit portfolio and discuss the pricing of CDO tranches for this model.

2 CDOs: Basic Concepts and Modeling Approaches

A *collateralized debt obligation* (CDO) is a structured product based on an underlying portfolio of reference entities subject to credit risk, such as corporate bonds, mortgages, loans or credit derivatives. Although several types of CDOs are traded in the market which mainly differ in the content of the portfolio and the cash flows between counterparties, the basic structure is the same. The originator (usually a bank) sells the assets of the portfolio to a so-called *special purpose vehicle* (SPV), a company which is set up only for the purpose of carrying out the securitization and the necessary transactions. The SPV does not need capital itself, instead it issues notes to finance the acquisition of the assets. Each note belongs to a certain loss piece or *tranche* after the portfolio has been divided into a number of them. Consequently the portfolio is no longer regarded as an asset pool but as a collateral pool. The tranches have different seniorities; the first loss piece or *equity tranche* has the lowest, followed by *junior mezzanine*, *mezzanine*, *senior* and finally *super-senior* tranches. The interest payments the SPV has to make to the buyer of a CDO tranche are financed from the cash flow generated by the collateral pool. Therefore the performance or the default risk of the portfolio is taken over by the investors. Since all liabilities of the SPV as a tranche seller are funded by proceeds from the portfolio, CDOs can be regarded as a subclass of so-called asset-backed securities. If the assets consist mainly of bonds resp. loans, the CDO is also called collateralized bond obligation (CBO) resp. collateralized loan obligation (CLO). For a *synthetic CDO* which we shall discuss in more detail below, the portfolio contains only credit default swaps. The motivation to build a CDO is given by economic reasons:

- By selling the assets to the SPV, the originator removes them from his balance sheet and therefore he is able to reduce his regulatory capital. The capital which is set free can then be used for new business opportunities.
- The proceeds from the sale of the CDO tranches are typically higher than the initial value of the asset portfolio because the risk-return profile of the tranches is more attractive for investors. This is both the result from and the reason for slicing the portfolio into tranches and the implicit collation and rebalancing hereby. Arbitrage CDOs are mainly set up to exploit this difference.

In general, CDO contracts can be quite sophisticated because there are no regulations for the compilation of the reference portfolio and its tranching or the payments to be made between the parties. The originator and the SPV can design the contract in a taylormade way, depending on the purposes they want to achieve. To avoid unnecessary complications, we concentrate in the following on synthetic CDOs which are based on a portfolio of credit default swaps.

2.1 Structure and Payoffs of CDSs and Synthetic CDOs

As mentioned before, the reference portfolio of a synthetic CDO consists entirely of *credit default swaps* (CDSs). These are insurance contracts protecting from losses caused by default of defaultable assets. The protection buyer A periodically pays a fixed premium to the protection seller B until a prespecified credit event occurs or the contract terminates. In turn, B makes a payment to A that covers his losses if the credit event has happened during the lifetime of the contract. Since there are many possibilities to specify the default event as well as the default payment, different types of CDSs are traded in the market, depending on the terms the counterparties have agreed on. The basic structure is shown in Fig. 1.

Throughout this article we will make the following assumptions: The reference entity of the CDS is a defaultable bond with nominal value L, and the credit event is the default of the bond issuer. If default has happened, B pays $(1-R)L$ to A where R denotes the recovery rate. On the other side A pays quarterly a fixed premium of $0.25 r_{CDS} L$ where r_{CDS} is the annualized fair CDS rate. To determine this rate explicitly, we fix some notation:

- r is the riskless interest rate, assumed to be constant over the lifetime $[0, T]$ of the CDS,
- $u(t)$ is the discounted value of all premiums paid up to time t when the annualized premium is standardized to 1,
- $G_1(t)$ is the distribution function of the default time T_1 with corresponding density $g_1(t)$ (its existence will be justified by the assumptions in subsequent sections).

The expected value of the discounted premiums (*premium leg*) can then be written as

$$PL(r_{CDS}) = r_{CDS} L \int_0^T u(t) g_1(t) \, dt + r_{CDS} L u(T)(1 - G_1(T)).$$

The expected discounted default payment (*default leg*) is given by

$$D = (1-R)L \int_0^T g_1(t) e^{-rt} \, dt.$$

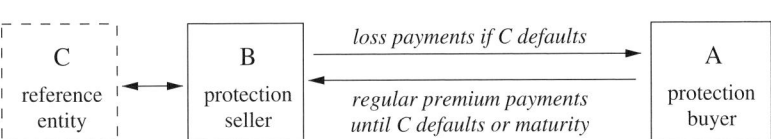

Fig. 1. Basic structure of a CDS

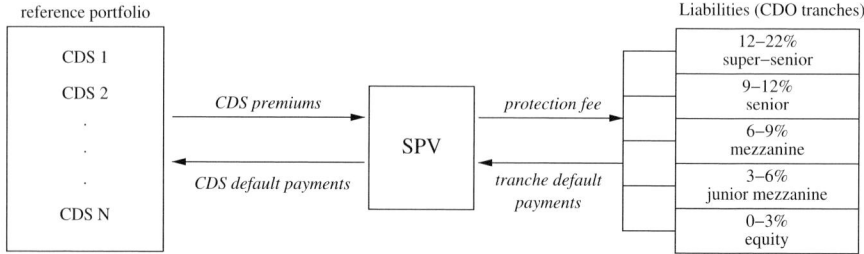

Fig. 2. Schematic representation of the payments in a synthetic CDO. The choice of the attachment points corresponds to DJ iTraxx Europe standard tranches

The no-arbitrage condition $PL(r_{CDS}) = D$ then implies

$$r_{CDS} = \frac{(1-R)\int_0^T g_1(t)e^{-rt}\,dt}{\int_0^T u(t)g_1(t)\,dt + u(T)(1-G_1(T))} = \frac{D}{PL(1)}. \qquad (1)$$

To explain the structure and the cash flows of a synthetic CDO assume that its reference portfolio consists of N different CDSs with the same notional value L. We divide this portfolio in subsequent tranches. Each tranche covers a certain range of percentage losses of the total portfolio value NL defined by lower and upper *attachment points* $K_l, K_u \leq 1$. The buyer of a tranche compensates as protection seller for all losses that exceed the amount of $K_l NL$ up to a maximum of $K_u NL$. On the other hand the SPV as protection buyer has to make quarterly payments of $0.25 r_c V_t$, where V_t is the notional value of the tranche at payment date t. Note that V_t starts with $NL(K_u - K_l)$ and is reduced by every default that hits the tranche. r_c is the fair tranche rate. See also Fig. 2.

In recent years a new and simplified way of buying and selling CDO tranches has become very popular, the trading of *single index tranches*. For this purpose standardized portfolios and tranches are defined. Two counterparties can agree to buy and sell protection on an individual tranche and exchange the cash flows shown in the right half of Fig. 2. The underlying CDS portfolio however is never physically created, it is merely a reference portfolio from which the cash flows are derived. So the left hand side of Fig. 2 vanishes in this case, and the SPV is replaced by the protection buyer. The portfolios for the two most traded indices, the Dow Jones CDX NA IG and the Dow Jones iTraxx Europe, are composed of 125 investment grade US and European firms respectively. The index itself is nothing but the weighted credit default swap spread of the reference portfolio. In Sects. 2.2 and 3.1 we shall derive the corresponding default probabilities. We will use market quotes for different iTraxx tranches and maturities to calibrate our models later in Sects. 3.2 and 4.2.

In the following we denote the attachment points by $0 = K_0 < K_1 < \cdots < K_m \leq 1$ such that the lower and upper attachment points of tranche i are

K_{i-1} and K_i respectively. Suppose for example that $(1-R)j = K_{i-1}N$ and $(1-R)k = K_i N$ for some $j < k$, $j,k \in \mathbb{N}$. Then the protection seller B of tranche i pays $(1-R)L$ if the $(j+1)^{st}$ reference entity in the portfolio defaults. For each of the following possible $k-j-1$ defaults the protection buyer receives the same amount from B. After the kth default occurred the outstanding notional of the tranche is zero and the contract terminates. However, the losses will usually not match the attachment points. In general, some of them are divided up between subsequent tranches: If $\frac{(j-1)(1-R)}{N} < K_i < \frac{j(1-R)}{N}$ for some $j \in \mathbb{N}$, then tranche i bears a loss of $NL\left(K_i - \frac{(j-1)(1-R)}{N}\right)$ (and is exhausted thereafter) if the jth default occurs. The overshoot is absorbed by the following tranche whose outstanding notional is reduced by $NL\left(\frac{j(1-R)}{N} - K_i\right)$. We use the following notation:

K_{i-1}, K_i are the lower/upper attachment points of tranche i,
  Z_t is the relative amount of CDSs which have defaulted up to time t, expressed as a fraction of the total number N,
  $L_t^i = \min[(1-R)Z_t, K_i] - \min[(1-R)Z_t, K_{i-1}]$ is the loss of tranche i up to time t, expressed as a fraction of the total notional value NL,
  r_i is the fair spread rate of tranche i,
$0 = t_0 < \cdots < t_n$ are the payment dates of protection buyer and seller,
$\beta(t_0, t_k)$ is the discount factor for time t_k.

Remark 1. Under the assumption of a constant riskless interest rate r we would have $\beta(t_0, t_k) = e^{-rt_k}$. Since this assumption is too restrictive one uses zero coupon bond prices for discounting instead. Therefore $\beta(t_0, t_k)$ will denote the price at time t_0 of a zero coupon bond with maturity t_k.

The assumption that all CDSs have the same notional value may seem somewhat artificial, but it is fulfilled for CDOs on standardized portfolios like the Dow Jones CDX or the iTraxx Europe.

With this notation the premium as well as the default leg of tranche i can be expressed as

$$PL_i(r_i) = \sum_{k=1}^{n}(t_k - t_{k-1})\beta(t_0, t_k)\, r_i\, \mathrm{E}\big[(K_i - K_{i-1} - L_{t_k}^i)NL\big], \qquad (2)$$

$$D_i = \sum_{k=1}^{n}\beta(t_0, t_k)\mathrm{E}\big[(L_{t_k}^i - L_{t_{k-1}}^i)NL\big],$$

where $\mathrm{E}[\cdot]$ denotes expectation. For the fair spread rate one obtains

$$r_i = \frac{\sum_{k=1}^{n}\beta(t_0, t_k)\big(\mathrm{E}[L_{t_k}^i] - \mathrm{E}[L_{t_{k-1}}^i]\big)}{\sum_{k=1}^{n}(t_k - t_{k-1})\beta(t_0, t_k)\big(K_i - K_{i-1} - \mathrm{E}[L_{t_k}^i]\big)}. \qquad (3)$$

Remark 2. To get arbitrage-free prices, all expectations above have to be taken under a risk neutral probability measure, which is assumed implicitly. One should be aware that risk neutral probabilities cannot be estimated from historical default data.

Since payment dates and attachment points are specified in the CDO contract and discount factors can be obtained from the market, the remaining task is to develop a realistic portfolio model from which the risk neutral distribution of Z_t can be derived, i.e. we need to model the joint distribution of the default times $T_1, \ldots, T_N$ of the reference entities.

2.2 Factor Models with Normal Distibutions

To construct this joint distribution, the first step is to define the marginal distributions $Q_i(t) = P(T_i \leq t)$. The standard approach, which was proposed in [21], is to assume that the default times T_i are exponentially distributed, that is, $Q_i(t) = 1 - e^{-\lambda_i t}$. The *default intensities* λ_i can be estimated from the clean spreads $r^i_{CDS}/(1-R)$ where r^i_{CDS} is the fair CDS spread of firm i which can be derived using formula (1). In fact, the relationship $\lambda_i \approx r^i_{CDS}/(1-R)$ is obtained directly from (1) by inserting the default density $g_1(t) = \lambda_i e^{-\lambda_i t}$ (see [22, section 9.3.3]).

As mentioned before, the CDX and iTraxx indices quote an average CDS spread for the whole portfolio in basis points (100bp = 1%), therefore the market convention is to set

$$\lambda_i \equiv \lambda_a = \frac{s_a}{(1-R)10000} \qquad (4)$$

where s_a is the average CDX or iTraxx spread in basis points. This implies that all firms in the portfolio have the same default probability. One can criticize this assumption from a theoretical point of view, but it simplifies and fastens the calculation of the loss distribution considerably as we will see below. Since λ_a is obtained from data of derivative markets, it can be considered as a risk neutral parameter and therefore the $Q_i(t)$ can be considered as risk neutral probability distributions.

The second step to obtain the joint distribution of the default times is to impose a suitable coupling between the marginals. Since all firms are subject to the same economic environment and many of them are linked by direct business relations, the assumption of independence of defaults between different firms obviously is not realistic. The empirically observed occurrence of disproportionally many defaults in certain time periods also contradicts the independence assumption. Therefore the main task in credit portfolio modeling is to implement a realistic dependence structure which generates loss distributions that are consistent with market observations. The following approach goes back to [32] and was motivated by the Merton model [25].

For each CDS in the CDO portfolio we define a random variable X_i as follows:

$$X_i := \sqrt{\rho}\, M + \sqrt{1-\rho}\, Z_i, \qquad 0 \leq \rho < 1, \quad i = 1, \ldots, N, \tag{5}$$

where $M, Z_1, \ldots, Z_N$ are independent and standard normally distributed. Obviously $X_i \sim N(0,1)$ and $\mathrm{Corr}(X_i, X_j) = \rho$, $i \neq j$. X_i can be interpreted as state variable for the firm that issued the bond which CDS i secures. The state is driven by two factors: the systematic factor M represents the macroeconomic environment to which all firms are exposed, whereas the idiosyncratic factor Z_i incorporates firm specific strengths or weaknesses.

To model the individual defaults, we define time-dependent thresholds by

$$d_i(t) := \Phi^{-1}(Q_i(t))$$

where $\Phi^{-1}(x)$ denotes the inverse of the standard normal distribution function or quantile function of $N(0,1)$. Observe that the $d_i(t)$ are increasing because so are Φ^{-1} and Q_i. Therefore we can define each default time T_i as the first time point at which the corresponding variable X_i is smaller than the threshold $d_i(t)$, that is

$$T_i := \inf\{t \geq 0 \,|\, X_i \leq d_i(t)\}, \qquad i = 1, \ldots, N. \tag{6}$$

This also ensures that the T_i have the desired distribution, because

$$P(T_i \leq t) = P\big(X_i \leq \Phi^{-1}(Q_i(t))\big) = P\big(\Phi(X_i) \leq Q_i(t)\big) = Q_i(t),$$

where the last equation follows from the fact that the random variable $\Phi(X_i)$ is uniformly distributed on the interval $[0,1]$. Moreover, the leftmost equation shows that $T_i \stackrel{d}{=} Q_i^{-1}(\Phi(X_i))$, so the default times inherit the dependence structure of the X_i. Since the latter are not observable, but serve only as auxiliary variables to construct dependence, such models are termed 'latent variable' models. Note that by (4) we have $Q_i(t) \equiv Q(t)$ and thus $d_i(t) \equiv d(t)$, therefore we omit the index i in the following.

Remark 3. Instead of inducing dependence by latent variables that are linked by the factor equation (5), one can also define the dependence structure of the default times more directly by inserting the marginal distribution functions into an appropriately chosen *copula*. We do not discuss this approach here further, but give some references at the end of Sect. 2.3.

To derive the loss distribution let A_k^t be the event that exactly k defaults have happened up to time t. From (6) and (5) we get

$$P(T_i < t \,|\, M) = P(X_i < d(t) \,|\, M) = \Phi\left(\frac{d(t) - \sqrt{\rho}\, M}{\sqrt{1-\rho}}\right).$$

Since the X_i are independent conditional on M, the conditional probability $P(A_k^t|M)$ equals the probability of a binomial distribution with parameters N and $p = P(T_i < t\,|M)$:

$$P(A_k^t|M) = \binom{N}{k} \Phi\left(\frac{d(t) - \sqrt{\rho}M}{\sqrt{1-\rho}}\right)^k \left(1 - \Phi\left(\frac{d(t) - \sqrt{\rho}M}{\sqrt{1-\rho}}\right)\right)^{N-k}.$$

The probability that at time t the relative number of defaults Z_t does not exceed q is

$$F_{Z_t}(q) = \sum_{k=0}^{[Nq]} P(A_k^t)$$

$$= \int_{-\infty}^{\infty} \sum_{k=0}^{[Nq]} \binom{N}{k} \Phi\left(\frac{d(t) - \sqrt{\rho}u}{\sqrt{1-\rho}}\right)^k \left(1 - \Phi\left(\frac{d(t) - \sqrt{\rho}u}{\sqrt{1-\rho}}\right)\right)^{N-k} dP_M(u).$$

If the portfolio is very large, one can simplify F_{Z_t} further using the following approximation which was introduced in [33] and which is known as large homogeneous portfolio (LHP) approximation. Let $p_t(M) := \Phi\left(\frac{d(t) - \sqrt{\rho}M}{\sqrt{1-\rho}}\right)$ and G_{p_t} be the corresponding distribution function, then we can rewrite F_{Z_t} in the following way:

$$F_{Z_t}(q) = \int_0^1 \sum_{k=0}^{[Nq]} \binom{N}{k} s^k (1-s)^{N-k} \, dG_{p_t}(s). \tag{7}$$

Applying the LHP approximation means that we have to determine the behaviour of the integrand for $N \to \infty$. For this purpose suppose that Y_i are independent and identically distributed (iid) Bernoulli variables with $P(Y_i = 1) = s = 1 - P(Y_i = 0)$. Then the strong law of large numbers states that $\bar{Y}_N = \frac{1}{N}\sum_{i=1}^N Y_i \to s$ almost surely which implies convergence of the distribution functions $F_{\bar{Y}_N}(x) \to 1\!\!1_{[0,x]}(s)$ pointwise on $\mathbb{R} \setminus \{s\}$. For all $q \neq s$ we thus have

$$\sum_{k=0}^{[Nq]} \binom{N}{k} s^k(1-s)^{N-k} = P\left(\sum_{i=1}^N Y_i \leq Nq\right) = P(\bar{Y}_N \leq q) \xrightarrow[N\to\infty]{} 1\!\!1_{[0,q]}(s).$$

Since the sum on the left hand side is bounded by 1, by Lebesgue's theorem we get from (7)

$$F_{Z_t}(q) \approx \int_0^1 1\!\!1_{[0,q]}(s) \, dG_{p_t}(s) = G_{p_t}(q) = P\left(-\frac{\sqrt{1-\rho}\,\Phi^{-1}(q) - d(t)}{\sqrt{\rho}} \leq M\right)$$

$$= \Phi\left(\frac{\sqrt{1-\rho}\,\Phi^{-1}(q) - d(t)}{\sqrt{\rho}}\right) \tag{8}$$

where in the last equation the symmetry relation $1 - \Phi(x) = \Phi(-x)$ has been used. This distribution is, together with the above assumptions, the current market standard for the calculation of CDO spreads according to equation (3). Since the relative portfolio loss up to time t is given by $(1-R)Z_t$, the expectations $\mathrm{E}[L_{t_k}^i]$ contained in (3) can be written as follows:

$$\mathrm{E}[L_{t_k}^i] = \int_{\frac{K_{i-1}}{1-R} \wedge 1}^{\frac{K_i}{1-R} \wedge 1} (1-R)\left(q - \tfrac{K_{i-1}}{1-R}\right) \mathrm{d}F_{Z_{t_k}}(q) + (K_i - K_{i-1})\left[1 - F_{Z_{t_k}}\left(\tfrac{K_i}{1-R} \wedge 1\right)\right]. \tag{9}$$

2.3 Deficiencies and Extensions

The pricing formula obtained from (3), (8) and (9) contains one unknown quantity: the correlation parameter ρ. This parameter has to be estimated before one can derive the fair rate of a CDO tranche. A priori it is not clear which data and which estimation procedure one could use to get ρ. In the Merton approach, defaults are driven by the evolution of the asset value of a firm. Consequently the dependence between defaults is derived from the dependence between asset values. The latter cannot be observed directly, therefore some practitioners have used equity correlations, which can be estimated from stock price data. A more direct and plausible alternative would be to infer correlations from historical default data, but since default is a rare event, this would require data sets over very long time periods which are usually not available.

With the development of a liquid market for single index tranches in the last years, a new source of correlation information has arisen: the *implied correlations* from index tranche prices. Similar to the determination of implied volatilities from option prices by inverting the Black–Scholes formula, one can invert the above pricing formula and solve numerically for the correlation parameter ρ which reproduces the quoted market price. This provides also a method to examine if the model and its assumptions are appropriate. If this is the case, the correlations derived from market prices of different tranches of the same index should coincide. However, in reality one observes a so-called *correlation smile*: the implied correlations of the equity and (super-)senior tranches are typically much higher than those of the mezzanine tranches. See Fig. 3 for an example. The smile indicates that the classical model is not flexible enough to generate realistic dependence structures. This is only partly due to the simplifications made by using the LHP approach. The deeper reason for this phenomenon lies in the fact that the model with normal factors strongly underestimates the probabilities of joint defaults. This has led to severe mispricings and inadequate risk forecasts in the past. The problem became evident in the so-called *correlation crisis* in May 2005: the factor model based on normal distributions was unable to follow the movement of market quotes occuring in reaction to the downgrading of Ford and General Motors to non-investment grade.

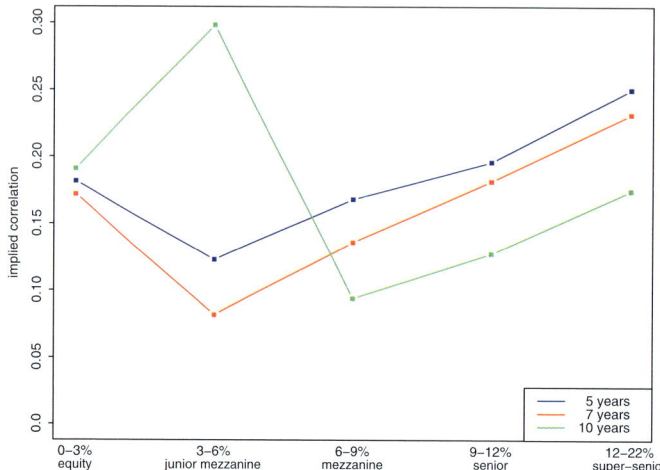

Fig. 3. Implied correlations calculated from the prices of DJ iTraxx Europe standard tranches at November 13, 2006, for different maturities T

A number of different approaches for dealing with this problem have been investigated. A rather intuitive extension to remedy the deficiencies of the normal factor model which we shall exploit in Sect. 3, is to allow for factor distributions which are much more flexible than the standard normal ones. Different factor distributions do not only change the shape of F_{Z_t}, but also have a great influence on the so-called *factor copula* implicitly contained in the joint distribution of the latent variables. In fact, the replacement of the normal distribution leads to a fundamental modification of the dependence structure which becomes much more complex and can even exhibit *tail-dependence*. A necessary condition for the latter to hold is that the distribution of the systematic factor M is heavy tailed. This fact was proven in [24]. The first paper in which alternative factor distributions are used is [17] where both factors are assumed to follow a Student t-distribution with 5 degrees of freedom. In [19], Normal Inverse Gaussian distributions are applied for pricing synthetic CDOs, and in [1] several models based on Gamma, Inverse Gaussian, Variance Gamma, Normal Inverse Gaussian and Meixner distributions are presented. In the last paper the systematic and idiosyncratic factors are represented by the values of a suitably scaled and shifted Lévy process at times ρ and $1 - \rho$.

Another way to extend the classical model is to implement *stochastic correlations* and *random factor loadings*. In the first approach which was developed in [15], the constant correlation parameter ρ in (5) is replaced by a random variable taking values in $[0, 1]$. The cumulative default distribution can then be derived similarly as before, but one has to condition on both, the systematic factor and the correlation variable. The concept of random factor loadings was first published in [2]. There the X_i are defined by

$X_i := m_i(M) + \sigma_i(M)Z_i$ with some deterministic functions m_i and σ_i. In the simplest case $X_i = m + (l 1\!\!\!1_{\{M<e\}} + h 1\!\!\!1_{\{M\geq e\}})M + \nu Z_i$ where $l, h, e \in \mathbb{R}$ are additional parameters and m, ν are constants chosen such that $\mathrm{E}[X_i] = 0$ and $\mathrm{Var}[X_i] = 1$. Further information and numerical details for the calibration of such models to market data can be found in [8].

As already mentioned in Remark 3, other approaches use copula models to define the dependence between the default times T_i. The concept of copulas was introduced in probability theory by Sklar [31]. A very useful and illustrative introduction to copulas and their application in risk management can be found in [22, chapter 5], for a thorough theoretical treatment, we refer to [26]. The first papers where copulas were used in credit risk models are [21] and [30]. A recent approach based on Archimedean copulas can be found in [5]. The pricing performance of models with Clayton and Marshall–Olkin copulas was investigated and compared with some other popular approaches in [7]. There the prices calculated from the Clayton copula model showed a slightly better fit to the market quotes, but they were still relatively close to those generated by the Gaussian model. The Marshall–Olkin copulas performed worse, since the deviations from market prices were greater than those of other models considered.

Alternatively, it is possible to come up with stochastic models for the dynamic evolution of the default state of the portfolio (instead of modeling just the distribution of the default times as seen from a given point in time t) and to look for dynamic models that can generate correlation skews. An example of this line of research is discussed in Sect. 4.

3 Calibration with Advanced Distributions

The factor distributions we implement to overcome the deficiencies mentioned above belong to the class of *generalized hyperbolic distributions* (GH) which was introduced in [4]. In the general case, their densities are given by

$$d_{GH(\lambda,\alpha,\beta,\delta,\mu)}(x) = a(\lambda,\alpha,\beta,\delta,\mu)\big(\delta^2 + (x-\mu)^2\big)^{(\lambda-\frac{1}{2})/2} e^{\beta(x-\mu)} \\ \times K_{\lambda-\frac{1}{2}}\big(\alpha\sqrt{\delta^2 + (x-\mu)^2}\big) \qquad (10)$$

with the norming constant

$$a(\lambda,\alpha,\beta,\delta,\mu) = \frac{(\alpha^2 - \beta^2)^{\frac{\lambda}{2}}}{\sqrt{2\pi}\,\alpha^{\lambda-\frac{1}{2}}\delta^\lambda K_\lambda(\delta\sqrt{\alpha^2 - \beta^2})}.$$

K_ν denotes the modified Bessel function of the third kind with index ν and $GH(\lambda,\alpha,\beta,\delta,\mu)$ the corresponding probability distribution. The influence of the parameters is as follows: $\alpha > 0$ determines the shape, $0 \leq |\beta| < \alpha$ the skewness, $\mu \in \mathbb{R}$ is a location parameter and $\delta > 0$ serves for scaling. $\lambda \in \mathbb{R}$ characterizes certain subclasses and has considerable influence on the size of

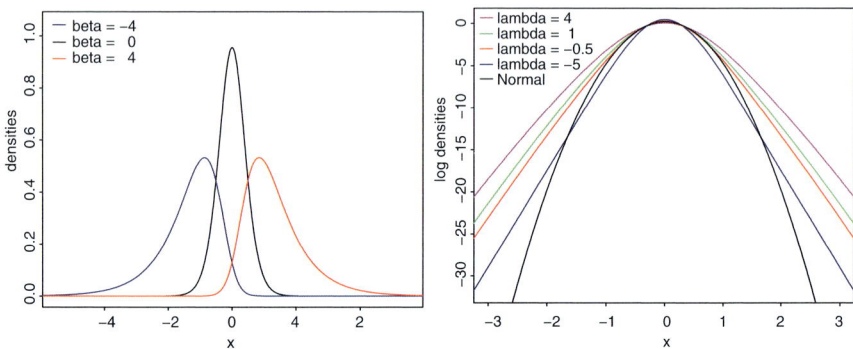

Fig. 4. Influence of the GH parameters β (*left*) and λ (*right*), where on the right hand side log densities are plotted

mass contained in the tails which can be seen from the asymptotic behaviour of the densities: $d_{GH(\lambda,\alpha,\beta,\delta,\mu)}(x) \sim |x|^{\lambda-1}e^{-\alpha|x|+\beta x}$ for $|x| \to \infty$. See also Fig. 4. Generalized hyperbolic distributions have already been shown to be a very useful tool in various fields of mathematical finance. An overview over different applications can be found in [9]. Let us mention some special subclasses and limiting cases which are of particular interest and which we will use later to calibrate the iTraxx data:

For $\lambda = -0.5$ one obtains the subclass of *Normal Inverse Gaussian distributions* (NIG) with densities

$$d_{NIG(\alpha,\beta,\delta,\mu)}(x) = \frac{\alpha\delta}{\pi} \frac{K_1\left(\alpha\sqrt{\delta^2+(x-\mu)^2}\right)}{\sqrt{\delta^2+(x-\mu)^2}} e^{\delta\sqrt{\alpha^2-\beta^2}+\beta(x-\mu)},$$

whereas $\lambda = 1$ characterizes the subclass of *hyperbolic distributions* (HYP) which was the first to be applied in finance in [10] and

$$d_{HYP(\alpha,\beta,\delta,\mu)}(x) = \frac{\sqrt{\alpha^2-\beta^2}}{2\alpha\delta K_1(\delta\sqrt{\alpha^2-\beta^2})} e^{-\alpha\sqrt{\delta^2+(x-\mu)^2}+\beta(x-\mu)}.$$

For positive λ, *Variance Gamma distributions* (VG), which were introduced in full generality in [23], can be obtained as weak limits of GH distributions. If $\lambda > 0$ and $\delta \to 0$, then the density (10) converges pointwise to

$$d_{VG(\lambda,\alpha,\beta,\mu)}(x) = \frac{(\alpha^2-\beta^2)^\lambda |x-\mu|^{\lambda-\frac{1}{2}}}{\sqrt{\pi}(2\alpha)^{\lambda-\frac{1}{2}}\Gamma(\lambda)} K_{\lambda-\frac{1}{2}}(\alpha|x-\mu|) e^{\beta(x-\mu)}.$$

However, if $\lambda < 0$ and $\alpha, \beta \to 0$, then (10) converges pointwise to the density of a scaled and shifted t-distribution with $f = -2\lambda$ degrees of freedom:

$$d_{t(\lambda,\delta,\mu)}(x) = \frac{\Gamma(-\lambda+\frac{1}{2})}{\delta\sqrt{\pi}\Gamma(-\lambda)} \left(1 + \frac{(x-\mu)^2}{\delta^2}\right)^{\lambda-\frac{1}{2}}, \text{ where } \Gamma(\lambda) = \int_0^\infty x^{\lambda-1} e^{-x} dx.$$

For a detailed derivation of these limits and their characteristic functions, we refer to [11].

Both the skewness and especially the heavier tails increase significantly the probability of joint defaults in the factor model

$$X_i = \sqrt{\rho}\, M + \sqrt{1-\rho}\, Z_i. \tag{11}$$

In the following we assume that $M, Z_1, \ldots, Z_N$ are independent as before, but $M \sim GH(\lambda_M, \alpha_M, \beta_M, \delta_M, \mu_M)$ and all Z_i are iid $\sim GH(\lambda_Z, \alpha_Z, \beta_Z, \delta_Z, \mu_Z)$ (including the above limiting cases). Thus the distribution functions of all X_i coincide. Denote the latter by F_X and the distribution functions of M and of the Z_i by F_M and F_Z, then one can derive the corresponding cumulative default distribution F_{Z_t} analogously as described in Sect. 2.2 and obtains

$$F_{Z_t}(q) \approx 1 - F_M\left(\frac{F_X^{-1}(Q(t)) - \sqrt{1-\rho}\, F_Z^{-1}(q)}{\sqrt{\rho}}\right). \tag{12}$$

Note that this expression cannot be simplified further as in equation (8) since the distribution of M is in general not symmetric.

Remark 4. As mentioned above, almost all densities of GH distributions possess exponentially decreasing tails, only the Student t limit distributions have a power tail. According to the results of [24], the joint distribution of the X_i will therefore show tail dependence if and only if the systematic factor M is Student t-distributed.

Further $GH(\lambda, \alpha, \beta, \delta, \mu) \xrightarrow{\mathcal{L}} N(\mu + \beta\sigma^2, \sigma^2)$ if $\alpha, \delta \to \infty$ and $\delta/\alpha \to \sigma^2$, so the normal factor model is included as a limit in our setting.

3.1 Factor Scaling and Calculation of Quantiles

To preserve the role of ρ as a correlation parameter, we have to standardize the factor distributions such that they have zero mean and unit variance. In the general case of GH distributions we fix shape, skewness and tail behaviour by specifying α, β, λ and then calculate $\bar{\delta}$ and $\bar{\mu}$ that scale and shift the density appropriately. For this purpose we first solve the equation

$$1 = \mathrm{Var}[GH(\lambda, \alpha, \beta, \delta, \mu)] = \frac{\delta^2}{\zeta}\frac{K_{\lambda+1}(\zeta)}{K_\lambda(\zeta)} + \beta^2 \frac{\delta^4}{\zeta^2}\left(\frac{K_{\lambda+2}(\zeta)}{K_\lambda(\zeta)} - \frac{K_{\lambda+1}^2(\zeta)}{K_\lambda^2(\zeta)}\right)$$

with $\zeta := \delta\sqrt{\alpha^2 - \beta^2}$ numerically to obtain $\bar{\delta}$ and then choose $\bar{\mu}$ such that

$$0 = \mathrm{E}[GH(\lambda, \alpha, \beta, \bar{\delta}, \bar{\mu})] = \bar{\mu} + \frac{\beta\bar{\delta}^2}{\bar{\zeta}}\frac{K_{\lambda+1}(\bar{\zeta})}{K_\lambda(\bar{\zeta})}, \quad \bar{\zeta} = \bar{\delta}\sqrt{\alpha^2 - \beta^2}.$$

Since the Bessel functions $K_{n+1/2}$, $n \geq 0$, can be expressed explicitly in closed forms, the calculations simplify considerably for the NIG subclass. We have

$$\mathrm{Var}[NIG(\alpha,\beta,\delta,\mu)] = \frac{\delta\alpha^2}{(\alpha^2-\beta^2)^{\frac{3}{2}}}, \quad \mathrm{E}[NIG(\alpha,\beta,\delta,\mu)] = \mu + \frac{\beta\delta}{\sqrt{\alpha^2-\beta^2}},$$

so the distribution can be standardized by choosing $\bar\delta = (\alpha^2-\beta^2)^{\frac{3}{2}}/\alpha^2$ and $\bar\mu = -\beta(\alpha^2-\beta^2)/\alpha^2$.

In the VG limiting case the variance is given by

$$\mathrm{Var}[VG(\lambda,\alpha,\beta,\mu)] = \frac{2\lambda}{\alpha^2-\beta^2} + \frac{4\lambda\beta^2}{(\alpha^2-\beta^2)^2} =: \sigma_{VG}^2,$$

so it would be tempting to use λ as a scaling parameter, but this would mean to change the tail behaviour which we want to keep fixed. Observing the fact that a VG distributed random variable X_{VG} equals in distribution the shifted sum of two Gamma variables, that is,

$$X_{VG} \stackrel{d}{=} \Gamma_{\lambda,\alpha-\beta} - \Gamma_{\lambda,\alpha+\beta} + \mu, \quad \text{where } d_{\Gamma_{\lambda,\sigma}}(x) = \frac{\sigma^\lambda}{\Gamma(\lambda)} x^{\lambda-1} e^{-\sigma x} 1\!\!1_{[0,\infty)}(x),$$

the correct scaling that preserves the shape is $\bar\alpha = \sigma_{VG}\alpha$, $\bar\beta = \sigma_{VG}\beta$. Then $\bar\mu$ has to fulfill

$$0 = \mathrm{E}[VG(\lambda,\bar\alpha,\bar\beta,\bar\mu)] = \bar\mu + \frac{2\lambda\bar\beta}{\bar\alpha^2-\bar\beta^2}.$$

The second moment of a Student t-distribution exists only if the number of degrees of freedom satisfies $f > 2$, so we have to impose the restriction $\lambda < -1$ in this case. Mean and variance are given by

$$\mathrm{Var}[t(\lambda,\delta,\mu)] = \frac{\delta^2}{-2\lambda-2} \quad \text{and} \quad \mathrm{E}[t(\lambda,\delta,\mu)] = \mu,$$

therefore one has to choose $\bar\delta = \sqrt{-2\lambda-2}$ and $\bar\mu = 0$.

We thus have a minimum number of three free parameters in our generalized factor model, namely λ_M, λ_Z and ρ if both M and Z_i are t-distributed, up to a maximum number of seven ($\lambda_M, \alpha_M, \beta_M, \lambda_Z, \alpha_Z, \beta_Z, \rho$) if both factors are GH or VG distibuted. If we restrict M and Z_i to certain GH subclasses by fixing λ_M and λ_Z, five free parameters are remaining.

Having scaled the factor distributions, the remaining problem is to compute the quantiles $F_X^{-1}(Q(t))$ which enter the default distribution F_{Z_t} by equation (12). Since the class of GH distributions is in general not closed under convolutions, the distribution function F_X is not known explicitly. Therefore one central task of the project was to develop a fast and stable algorithm for the numerical calculation of the quantiles of X_i, because simulation techniques had to be ruled out from the very beginning for two reasons: The default probabilities $Q(t)$ are very small, so one would have to generate a very large data set to get reasonable quantile estimates, and the simulation would have to be restarted whenever at least one model parameter has been modified. Since the pricing formula is evaluated thousands of times with different parameters during calibration, this procedure would be too time-consuming. Further, the routine used to calibrate the models tries to find an extremal point by searching the direction of the steepest ascend within the parameter space in each optimization step. This can be done successfully only if the model prices depend exclusively on the parameters and not additionally on random effects. In the latter case the optimizer may behave erratically and will never reach an extremum.

We obtain the quantiles of X_i by Fourier inversion. Let $\hat{P}_X$, $\hat{P}_M$ and $\hat{P}_Z$ denote the characteristic functions of X_i, M and Z_i, then equation (11) and the independence of the factors yield

$$\hat{P}_X(t) = \hat{P}_M(\sqrt{\rho}\,t) \cdot \hat{P}_Z(\sqrt{1-\rho}\,t).$$

With the help of the inversion formula we get a quite accurate approximation of F_X from which the quantiles $F_X^{-1}(Q(t))$ can be derived. For all possible factor distributions mentioned above, the characteristic functions $\hat{P}_M$ and $\hat{P}_Z$ are well known; see [11] for a derivation and explicit formulas.

In contrast to this approach there are two special settings in which the quantiles of X_i can be calculated directly. The first one relies on the following convolution property of the NIG subclass,

$$NIG(\alpha, \beta, \delta_1, \mu_1) * NIG(\alpha, \beta, \delta_2, \mu_2) = NIG(\alpha, \beta, \delta_1 + \delta_2, \mu_1 + \mu_2),$$

and the fact that if $Y \sim NIG(\alpha, \beta, \delta, \mu)$, then $aY \sim NIG\big(\frac{\alpha}{|a|}, \frac{\beta}{a}, \delta|a|, \mu a\big)$. Thus if both M and Z_i are NIG distributed and the distribution parameters of the latter are defined by $\alpha_Z := \alpha_M \sqrt{1-\rho}/\sqrt{\rho}$ and $\beta_Z = \beta_M \sqrt{1-\rho}/\sqrt{\rho}$, then it follows together with equation (11) that $X_i \sim NIG\big(\frac{\alpha_M}{\sqrt{\rho}}, \frac{\beta_M}{\sqrt{\rho}}, \frac{\bar{\delta}_M}{\sqrt{\rho}}, \frac{\bar{\mu}_M}{\sqrt{\rho}}\big)$, where $\bar{\delta}_M$ and $\bar{\mu}_M$ are the parameters of the standardized distribution of M as described before.

In the VG limiting case the parameters α, β and μ behave as above under scaling, and the corresponding convolution property is

$$VG(\lambda_1, \alpha, \beta, \mu_1) * VG(\lambda_2, \alpha, \beta, \mu_2) = VG(\lambda_1 + \lambda_2, \alpha, \beta, \mu_1 + \mu_2).$$

Consequently if both factors are VG distributed and the free parameters of the idiosyncratic factor are chosen as follows, $\lambda_Z = \lambda_M(1-\rho)/\rho$, $\alpha_Z = \alpha_M$, $\beta_Z = \beta_M$, then $X_i \sim VG\big(\frac{\lambda_M}{\rho}, \frac{\bar{\alpha}_M}{\sqrt{\rho}}, \frac{\bar{\beta}_M}{\sqrt{\rho}}, \frac{\bar{\mu}_M}{\sqrt{\rho}}\big)$.

This stability under convolutions, together with the appropriate parameter choices for the idiosyncratic factor, was used in [19] and all models considered in [1]. We do not use this approach here because it reduces the number of free parameters and therefore the flexibility of the factor model. Moreover, in such a setting the distribution of the idiosyncratic factor is uniquely determined by the systematic factor, which contradicts the intuitive idea behind the factor model and lacks an economic interpretation.

3.2 Calibration Results for the Dj iTraxx Europe

We calibrate our generalized factor model with market quotes of DJ iTraxx Europe standard tranches. As mentioned before, the iTraxx Europe index is based on a reference portfolio of 125 European investment grade firms and quotes its average credit spread which can be used to estimate the default intensity of all constituents according to equation (4). The diversification of the portfolio always remains the same. It contains CDSs of 10 firms from automotive industry, 30 consumers, 20 energy firms, 20 industrials, 20 TMTs (technology, media and telecommunication companies) and 25 financials. In each sector, the firms with the highest liquidity and volume of trade with respect to their defaultable assets (bonds and CDSs) are selected. The iTraxx portfolio is reviewed and updated quarterly. Not only companies that have defaulted in between are replaced by new ones, but also those which no longer fulfill the liquidity and trading demands. Of course, the recomposition affects future deals only. Once two counterparties have agreed to buy and sell protection on a certain iTraxx tranche, the current portfolio is kept fixed for them in order to determine the corresponding cash flows described in Sect. 2.1. The names and attachment points of the five iTraxx standard tranches are given in Figs. 2 and 3. For each of them four contracts with different maturities (3, 5, 7 and 10 years) are available.

The settlement date of the sixth iTraxx series was December 20, 2006, so the 5, 7, and 10 year contracts mature on December 20, 2011 resp. 2013 and 2016. We consider the market prices of the latter on all standard tranches at November 13, 2006. For the mezzanine and senior tranches, these equal the annualized fair spreads r_i which can be obtained from equation (3) and are also termed *running spreads*. However, the market convention for pricing the equity tranche is somewhat different: In this case the protection buyer has to pay a certain percentage s_1 of the notional value $K_1 NL$ as an *up-front fee* at the starting time t_0 of the contract and a fixed spread of 500bp on the outstanding notional at $t_1, \ldots, t_n$. Therefore the premium leg for the equity tranche is given by

$$PL_1(s_1) = s_1 K_1 NL + 0.05 \sum_{k=1}^{n} (t_k - t_{k-1}) \beta(t_0, t_k) \mathrm{E}\big[(K_1 - L^1_{t_k}) NL\big],$$

and the no-arbitrage condition $PL_1(s_1) = D_1$ then implies

$$s_1 = \frac{\sum_{k=1}^{n} \beta(t_0, t_k) \Big(\mathrm{E}\big[L^1_{t_k}\big] - \mathrm{E}\big[L^1_{t_{k-1}}\big] - 0.05(t_k - t_{k-1})\big(K_1 - \mathrm{E}\big[L^1_{t_k}\big]\big)\Big)}{K_1}. \tag{13}$$

Since the running spread is set to a constant of 500 bp, the varying market price quoted for the equity tranche is the percentage s_1 defining the magnitude of the up-front fee.

We calibrate our generalized factor model by least squares optimization, that is, we first specify to which subclass of the GH family the distributions F_M and F_Z belong and then determine the correlation and distribution parameters numerically which minimize the sum of the squared differences between model and market prices over all tranches. Although our algorithm for computing the quantiles $F_X^{-1}(Q(t))$ allows us to combine factor distributions of different GH subclasses, we restrict both factors to the same subclass for simplicity reasons. Therefore in the following table and figures the expression VG, for example, denotes a factor model where M and the Z_i are variance gamma distributed. The model prices are calculated from equations (3) and (13), using the cumulative default distribution (12) resp. (8) for the normal factor model which serves as a benchmark. The recovery rate R which has a great influence on the expected losses $\mathrm{E}[L_{t_k}^i]$ according to equation (9) is always set to 40 %; this is the common market assumption for the iTraxx portfolio.

One should observe that the prices of the equity tranches are usually given in percent, whereas the spreads of all other tranches are quoted in basis points. In order to use the same units for all tranches in the objective function to be minimized, the equity prices are transformed into basis points within the optimization algorithm. Thus they are much higher than the mezzanine and senior spreads and therefore react to parameter changes in a more sensitive way, which amounts to an increased weighting of the equity tranche in the calibration procedure. This is also desirable from an economical point of view since the costs for mispricing the equity tranche are typically greater than for all other tranches.

Remark 5. For the same reason, the normal factor model is usually calibrated by determining the implied correlation of the equity tranche first and then using this to calculate the fair spreads of the other tranches. This ensures that at least the equity price is matched perfectly. To provide a better comparison with our model, we give up this convention and also use least squares estimation in this case. Therefore the fit of the equity tranche is sometimes less accurate, but the distance between model and market prices is smaller for the higher tranches instead.

Our calibration results for the 5 and 7 year iTraxx tranches are shown in Figs. 5 and 6. The normal benchmark model performs worst in all cases. The performance of the t model is comparable with the NIG and HYP models, whereas the VG model provides the best fit for both maturities. Since the t model is the only one exhibiting tail dependence (confer Remark 4) but does not outperform the NIG, HYP and VG models, one may conclude that this property is negligible in the presence of more flexible factor distributions. This may also be confirmed by the fact that all estimated GH parameters β_M and β_Z are different from zero which implies skewness of the factor distributions. Furthermore the parameter ρ is usually higher in the GH factor models than in the normal benchmak model, indicating that correlation is still of some importance, but has a different impact on the pricing formula because of the more complex dependence structure.

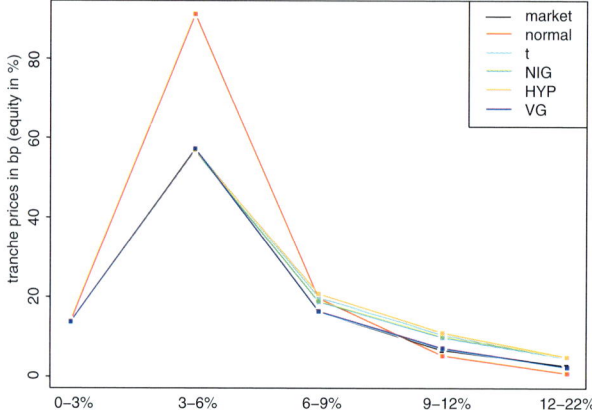

Fig. 5. Comparison of calibrated model prices and market prices of the 5 year iTraxx contracts

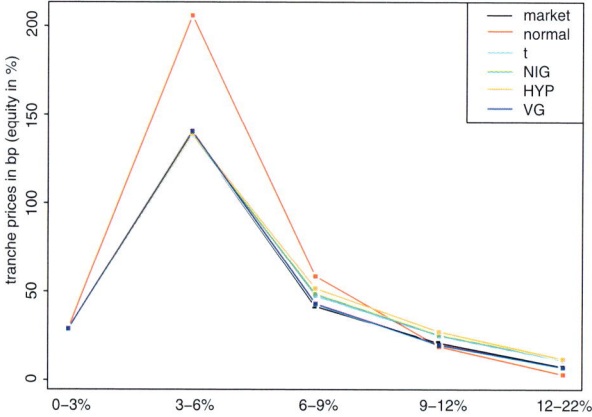

Fig. 6. Comparison of calibrated model prices and market prices of the 7 year iTraxx contracts

The VG model even has the potential to fit the market prices of all tranches and maturities simultaneously with high accuracy, which we shall show below. However, before that we want to point out that the calibration over different maturities requires some additional care to avoid inconsistencies when calculating the default probabilities. As can be seen from Fig. 7, the average iTraxx spreads s_a are increasing in maturity and by equation (4) so do the default intensities λ_a. This means that the estimated default probabilities $Q(t) = 1 - e^{-\lambda_a\,t}$ of a CDO with a longer lifetime are always greater than those of a CDO with a shorter maturity. While this can be neglected when concentrating on just one maturity, this fact has to be taken into account

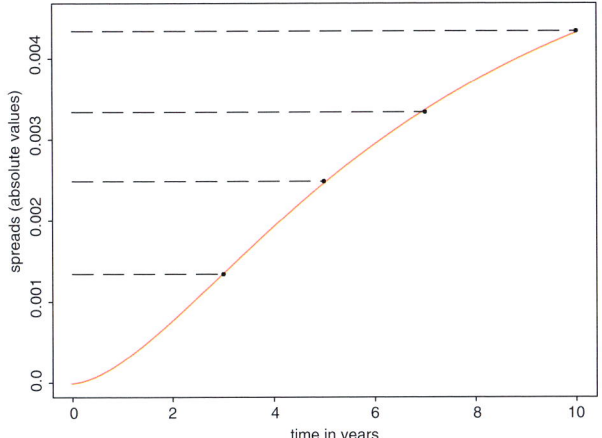

Fig. 7. Constant iTraxx spreads of November 13, 2006, and fitted Nelson–Siegel curve $\hat{r}_{NS}$ with parameters $\hat{\beta}_0 = 0.0072$, $\hat{\beta}_1 = -0.0072$, $\hat{\beta}_2 = -0.0069$, $\hat{\tau}_1 = 2.0950$

when considering iTraxx CDOs of different maturities together. Since the underlying portfolio is the same, the default probabilities should coincide during the common lifetime.

To avoid these problems we now assume that the average spreads $s_a = s(t)$ are time-dependent and follow a Nelson–Siegel curve. This parametric family of functions has been introduced in [27] and has become very popular in interest rate theory for the modeling of yield curves where the task is the following: Let $\beta(0, t_k)$ denote today's price of a zero coupon bond with maturity t_k as before, then one has to find a function f (*instantaneous forward rates*) such that the model prices $\beta(0, t_k) = \exp\bigl(-\int_0^{t_k} f(t)\,dt\bigr)$ approximate the market prices reasonably well for all maturities t_k. Since instantaneous forward rates cannot be observed directly in the market, one often uses an equivalent expression in terms of *spot rates*: $\beta(0, t_k) = \exp(-r(t_k) t_k)$, where the spot rate is given by $r(t_k) = \frac{1}{t_k} \int_0^{t_k} f(t)\,dt$. Nelson and Siegel suggested to model the forward rates by

$$f_{NS(\beta_0,\beta_1,\beta_2,\tau_1)}(t) = \beta_0 + \beta_1 e^{-\frac{t}{\tau_1}} + \beta_2 \frac{t}{\tau_1} e^{-\frac{t}{\tau_1}}.$$

The corresponding spot rates are given by

$$r_{NS(\beta_0,\beta_1,\beta_2,\tau_1)}(t) = \beta_0 + (\beta_1 + \beta_2)\frac{\tau_1}{t}\left(1 - e^{-\frac{t}{\tau_1}}\right) - \beta_2 e^{-\frac{t}{\tau_1}}. \quad (14)$$

In order to obtain time-consistent default probabilities resp. intensities we replace s_a in equation (4) by a Nelson–Siegel spot rate curve (14) that has been fitted to the four quoted average iTraxx spreads, that is,

$$\lambda_a = \lambda(t) = \frac{\hat{r}_{NS}(t)}{(1 - R)10000}, \quad (15)$$

and $Q(t) := 1 - e^{-\lambda(t)\,t}$. The Nelson–Siegel curve estimated from the iTraxx spreads of November 13, 2006, is shown in Fig. 7. At first glance the differences between constant and time-varying spreads seem to be fairly large, but one should observe that these are the absolute values which have already been divided by 10000 and therefore range from 0 to 0.004338, so the differences in the default probabilities are almost negligible.

Under the additional assumption (15), we have calibrated a model with VG distributed factors to the tranche prices of all maturities simultaneously. The results are summarized in Table 1 and visualized in Fig. 8. The fit is ex-

Table 1. Results of the VG model calibration simultaneously over all maturities. Estimated parameters are as follows: $\lambda_M = 0.920$, $\alpha_M = 5.553$, $\beta_M = 1.157$, $\lambda_Z = 2.080$, $\alpha_Z = 2.306$, $\beta_Z = -0.753$, $\rho = 0.321$.

Tranches	Market	VG	Market	VG	Market	VG
	5Y		7Y		10Y	
0–3 %	13.60 %	13.60 %	28.71 %	28.72 %	42.67 %	42.67 %
3–6 %	57.16bp	53.30bp	140.27bp	132.27bp	360.34bp	357.60bp
6–9 %	16.31bp	17.19bp	41.64bp	41.83bp	105.08bp	111.17bp
9–12 %	6.65bp	8.23bp	21.05bp	19.90bp	43.33bp	52.00bp
12–22 %	2.67bp	3.05bp	7.43bp	7.34bp	13.52bp	18.97bp

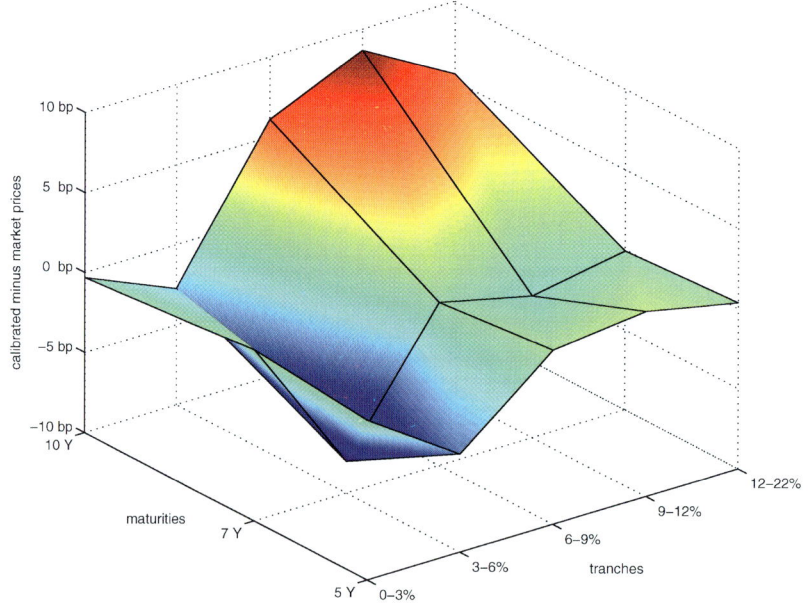

Fig. 8. Graphical representation of the differences between model and market prices obtained from the simultaneous VG calibration

cellent. The maximal absolute pricing error is less than 9bp, and for the 5 and 7 year maturities the errors are, apart from the junior mezzanine tranches, almost as small as in the previous calibrations. The junior mezzanine tranche is underpriced for all maturities, but it is difficult to say whether this is caused by model or by market imperfections. Nevertheless the overall pricing performance of the extended VG model is comparable or better than the performance of the models considered in [1, 7, 19], although the latter were only calibrated to tranche quotes of a single maturity.

Also note that this model admits a flat correlation structure not only over all tranches, but also over different maturities: all model prices contained in Table 1 were calculated using the same parameter ρ. Thus the correlation smiles shown in Fig. 3 which in some sense question the factor equation (5) are completely eliminated. Therefore the intuitive idea of the factor approach is preserved, but one should keep in mind that in the case of GH distributed factors the dependence structure of the joint distribution of the X_i is more complex and cannot be described by correlation alone.

4 A Dynamic Markov Chain Model

In this section we discuss an entirely different approach to explain observed CDO spreads, rooted more in the theory of stochastic processes. Our exposition summarizes results from [13].

4.1 The Model

We begin with some notation. Given some probability space $(\Omega, \mathcal{F}, Q)$, Q the risk-neutral measure used for pricing, we define the *default indicator* of firm i at time t by $Y_{t,i} = \mathbb{1}_{\{T_i \leq t\}}$. Note that the default indicator process $Y_i = (Y_{t,i})_{t \geq 0}$ is a right continuous process which jumps from 0 to 1 at the default of firm i. The evolution of the default state of the portfolio is then described by the process $Y = (Y_{t,1}, \ldots, Y_{t,N})_{t \geq 0}$; obviously, $Y_t \in S^Y := \{0,1\}^N$. We use the following notation for flipping the ith coordinate of a default state: given $y \in S^Y$ we define $y^i \in S^Y$ by

$$y_i^i := 1 - y_i \text{ and } y_j^i := y_j, \ j \in \{1, \ldots, N\} \setminus \{i\}. \tag{16}$$

The *default history* (the internal filtration of the process Y) is denoted by $(\mathcal{H}_t)$, i.e. $\mathcal{H}_t = \sigma(Y_s : s \leq t)$. An $(\mathcal{H}_t)$-adapted process $(\lambda_{t,i})$ is called the *default intensity* of T_i (with respect to $(\mathcal{H}_t)$) if

$$Y_{t,i} - \int_0^{T_i \wedge t} \lambda_{s,i} \, ds \text{ is an } (\mathcal{H}_t)\text{-martingale.}$$

Intuitively, $\lambda_{t,i}$ gives the instantaneous chance of default of a non-defaulted firm i given the default history up to time t. It is well-known that the default intensities determine the law of the marked point process (Y_t); see for instance [6] for a detailed account of the mathematics of marked point processes.

Modeling the Dynamics of Y

We assume that the default intensity of a non-defaulted firm i at time t is given by a function $\lambda_i(t, Y_t)$ of time and of the current default state Y_t. Hence the default intensity of a firm may change if there is a change in the default state of other firms in the portfolio; in this way dependence between default events can be modeled explicitly. Formally, we model the default indicator process by a time-inhomogeneous Markov chain with state space S^Y. The next assumption summarizes the mathematical properties of Y.

Assumption 1 (Markov family). *Consider bounded and measurable functions $\lambda_i : [0, \infty) \times S^Y \to \mathbb{R}_+$, $1 \leq i \leq N$. There is a family $Q_{(t,y)}$, $(t, y) \in [0, \infty) \times S^Y$, of probability measures on $(\Omega, \mathcal{F}, (\mathcal{H}_t))$ such that $Q_{(t,y)}(Y_t = y) = 1$ and such that $(Y_s)_{s \geq t}$ is a finite-state Markov chain with state space S^Y and transition rates $\lambda(s, y_1, y_2)$ given by*

$$\lambda(s, y_1, y_2) = \begin{cases} (1 - y_{1,i}) \lambda_i(s, y_1), & \text{if } y_2 = y_1^i \text{ for some } i \in \{1, \ldots, N\}, \\ 0 & \text{else.} \end{cases} \quad (17)$$

Relation (17) has the following interpretation: In t the chain can jump only to the set of neighbors of the current state Y_t that differ from Y_t by exactly one default; in particular there are no joint defaults. The probability that firm i defaults in the small time interval $[t, t+h)$ thus corresponds to the probability that the chain jumps to the neighboring state $(Y_t)^i$ in this time period. Since such a transition occurs with rate $\lambda_i(t, Y_t)$, it is intuitively obvious that $\lambda_i(t, Y_t)$ is the default intensity of firm i at time t; a formal argument is given in [13].

The *numerical treatment* of the model can be based on Monte Carlo simulation or on the Kolmogorov forward and backward equation for the transition probabilities; see again [13] for further information. An excellent introduction to continuous-time Markov chains is given in [28].

Modeling Default Intensities

The default intensities $\lambda_i(t, Y_t)$ are crucial ingredients of the model. If the portfolio size N is large – such as in the pricing of typical synthetic CDO tranches – it is natural to assume that the portfolio has a homogeneous group structure. This assumption gives rise to intuitive parameterizations for the default intensities; moreover, the homogeneous group structure leads to a substantial reduction in the size of the state space of the model. Here we concentrate on the extreme case where the entire portfolio forms a single homogeneous group so that the processes Y_i are *exchangeable*; this simplifying assumption is made in most CDO pricing models; see also Sect. 2. Denote the number of defaulted firms at time t by

$$M_t := \sum_{i=1}^{N} Y_{t,i}.$$

As discussed in [13], in a homogeneous model default intensities are necessarily of the form

$$\lambda_i(t, Y_t) = h(t, M_t) \text{ for some } h : [0, \infty) \times \{0, \ldots, N\} \to \mathbb{R}_+ . \tag{18}$$

Note that the assumption that default intensities depend on Y_t via the number of defaulted firms M_t makes sense also from an economic viewpoint, as unusually many defaults might have a negative impact on the liquidity of credit markets or on the business climate in general. This point is discussed further in [12] and [16].

The simplest exchangeable model is the *linear counterparty risk* model. Here

$$h(t, l) = \lambda_0 + \lambda_1 l, \quad \lambda_0 > 0, \lambda_1 \geq 0, l \in \{0, \ldots, N\} . \tag{19}$$

The interpretation of (19) is straightforward: upon default of some firm the default intensity of the surviving firms increases by the constant amount λ_1 so that default dependence increases with λ_1; for $\lambda_1 = 0$ defaults are independent. Model (19) is the homogeneous version of the so-called looping-defaults model of [18].

The next model generalizes the linear counterparty risk model in two important ways: first, we introduce time-dependence and assume that a default event at time t increases the default intensity of surviving firms only if M_t exceeds some deterministic threshold $\mu(t)$ measuring the expected number of defaulted firms up to time t; second, we assume that on $\{l > \mu(t)\}$ the function $h(t, \cdot)$ is strictly *convex*. Convexity of h implies that large values of M_t lead to very high values of the default intensities, thus triggering a cascade of further defaults. This will be important in explaining properties of observed CDO prices below. The following specific model with the above features will be particularly useful:

$$h(t, l) = \lambda_0 + \frac{\lambda_1}{\lambda_2}\left(\exp\left(\lambda_2 \frac{(l - \mu(t))^+}{N}\right) - 1\right), \quad \lambda_0 > 0, \lambda_1, \lambda_2 \geq 0; \tag{20}$$

in the sequel we call (20) *convex counterparty risk* model. In (20) λ_0 is a level parameter that mainly influences credit quality. λ_1 gives the slope of $h(t, l)$ at $\mu(t)$; intuitively this parameter models the strength of default interaction for "normal" realisations of M_t. The parameter λ_2 controls the degree of convexity of h and hence the tendency of the model to generate default cascades; note that for $\lambda_2 \to 0$ (and $\mu(t) \equiv 0$) (20) reduces to the linear model (19).

The Markov Property of M

It is straightforward that for default intensities of the form (18) the process $M = (M_t)_{t \geq 0}$ is itself a Markov chain with generator given by

$$G^M_{[t]} f(l) = (N - l) h(t, l) \big(f(l+1) - f(l)\big). \tag{21}$$

In fact, since Assumption 1 excludes joint defaults, M_t can jump only to M_t+1. The intensity of such a transition is proportional to the number $N - M_t$ of surviving firms at time t as well as to their default intensity $h(t, M_t)$. This is important: since the portfolio loss satisfies $L_t = (1 - R)M_t/N$, the loss processes L^i of the individual tranches (and of course the overall portfolio loss) are given by functions of M_t, so that we may concentrate on the process M. As shown in [13] this considerably simplifies the numerical analysis of the model. Similar modeling ideas have independently been put forward in [3].

4.2 Analysis of CDO Tranches

Next we turn to an analysis of synthetic CDO tranches in the context of the Markov chain model; in particular, we are interested in modeling the well-known implied correlation skew described in Sect. 2.3. Recall that according to equation (3), the computation of fair tranche spreads r_i boils down to evaluating the distribution of L_t^i – and hence the distribution of M_t – at the premium payment dates. The latter can be computed efficiently using the *Kolmogorov forward equations* or by simulation; see [13] for details.

The basic idea for generating correlation skews in the context of the convex counterparty risk model (20) is simple: by increasing λ_2 we can generate occasional large clusters of defaults without affecting the left tail of the distribution of L_t too much; in this way we can reproduce the spread of the mezzanine and senior CDO tranches in a way which is consistent with the observed spread of the equity tranche. In order to confirm this intuition we consider a numerical example with spread data from [17]. In Table 2 we give the CDO spreads if the convexity parameter λ_2 is varied; λ_0 and λ_1 were calibrated to the index level

Table 2. CDO spreads in the convex counterparty risk model (20) for varying λ_2. λ_0 and λ_1 were calibrated to the index level of 42bp and the market quote for the equity tranche, assuming $\delta = 0.6$. For $\lambda_2 \in [8, 10]$ the qualitative properties of the model-generated CDO spreads resemble closely the behaviour of the market spreads; with state-dependent LGD the fit is almost perfect

tranches	[0,3]	[3,6]	[6,9]	[9,12]	[12,22]	
market spreads	27.6%	168.0bp	70.0bp	43.0bp	20.0bp	
model spreads						$\sum$ abs. err.
$\lambda_2 = 0$	27.6%	223.1bp	114.5bp	61.1bp	16.9bp	120.8bp
$\lambda_2 = 5$	27.6%	194.2bp	95.7bp	54.9bp	23.3bp	67.1bp
$\lambda_2 = 8$	27.6%	172.1bp	80.0bp	46.7bp	23.7bp	21.5bp
$\lambda_2 = 8.54$	27.6%	168.0bp	77.1bp	45.1bp	23.5bp	12.7bp
$\lambda_2 = 10$	27.6%	156.9bp	69.4bp	40.7bp	22.7bp	16.7bp
state-dependent LGD $\delta_0 = 0.5; \delta_1 = 7.5$	27.6%	168.0bp	71.2bp	39.3bp	19.6bp	5.3bp

and the observed market quote of the equity tranche. The results show that for appropriate values of λ_2 the model can reproduce the qualitative behavior of the observed tranche spreads in a very satisfactory way. This observation is interesting, as it provides an explanation of correlation skews of CDOs in terms of the *dynamics* of the default indicator process. Similarly as in [2], the model fit can be improved further by considering a state-dependent loss given default of the form $\delta_t = \delta_0 + \delta_1 M_t$ with $\delta_0, \delta_1 > 0$; see again Table 2.

Comments

Implied correlations for CDO tranches on the iTraxx Europe have changed substantially since August 2004. More importantly, the analysis presented in Table 2 presents only a "snapshot" of the CDO market at a single day. For these reasons in [13] the convex counterparty risk model (20) was recalibrated to 6 months of observed 5 year tranche spreads on the iTraxx Europe in the period 23.9.2005–03.03.2006. It turned out that the resulting parameters were quite stable over time.

In this paper we have calibrated a parametric version of the model (20) to observed CDO spreads. For an interesting nonparametric calibration procedure based on the Kolmogorov forward equation we refer to [3, 20, 29].

Dynamic Markov chain models are very useful tools for studying the practically relevant risk management of CDO tranches via dynamic hedging; we refer to the recent papers [14] and [20] for details.

References

1. H. Albrecher, S. A. Ladoucette, and W. Schoutens, *A generic one-factor Lévy model for pricing synthetic CDOs*. In: M. C. Fu, R. A. Jarrow, J.-Y. Yen, and R. J. Elliott (eds.), Advances in Mathematical Finance, Birkhäuser, Boston, 259–277 (2007)
2. L. Andersen, J. Sidenius, *Extensions to the Gaussian copula: random recovery and random factor loadings*. Journal of Credit Risk 1, 29–70 (2005)
3. M. Arnsdorf, I. Halperin, *BSLP: Markovian bivariate spread-loss model for portfolio credit derivatives*. Working paper, JP Morgan (2007)
4. O. E. Barndorff-Nielsen, *Exponentially decreasing distributions for the logarithm of particle size*. Proceedings of the Royal Society London, Series A 353, 401–419 (1977)
5. T. Berrada, D. Dupuis, E. Jacquier, N. Papageorgiou, and B. Rémillard, *Credit migration and basket derivatives pricing with copulas*. Journal of Computational Finance 10, 43–68 (2006)
6. P. Brémaud, *Point Processes and Queues: Martingale Dynamics*. Springer, New York (1981)
7. X. Burtschell, J. Gregory, and J.-P. Laurent, *A comparative analysis of CDO pricing models*. Working paper, BNP Paribas (2005)
8. X. Burtschell, J. Gregory, and J.-P. Laurent, *Beyond the Gaussian copula: stochastic and local correlation*. Journal of Credit Risk 3, 31–62 (2007)

9. E. Eberlein, *Application of generalized hyperbolic Lévy motions to finance*. In: O. E. Barndorff-Nielsen, T. Mikosch, and S. Resnick (eds.), Lévy processes: Theory and Applications, Birkhäuser, Boston, 319–337 (2001)
10. E. Eberlein, U. Keller, *Hyperbolic distributions in finance*. Bernoulli 1, 281–299 (1995)
11. E. Eberlein, E. A. von Hammerstein, *Generalized hyperbolic and inverse Gaussian distributions: limiting cases and approximation of processes*. In: R. C. Dalang, M. Dozzi, and F. Russo (eds.), Seminar on Stochastic Analysis, Random Fields and Applications IV, Progress in Probability 58, Birkhäuser, Basel, 221–264 (2004)
12. R. Frey, *A mean-field model for interacting defaults and counterparty risk*. Bulletin of the International Statistical Institute (2003)
13. R. Frey, J. Backhaus, *Credit derivatives in models with interacting default intensities: a Markovian approach*. Preprint, Universität Leipzig (2006), available from www.math.uni-leipzig.de/~frey/publications-frey.html.
14. R. Frey, J. Backhaus, *Dynamic hedging of synthetic CDO tranches with spread- and contagion risk*. Preprint, Universität Leipzig (2007), available from www.math.uni-leipzig.de/~frey/publications-frey.html.
15. J. Gregory, J.-P. Laurent, *In the core of correlation*. Risk, October, 87–91 (2004)
16. U. Horst, *Stochastic cascades, credit contagion and large portfolio losses*. Journal of Economic Behavior & Organization 63, 25–54 (2007)
17. J. Hull, A. White, *Valuation of a CDO and an nth to default CDS without Monte Carlo simulation*. Journal of Derivatives 12, 8–23 (2004)
18. R. Jarrow, F. Yu, *Counterparty risk and the pricing of defaultable securities*. Journal of Finance 53, 2225–2243 (2001)
19. A. Kalemanova, B. Schmid, and R. Werner, *The normal inverse Gaussian distribution for synthetic CDO pricing*. Working paper, risklab germany (2005)
20. J.-P. Laurent, A. Cousin, and J. D. Fermanian, *Hedging default risk of CDOs in Markovian contagion models*. Workinp paper, ISFA Actuarial School, Université de Lyon (2007)
21. D. Li, *On default correlation: a copula function approach*. Journal of Fixed Income 9, 43–54 (2000)
22. A. J. McNeil, R. Frey, and P. Embrechts, *Quantitative Risk Management*. Princeton Series in Finance, Princeton University Press, Princeton (2005)
23. D. B. Madan, P. Carr, and E. C. Chang, *The variance gamma process and option pricing*. European Financial Review 2, 79–105 (1998)
24. Y. Malevergne, D. Sornette, *How to account for extreme co-movements between individual stocks and the market*. Journal of Risk 6, 71–116 (2004)
25. R. C. Merton, *On the pricing of corporate debt: the risk structure of interest rates*. Journal of Finance 29, 449–470 (1974)
26. R. B. Nelsen, *An Introduction to Copulas*. Lecture Notes in Statistics 139, Springer, New York (1999)
27. C. R. Nelson, A. F. Siegel, *Parsimonious modeling of yield curves*. Journal of Business 60, 473–489 (1986)
28. J. R. Norris, *Markov Chains*. Cambridge University Press, Cambridge (1997)
29. P. J. Schönbucher, *Portfolio loss and the term-structure of loss transition rates: a new methodology for the pricing of portfolio credit derivatives*. Working paper, ETH Zürich (2006)
30. P. J. Schönbucher, D. Schubert, *Copula-dependent default risk in intensity models*. Preprint, ETH Zürich and Universität Bonn (2001)

31. A. Sklar, *Fonctions de répartition à n dimensions et leurs marges*. Publications de l'Institute de Statistique de l'Université de Paris 8, 229–231 (1959)
32. O. A. Vasiček, *Probability of loss on loan portfolio*. Mimeo, KMV Corporation (1987)
33. O. A. Vasiček, *Limiting loan loss probability distribution*. Mimeo, KMV Corporation (1991)

Contributions to Multivariate Structural Approaches in Credit Risk Modeling

Swantje Becker*, Stefanie Kammer*, and Ludger Overbeck

Universität Gießen, Arndtstraße 2, 35392 Gießen, Germany
{swantje.becker,stefanie.kammer,ludger.overbeck}@math.uni-giessen.de

Summary. Credit risk models are usually differentiated into reduced form models and structural models. The latter are usually more powerful if many credits are to be modelled, more precisely if the focus stays with the dependency structure of credits, whereas reduced form models are more adequate if single credits, like term structure of credit spreads, are considered. This paper has two objectives the first one is to analyze the credit spread dynamcis of a wide class of structural models and the second one to understand the dependency structure if the multivariate asset value model is assumed to be a shot-noise process.

1 Introduction

Credit risk and the capital markets products transferring and structuring credit risk have evolved dramatically over the last decade. The first liquidly traded credit derivative was the simple credit default swap, which transfer the credit risk of an underlying bond synthetically to an investor, without the necessity that the investor buys the bond.

If the risk out of entire portfolios of bonds, loans or other credit risky instruments is transferred to the capital market, one usually employs a securitization or other structured credit products. As an example: Synthetic transaction based on credit default swaps (CSO).

Fig. 1. Credit Default Swap (CDS)

---

* Supported by BMBF-grant 03OVNHF

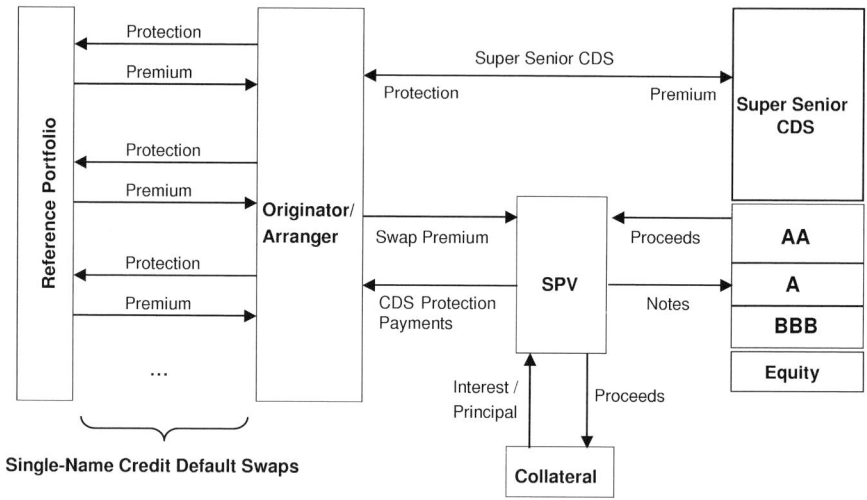

Fig. 2. Collaterized Swap Obligation (CSO)

Losses in the underlying portfolio, are first covered by the lowest tranche, i.e. in a synthetic structure, if the first loss in the portfolio on the left hand side happens, of an amount of l_1, then the holder of the lowest tranche has to pay the amount of l_1 the swap counterparty.

This increasing market made it indispensible to model the underlying risk properly. There are two classes of models. The first one are made up by the reduced form models, which are reduced to the modeling of the default probabilities themselves, without asking how defaults happens, cf. e.g. [10].

In contrast, the basic feature of a structural model lies in the attempt to model a causal structure, which finally leads to default. In the simplest case the structure consists in a stochastic process $Y = (Y_t)$ determining, whether a default event has happened up to a given time $T > 0$. In that case

$$D_T = \{Y_T \leq K\} \tag{1}$$

with a default threshold K_T. A more dynamic approach, the exit or hitting time model, constructs the default time τ as the first hitting time of K,

$$\tau = \inf\{s \geq 0 : Y_s < K\}. \tag{2}$$

In Sect. 2 we will present the one-dimensional dynamics of the exit time model in the case where X is a Brownian motion, and a Brownian motion with deterministic or stochastic time change. The deterministic time change provides an exit time model matching any term structure of default probabilities. In the context of time changes, we will present for any distribution function $p(t)$

on $[0,\infty)$, the distribution of the default time implied by the term structure of default probabilities, a time change $\tilde{Y}$ of a Brownian motion Y such that
$$P[\inf\{s|\tilde{Y}_s \leq K\} \leq t] = p(t).$$
This model yields then a stochastic process $s(t)$ for spreads, basically the dynamics of future default probabilities. More generally also process $\tilde{Y}_s = Y_{g_s}$ with a stochastic time change (g_s) are introduced and their implied spread dynamics in an exit time model.

In Sect. 3 we will extend the exit time models based on Brownian motion towards a shot-noise process. Roughly, a shot-noise process is a Brownian motion which additional Poisson distributed jump times and arbitrary jump distribution. The "shot" character comes from the fast mean reversion of the process, i.e. the reversion to the state just before the jump is done along a determinstic exponential function.

We mainly focus on the impact of the shots to the joint default probabilities and in a simulation study we are going to present some results on the valuation of CDOs.

2 Exit Times of Time Changed Processes

In a structural first-passage approach a default event happens the first time the asset-value or ability-to-pay process falls below some pre-specified default boundary K depending on the firm's debt. The structural model thus links equity and debt of a firm's capital structure. The default time of an entity was defined in (2). We assume that we observe the underlying asset-value process and know whether a default happened. Mathematically spoken we assume the information is given by the filtration
$$\mathcal{F}_t^Y = \sigma(Y_s : s \leq t),$$
such that τ is a $\mathbb{F}^Y$-stopping time. A default event can e.g. indicate a credit-rating downgrade or a total default; for us this is not important. We call $\mathbb{P}(\tau \leq t)$ the default probability, and for $0 \leq t \leq T$ define the survival probability conditional on the information $\mathcal{F}_t^Y$ as follows:
$$Q(t,T) := \mathbb{P}\left(\tau > T \mid \mathcal{F}_t^Y\right).$$
We now describe the credit default swap (CDS, see Fig. 1) and the CDS credit spread:

A CDS is a contract between a protection seller and a protection buyer. The protection seller offers protection against default of a reference entity during a certain time period, say between t and maturity T. Therefore the protection buyer regularly pays an insurance fee, the credit spread $s(t,T)$, but only as long as the entity is not defaulted. In case of a default before maturity of the CDS the protection seller pays the claim amount $1 - R$ to the protection buyer, where R denotes the recovery rate of the entity. We assume that there is no counterparty default risk, and for simplicity zero interest rates. Under no default until t, the fair continuously-paid credit spread is given by

$$s(t,T) = \frac{(1-R)(1-Q(t,T))}{\int_t^T Q(t,u)\,\mathrm{d}u}. \tag{3}$$

Usually a CDS spread is considered in terms of time to maturity M, then we have to insert $T = t + M$ into formula (3). We want to describe the dynamics of the credit-spread via a stochastic differential equation (SDE), a local volatility model:

$$\mathrm{d}s_t = s(t)\left[\mu(t,s_t)\,\mathrm{d}t + \sigma(t,s_t)\,\mathrm{d}W_t\right], \tag{4}$$

where W denotes a Brownian motion, μ can be interpreted as drift and σ as volatility of credit spread. We are especially interested in the credit-spread volatility. Fig. 3 shows credit-spread volatility of IBM versus time and credit spread, respectively. Certainly credit-spread volatility changes with time and seems to depend on the spread level.

Before we introduce the deterministic and the stochastic time-change model we shortly summarize the history of the time change and give its definition:

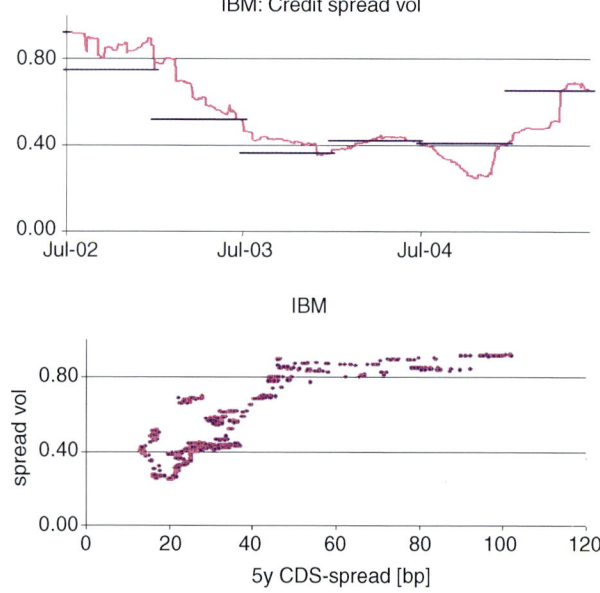

Fig. 3. Credit-spread volatility estimates versus time (*first plot*) resp. versus credit spread (*second plot*)

BOCHNER [5] first introduced a time-changed Brownian motion. FELLER [12] first presented subordinators as a time change to Markov processes. CLARK [8] introduced Brownian motion with an independent time change as a price process in finance. MONROE [18] showed that a very general semi-martingale can be embedded in Brownian motion via a time change. IKEDA & WATANABE [15] studied time-change models for solving SDEs. Øksendal [19] studied when a stochastic integral can be represented as a time change of a diffusion. GEMAN, MADAN & YOR [13] and CARR & WU [7] introduced subordinated Lévy process. SCHOUTENS [22] used these in derivative pricing.

Definition 2.1 (Time change). *A deterministic time change is a positive, non-decreasing function and a stochastic time change a positive, non-decreasing stochastic process.*

The non-decreasing property of the time transformation can be understood such that information that has been obtained once, will never be lost. A stochastic time change adds stochastic volatility to a process. The original clock will sometimes be called normal clock and the new clock will be called business clock. The time change may be interpreted as experienced time, that runs faster when the information flow is bigger (or speeds up). In other words, experienced time is a measure of the amount of information arrival.

2.1 Deterministic Time-Change Model

The difficulty of the structural model is its calibration to a market-given term structure of default probabilities F. The deterministic time-change model by OVERBECK & SCHMIDT [20],

$$\tau = \inf\{s \geq 0 : W_{\mathcal{T}_s} < K\}, \quad \mathcal{T}_t = \left[\frac{K}{\Phi^{-1}\left(\frac{F(t)}{2}\right)}\right]^2,$$

perfectly matches any default probability curve: $F(t) = \mathbb{P}(\tau \leq t) \ \forall\ t \geq 0$. The model leaves no degrees of freedom to influence the credit-spread dynamics: for a given asset values $Y_t = W_{\mathcal{T}_t}$ and a corresponding credit spread the credit-spread volatility is fixed. Figure 4 shows survival probabilities corresponding to various asset values $W_{\mathcal{T}_t}$, and credit-spread volatilities corresponding to various spread values s_t.

In order to be able to influence the credit-spead dynamics we introduce a stochastic time-change model in the next subsection.

2.2 Stochastic Time-Change Model

Our general stochastic time-change process is given by

$$Y_t = \sigma W_{G_t} + \mu\, G_t, \tag{5}$$

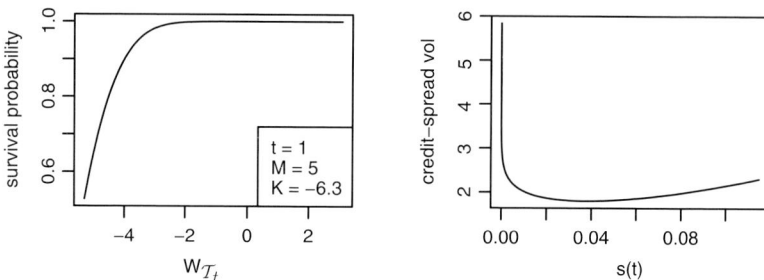

Fig. 4. Survival probabilities corresponding to various asset values W_{T_t}, and credit-spread volatilities corresponding to various spread values s_t

with drift parameter μ and volatility parameter σ_i. We make the following assumption about the time change:

Assumption 2.2 (Continuous time change). *G be a non-decreasing process with continuous paths and starting value $G_0 = g \geq 0$, furthermore G be independent of the Wiener process W.*

Herewith the survival probability can be determined:

$$Q(t,T) = P[\tau > t | \mathcal{F}_t] \qquad (6)$$

$$= 1 - \int_0^\infty \left[\Phi\left(\frac{K - Y_t}{\sigma\sqrt{z}} - \frac{\mu}{\sigma}\sqrt{z}\right) \right.$$

$$\left. + e^{2\frac{\mu}{\sigma^2}(K - Y_t)} \Phi\left(\frac{K - Y_t}{\sigma\sqrt{z}} + \frac{\mu}{\sigma}\sqrt{z}\right) \right] \mathbb{P}\left(G_T - G_t \in dz \mid \mathcal{F}_t^Y\right),$$

and thus, with (3), also the credit spread as a function of t and Y_t, i.e.

$$s(t) = f(t, Y_t). \qquad (7)$$

The formulas are analytical whenever the conditional densitiy of the time-change increment, $\mathbb{P}(G_T - G_t \in dz \mid \mathcal{F}_t^Y)$, is analytical. This is for example the case for the so-called *simple time change*:

$$G_t = g + \hat{\sigma}^2 \int_0^t B_s^2 \, ds , \qquad (8)$$

where B is a Brownian motion independent of W.

2.3 Credit-Spread Dynamics

Applying Itô's rule on the spread formula (3) in the representation indicated by (7) yields the following spread dynamics:

$$ds_t = s_t \left[\mu^s(t, Y_t) \, dt - \sigma^s(t, Y_t) \, dY_t\right] ,$$

with drift $\mu^s := \frac{f_t + \frac{1}{2}f_{yy}}{s}$ and $\sigma^s := -\frac{f_y}{s}$, where f_t, f_y, f_{yy} are the partial derivatives of f w.r.t. t and y.

We want to determine credit-spread dynamics, under the model without drift $Y_t = W_{G_t}$, that follow a SDE (4). Therefor we make an additional assumption:

Assumption 2.3 (Absolute continuity). *There is a stochastic process (g_t) with $\mathbb{E}[\int_0^t g_s^2 \, ds] < \infty$ for all $t \geq 0$ such that*

$$G_t = g_0^2 + \int_0^t g_s^2 \, ds \ ,$$

that is G is absolutely continuous.

Then $Y_t = W_{G_t}$ is equivalent in distribution to the stochastic integral,

$$Y_t \stackrel{\mathcal{L}}{=} Y_0 + \int_0^t g_s \, dW_s \ .$$

This leads to the following credit-spread dynamics:

$$ds_t \stackrel{\mathcal{L}}{=} s_t \left[\mu^s(t, Y_t) \, dt - \sigma^s(t, Y_t) g_t \, dW_t \right] \ ,$$

with credit-spread volatility (for $T = t + M$)

$$\sigma^s \cdot g_t = \frac{2(1-R)}{s_t} g_t \cdot$$

$$\left\{ \frac{\int_0^\infty \frac{1}{\sqrt{z}} \phi\left(\frac{K-Y_t}{\sqrt{z}}\right) \mathbb{P}\left(G_{t+M} - G_t \in dz \mid \mathcal{F}_t^Y\right)}{\alpha_t} \right.$$

$$\left. +2 \frac{\int_0^\infty \Phi\left(\frac{K-Y_t}{\sqrt{z}}\right) \mathbb{P}\left(G_{t+M} - G_t \in dz \mid \mathcal{F}_t^Y\right)}{\alpha_t^2} \beta_t \right\} \ ,$$

and abbreviations

$$\alpha_t = M - 2 \int_t^{t+M} \int_0^\infty \Phi\left(\frac{K-Y_t}{\sqrt{z}}\right) \mathbb{P}\left(G_u - G_t \in dz \mid \mathcal{F}_t^Y\right)$$

$$\beta_t = \int_t^{t+M} \int_0^\infty \frac{1}{\sqrt{z}} \phi\left(\frac{K-Y_t}{\sqrt{z}}\right) \mathbb{P}(G_u - G_t \in dz \mid \mathcal{F}_t^Y) \, du \ .$$

Again credit-spread dynamics are analytical whenever this holds for $\mathbb{P}(G_T - G_t \in dz \mid \mathcal{F}_t^Y)$.

As an example we consider the simple time change (8). It has three degrees of freedom for calibration: in $\frac{K-Y_0}{\sigma \hat{\sigma}}$, $\frac{\mu \hat{\sigma}}{\sigma}$, and g. The first two can be used for calibration to the default-probability cirve and one to influence the credit spread volatility. Figure 5 shows possible calibrations to an average CCC default-probability curve (from the years 1981 to 2002) of Standard & Poor's, and Fig. 6 illustrates the influence of $G_0 = g$ on default-probability and credit-spread curves. Especially $g > 0$ leads to non-zero instantaneous credit spreads.

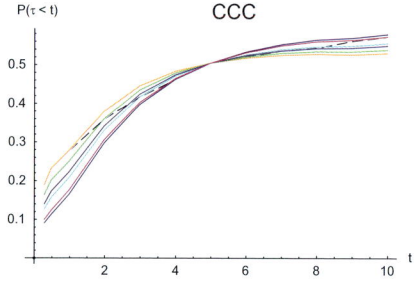

Fig. 5. Possible calibrations to an average CCC default-probability curve of S&P

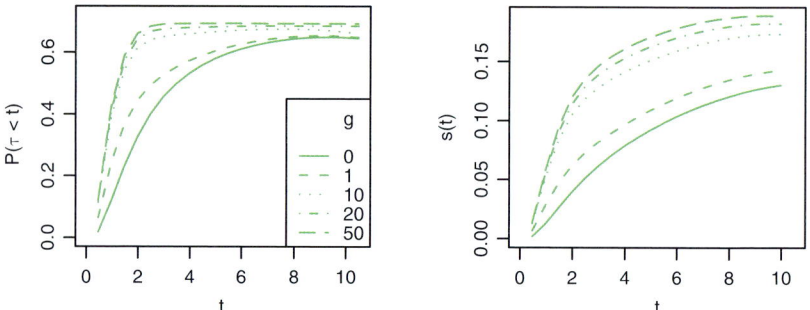

Fig. 6. Influence of the starting value $G_0 = g$ on default-probability and credit-spread curves

2.4 The Multivariate Model

The extension of (5) to a multivariate model is straightforward:

$$Y_t^i = \sigma_i W_{G_t}^i + \mu_i G_t, \quad i = 1, \ldots, m,$$

where each asset-value process has individual drift and volatility parameter, Brownian motion and if wanted also an individual time change G^i. We focus on the two-dimensional model under a joint time change G, given by the simple time change (8), and allow for correaltion between the Brownian motions, $\rho = \mathrm{Corr}(W^1, W^2)$. Figure 7 shows joint survival probabilities (JSP) curves, for fixed parameters $g = 0$, $Y_0 = 0$, $\sigma = 1$, when varying the time-change parameter $\hat{\sigma}$ the threshold level K, and the Brownian correlation ρ. A higher correlation yields higher JSPs, and a higher time change volatility (influenced by $\hat{\sigma}$) yields steeper JSP curves. Note that the time change and the Brownian correlation parameter ρ lead to different dependence structures.

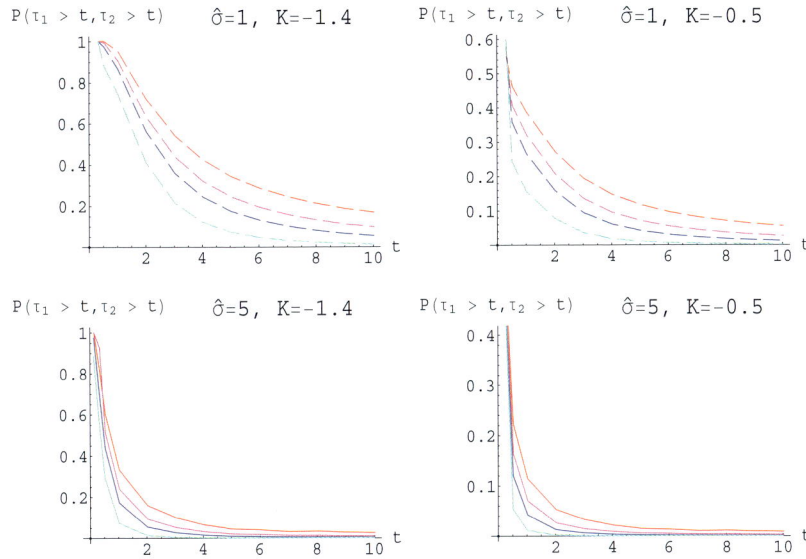

Fig. 7. Joint survival-probability curves for $\rho = 0.9$ (*uppermost curves*), 0.5, 0.1 and -0.5 (*lowest curves*)

3 Asset-Value Processes with Shot-Noise

We define the asset-value process with shot-noise via

$$V_t^i = V_0^i e^{\left(\mu_i - \frac{\sigma_i^2}{2}\right)t + \sigma_i\left(\rho B_t^M + \sqrt{1-\rho^2}B_t^i\right) + \sum_{j=1}^{N_t}(Y_j - sgn(Y_j)\cdot \min(|\alpha\cdot(t-\tau_j)|,|Y_j|))}, \quad (9)$$

where μ_i is a constant drift and σ_i a positiv volatility. B_t^M is the brownian motion of the market and B_t^i the asset specific brownian motion. All assets have the same jump parameters. We assume that the jump amplitude Y_j is i.i.d. $N(\mu_Y, \sigma_Y^2)$. $(N_t)_{t\geq 0}$ is a poisson process with intensity λ. $\tau_j = \sum_{k=1}^{j} E_k$ are the jump times, with E_k i.i.d. $Exp(\lambda)$. $\alpha > 0$ is our shot-noise parameter.

In the classical approach by Merton [17] a firm defaults if

$$V_T^i \leq K_i.$$

In the first passage approach a default appears if

$$\min_{0\leq s\leq T} V_s^i \leq K_i.$$

3.1 Some Parameter Sensitivities for the Joint Default Probability

We consider two firms and simulate the default probabilities in the classical approach. We fix the single firm first passage default probability at 2% and

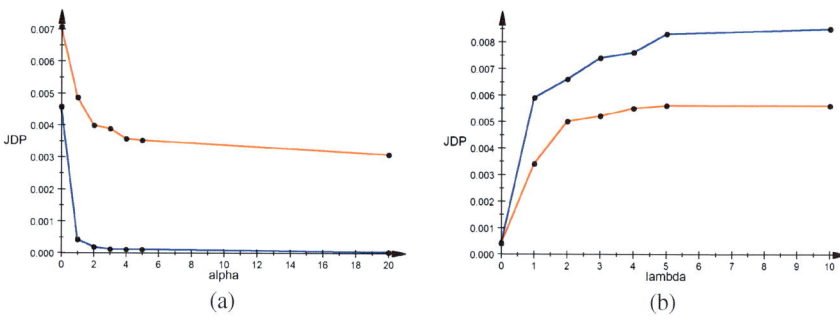

Fig. 8. (**a**) Impact of α on the JDP (*blue:* JDP in T, *red:* first passage JDP). (**b**) Impact of λ on the first passage JDP (*blue:* $\alpha = 0$, *red:* $\alpha = 1$)

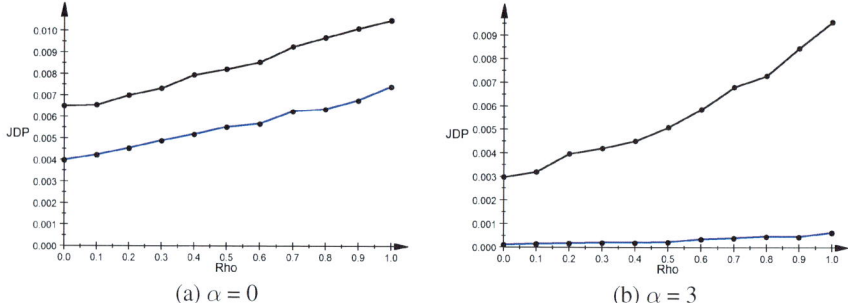

Fig. 9a,b. JDP for different ρ (*blue:* JDP in T, *black:* first passage JDP)

1% (therefor we adapt the default thresholds K_1 and K_2). Figure 8a shows the impact of α on the joint default probability (JDP). The blue curve is the JDP in maturity T, the red curve is the first passage JDP. The bigger the α the smaller the JDP. Figure 8b shows the impact of the intensity λ for different α. The blue curve is a jump diffusion model ($\alpha = 0$) the red curve is a shot-noise model ($\alpha = 1$). The curves are increasing in λ (as expected) and the shot-noise curve is lower than the jump diffusion curve.

Figure 9 shows the effect of different correlations. The JDP is increasing in ρ.

3.2 Valuation Of CDOs

We simulate 125 firms and evaluate the spreads for CDO tranches with attachement points at 0%, 3%, 6%, 9%, 12% and 22%. For definitions, terminology and valuations for CDOs we refer e.g. to [3] or [11]. The single firm default probabilities are fixed, we consider two kinds of firms.

First, we check the influence of ρ. As we can see in Fig. 10a the effect is not the same for all tranches. For the equity tranche a higher correlation implies a higher probability for no defaults, less risk and a lower spread. For the senior

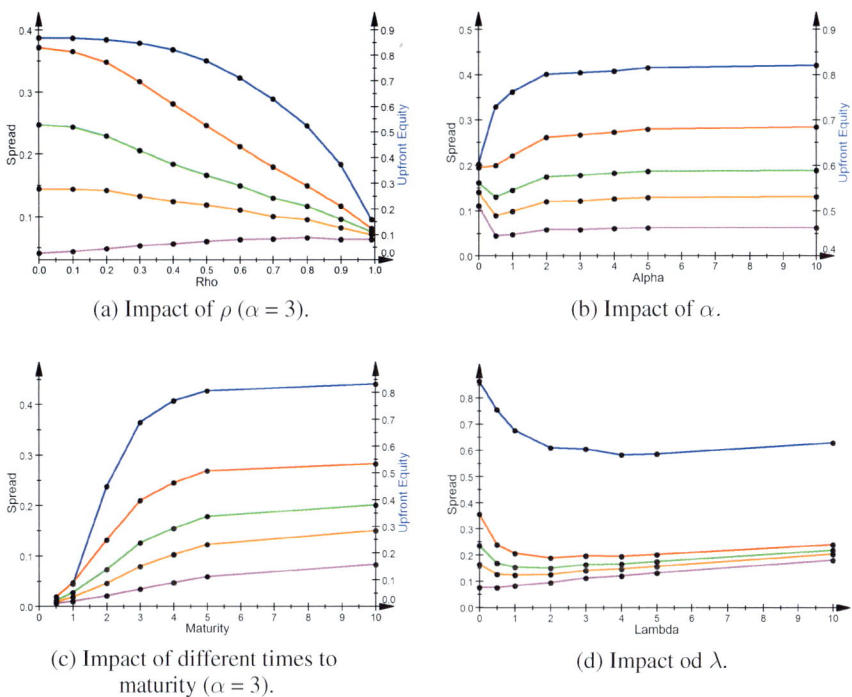

Fig. 10a–d. *blue:* equity, *red:* first mezzanine, *green:* second mezzanine, *orange:* third mezzanine, *pink:* senior

tranche a higher correlation implies a higher probability for many defaults, more risk and a higher spread. For the mezzanine tranches both behaviours are possible.

For increasing α the behaviour of the tranches differs, too (see Fig. 10b). The JDP is decreasing in α. A smaller JDP is like a smaller correlation, it implies a higher equity spread and a lower senior spread.

Figure 10c shows, that the effect of different times to maturity is as expected (long time $\to$ big risk $\to$ high spread). The first tranches react more sensitive to time.

Increasing λ causes an increasing JDP-curve which provokes different changes of the spreads of the different tranches (see Fig. 10d).

Calibration to the Market

We calibrate our model to the market data (5-year-iTraxx-Data, October 2006). Table 1 shows the results for normal distributed jumps and for exponential distributed jumps. The sum of the differences is 0.538% in the normal case and 0.1923% in the exponential case. In both cases we have a good fit, but the exponential case fits a bit better. In the normal case, the first

Table 1. 5-year-CDO

tranche	market price	normal jumps	exponential jumps
0%–3%	13.0%	13.24%	12.97%
3%–6%	59.25 bp	82.09 bp	47.42 bp
6%–9%	13.25 bp	12.03 bp	10.92 bp
9%–12%	5.25 bp	1.96 bp	4.22 bp
12%–22%	2.6 bp	0.16 bp	1.56 bp

mezzanine spread ist overestimated, the higher tranches are underestimated. In the exponential case, the spreads for the tranches higher than equity are underestimated.

Implied Correlation

Comparable to the implied volatility for options in the model of Black and Scholes [2] we can evaluate the implied correlation for CDOs. Generally the market observes different implied correlations for the different tranches, the so called correlation smile (cf. volatility smile). We show that our model produces a correlation smile, too.

We calculate the spreads for different fictive portfolios, assume that the spreads are the real ones and evaluate the implied correlations via the Vasicek-model (see [23]). The first two portfolios are the resulting models of the calibration section, model 1 is the one with normal jumps, model 2 with exponential jumps. Model 3 is a pure diffusion model ($\lambda = 0$). The other portfolios have normal distributed jumps. Model 4 is a jump diffusion model ($\lambda = 4$, $\alpha = 0$)

Table 2. Models with normal and exponential distributed jumps

Tranche	Model 1 CDO	ρ	Model 2 CDO	ρ
0%–3%	13.24%	0.657	12.97%	0.726
3%–6%	82.09 bp	0.361	47.42 bp	0.272
6%–9%	12.03 bp	0.373	10.92 bp	0.348
9%–12%	1.96 bp	0.369	4.22 bp	0.389
12%–22%	0.16 bp	0.373	1.56 bp	0.449

Table 3. Diffusion-, Jump-Diffusion- and Shot-Noise-Model

Tranche	Model 3 CDO	ρ	Model 4 CDO	ρ	Model 5 CDO	ρ
0%–3%	3.05%	0.822	32.28%	0.880	16.18%	0.573
3%–6%	6.93 bp	0.269	860.07 bp	0.814	59.51 bp	0.321
6%–9%	0.089 bp	0.250	646.74 bp	0.813	6.50 bp	0.334
9%–12%	n.a.	n.a.	512.68 bp	0.343	0.81 bp	0.335
12%–22%	n.a.	n.a.	341.83 bp	0.754	0.06 bp	0.342

and model 5 a shot-noise model ($\lambda = 4$, $\alpha = 3$). The results are shown in Tables 2 and 3.

The implied correlation is higher for the equity and the senior tranche. Table 3 shows that adding jumps causes higher implied correlations, a shot-noise effect reduces the implied correlations.

4 Summary

The asset-value model is extended in two directions.

First time changed process, which can be viewed as time dependent or stochastic volatility models, provide a flexible tool. They provide the possibility to calibrate the model to several points in the term structure of defaults or in the CDS curve. Additionally, features of the spread dynamic could be reproduced. But, currently no special features of the spread dynamics are available, to test the model on real data. Also in terms of dependency modeling they deliver an additional degree of freedom going beyond the standard modeling based on correlated Brownian motions. The common time change, or the dependency structure in the multivariate time changes enables the introduction of these additional dependency features.

This is also true for shot-noise process. Here common jumps in the asset value process yield to an additional joint behavior of default going beyond correlations. The sensitivity of the Joint Default Probabilities to the different kind of dependency parameters, like correlation and common jump intensity is analysed in detail. This additional degree of freedom enables also a consistent way to price all tranches in a CDO structure properly. Therefore shot-noise processes are able to reproduce the so-called correlation skew observed in the liquidly traded market of standard single tranche CDOs.

References

1. F. Black and J.C. Cox (1976). Valuing corporate securities: Some effects of bond indenture provisions, *Journal of Finance* 31, 351–367.
2. F. Black and M. Scholes (1973). The pricing of options and corporate liabilities. *Journal of Political Economy* 81, 637–653.
3. C. Bluhm and L. Overbeck (2006). Structured Credit Portfolio Analysis, Baskets & CDOs. *Chapman & Hall/CRC Financial Mathematics Series*; CRC Press.
4. C. Bluhm, L. Overbeck and C. Wagner (2003). An Introduction to Credit Risk Modeling. *Chapman & Hall/CRC Financial Mathematics Series*; 2nd Reprint; CRC Press.
5. S. Bochner (1949). Diffusion equation and stochastic processes. *Journal of Mathematics* 35, 368–370.
6. A. Borodin and P. Salminen (2002). *Handbook of Brownian Motion – Facts and Formulae*, 2nd edition. Birkhäuser.

7. P. Carr and L. Wu (2003). Time-changed Lévy processes and option pricing. *Journal of Financial Economics* 71, 113–141.
8. P.K. Clark (1973). A subordinated stochastic process model with finite variance for speculative prices. *Econometrica* 41, 135–155.
9. J.C. Cox, J. Ingersoll and S. Ross (1985). A theory of the term structure of interest rates. *Econometrica* 53, 385–407.
10. D. Duffie & K. Singleton. Princeton (2003).
11. D. Duffie and N. Garleanu (2001). Risk and valuation of collateralized debt obligations, *Financial Analysts Journal* 57 (1), 41–59.
12. W. Feller (1966). Infinitely divisible distributions and Bessel functions associated with random walks. *SIAM Journal of Applied Mathematics* 14, 864–875.
13. H. Geman, D. Madan and M. Yor (2000). Time changes for Lévy processes. *Mathematical Finance* 11, 79–96.
14. J.M. Harrison (1985). *Brownian Motion and Stochastic Flow Systems*. Wiley.
15. N. Ikeda and S. Watanabe (1981). *Stochastic Differential Equations and Diffusion Processes*. North-Holland and Kodansha.
16. R. A. Jarrow, D. Lando and S. M. Turnbull (1997). A Markov Model for the Term Structure of Credit Risk Spreads. *Review of Financial Studies* 10, 481–523.
17. R.C. Merton (1974). On the pricing of corporate debt: the risk structure of interest rates. *Journal of Finance* 29, 449–470.
18. I. Monroe (1978). Processes that can be embedded in Brownian Motion, *The Annals of Probability* 6(1), 42–56.
19. B. Øksendal (1990). When is a stochastic integral a time change of a diffusion? *Journal of Theoretical Probability* 3(2), 207–226.
20. L. Overbeck and W. Schmidt (2005). Modeling default dependence with threshold models. *Journal of Derivatives* 12(4), 10–19.
21. J.A. Rebholz (1994). *Planar diffusions with applications to Mathematical finance*. Thesis for Ph.D., University of California, Berkeley.
22. W. Schoutens (2003). *Lévy processes in Finance*. Wiley.
23. O. Vasicek (1987). Probability of loss on loan portfolio, *KMV Corporation*, [http://www.moodyskmv.com/research/whitepaper/Probability_of_Loss_on_Loan_Portfolio.pdf].
24. C. Zhou (2001). An analysis of default correlations and multiple defaults. *Review of Financial Studies* 14(2), 555–576.

Economic Capital Modelling and Basel II Compliance in the Banking Industry

Klaus Böcker[1] and Claudia Klüppelberg[2]

[1] Risk Integration, Reporting & Policies – Risk Analytics and Methods – UniCredit Group, Munich Branch, `klaus.boecker@hvb.de`
[2] Center for Mathematical Sciences, Munich University of Technology, 85747 Garching bei München, Germany, `cklu@ma.tum.de`

1 Introduction

> *It would be a mistake to conclude that the only way to succeed in banking is through ever-greater size and diversity. Indeed, better risk management may be the only truly necessary element of success in banking.*
> Alan Greenspan, Speach to the American Bankers Association, 10/5/2004.

Risk is an inevitable part of every financial institution, above all banks and insurance companies. Risks are implicitly accepted when such institutions provide their financial services to customers and explicitly when they take risk positions that offer profitable, above-average returns. There is no unique view on risk and usually it is considered in certain sub-classes such as market risk, credit risk and operational risk, also interest rate risk and liquidity risk. Market risk is associated with trading activities; it is defined as the potential loss arising from adverse price changes of a bank's positions in financial markets and encompasses interest rate, foreign exchange, equity and credit-spread risk. Credit risk is defined as potential losses arising from a customer's default or loss of credit rating. Such risks usually include loan default risk, counterparty risk, issuer risk and country risk. Finally, operational risk is due to losses resulting from inadequate or failed internal processes, human errors, technological breakdowns, or from external events.

Moreover, risk can be distinguished by the negative effects and potential hazards it has on different kinds of stakeholders, e.g risks may seriously threaten the firm's market value (shareholders' perspective), create losses to their lenders (debtholders' perspective), or jeopardizing the stability of the financial system (regulators' perspective). Though the individual interests of these groups may be rather diverse, all parties are interested in an continued existence of the institution. Hence, a bank needs a certain amount of capital relative to its risk as a buffer against future potential losses. This capital base must be sufficient so that also very unlikely losses, measured at a high confidence level, can be absorbed.

The growing awareness of risk inherent in banking industry is partially owing to spectacular crunches like the Saving & Loans crisis in the 1970s or the Japanese banking crisis in the 1990s and led to an increasing demand for banking supervision at the international level, finally resulting in the Basel Committee of Banking Supervision under the auspices of the Bank for International Settlement (BIS) in Basel. The basic idea underlying modern banking regulation is pretty simple, namely that banks should quantify their risks and then are required to keep a certain amount of equity capital (the so-called "capital charge") as a buffer against it. For instance, the minimum capital ratio according to the "Basel Accord" should be 8 % of the so-called "risk-weighted assets", although some regulators set different target levels for individual banks, which may be substantially higher than 8 %.

The first important proposal of the Committee was the "1988 Accord", and even though it was primarily dealing with rather crude methods for assessing credit risk, "Basel I" was a major step towards a common framework for calculating minimum capital standards for international banks. In 1996 the Committee then released an amendment to the Basel I Accord where banks were allowed to build sophisticated internal models for calculating capital charges for their market risk exposures.

The new Basel Accord "Basel II" [BII04], which should be fully implemented by year-end 2007, describes a more comprehensive risk measure and minimum standard for capital adequacy and is structured in three Pillars. Pillar I imposes new methodologies of calculating regulatory capital, thereby mainly focusing on credit risk and operational risk. For the latter, banks can then use – similar as it is already the case for market risk – their own internal modelling techniques (commonly referred to as advanced measurement approaches (AMA)) to determine capital charges, and we consider this subject again in Sect. 2.

Pillar II then introduces the so-called Internal Capital Adequacy Assessment Process (ICAAP) and contains guidance to supervisors on how they should review an institution's ICAAP. Besides the treatment of so-called "other" risks that are not covered under Pillar I such as interest rate risk or credit concentration risk, it deals with an institution's overall risk exposure. According to the Committee of European Banking Supervisors [CEBS], banks should calculate an "overall capital number" as an integral part of their ICAAP. This single-number metric should encompass all risks related to different businesses and risk types. Above all, regulators want to understand the extent to which the institution has introduced diversification and correlation effects when aggregating different risk types. A particularly important example of this issue is considered in Sect. 3 where the inter-risk correlation between credit and market risk is investigated.

A milestone in mathematical finance was the idea of dynamic replication introduced in 1973 by Fischer Black, Myron Scholes and Robert C.

Merton [BS73], revolutionizing the theory of pricing and hedging of financial derivatives completely. Then, since the introduction of internal market risk models in 1996, quantitative risk management has become an interesting and fruitful research area for mathematicians and statisticians; cf. Föllmer & Klüppelberg [FK02].

Although our project focussed at the beginning on credit risk problems alone with results documented in Hillebrand [H06], our industry partner was interested in further collaboration in operational risk and aggregation of different risk types, more precisely in aggregation of market and credit risk. As these are new areas with many interesting open problems, we henceforth concentrate on these cutting-edge topics.

Our paper is organised as follows. In Sect. 2 we suggest a novel method for calculating operational risk at a high confidence level by using the new concept of Lévy copulas. Our results can be used as an approximation for operational Value-at-Risk and deliver important insights into extremal dependence modelling in general. In Sect. 3 we then investigate the interaction between a credit portfolio and another risk type, which can be thought of as market risk. Combining Merton-like factor models for credit risk with linear factor models for market risk, we analytically calculate their inter-risk correlation and show how inter-risk correlation bounds can be derived. For known inter-risk correlation the total aggregated credit and market risk can be approximated (cf. (20) below). We conclude with a discussion of possible overlapping risk and indicate the assignment problem of a simple financial instrument to one specific risk like operational, credit or market risk.

2 Analytical Approximation of Operational Risk

One of the determinants of Basel II is Operational Risk, defined as losses resulting from inadequate or failed internal processes, human errors, technological breakdowns, or from external events. Risk in all categories of Basel II is defined as Value-at-Risk (VAR) of the total loss (per year) at a certain confidence level κ near 1. If we denote by S this total loss, then VAR(κ) is the capital amount such that total losses remain below VAR with at least probability κ. This is a rather simplistic risk measure; it only becomes nontrivial because the total loss S is not a straightforward quantity. Below we concentrate on the advanced measurement approach (AMA) and indicate the problems involved for obtaining VAR(κ). It is important to note that the Basel Committee specifies as quantitative standards a confidence level of $\kappa = 0.999$ and only models, which capture potentially severe tail loss events.

2.1 The Loss Distribution Approach

A required feature of AMA for measuring operational risk in the context of Pillar II is that it allows for explicit correlations between different operational risks, usually classified according to an event type/business line matrix consisting of eight business lines and seven loss event types. The core problem here is the multivariate modelling and how the dependence structure between different matrix cells affects a bank's total operational risk. The prototypical loss distribution approach (LDA) assumes that, for each cell $i = 1, \ldots, d$, the cumulated operational loss $S_i(t)$ up to time t is described by an aggregate loss process

$$S_i(t) = \sum_{k=1}^{N_i(t)} X_k^i, \quad t \geq 0, \tag{1}$$

where for each i the sequence $(X_k^i)_{k \in \mathbb{N}}$ are independent and identically distributed (iid) positive random variables with distribution function F_i describing the magnitude of each loss event (loss severity), and $(N_i(t))_{t \geq 0}$ counts the number of losses in the time interval $[0, t]$ (called frequency), independent of $(X_k^i)_{k \in \mathbb{N}}$. For regulatory capital and economic capital purposes, the time horizon is usually fixed to $t = 1$ year. The bank's total operational risk is then given as

$$S^+(t) := S_1(t) + S_2(t) + \cdots + S_d(t), \quad t \geq 0. \tag{2}$$

The present literature suggests to model dependence between different operational risk cells by means of different concepts, which basically split into models for frequency dependence on the one hand and for severity dependence on the other hand.

Here we suggest a model based on the new concept of Lévy copulas (see e.g. Cont & Tankov [CT04]), which models dependence in frequency and severity simultaneously, yielding a model with comparably few parameters. Moreover, our model has the same advantage as a distributional copula: the dependence structure between different cells can be separated from the marginal processes S_i for $i = 1, \ldots, d$. This approach allows for closed-form approximations for operational VAR (OpVAR).

2.2 Dependent Operational Risks and Lévy Copulas

In accordance with a recent survey of the Basel Committee on Banking Supervision about AMA practices at financial services firms, we assume that the loss frequency processes N_i in (1) follows a homogeneous Poisson process with rate $\lambda_i > 0$. Then the aggregate loss (1) constitutes a compound Poisson process and is therefore a Lévy process.

A key element in the theory of Lévy processes is the notion of the so-called Lévy measure. A Lévy measure controls the jump behaviour of a Lévy process and, therefore, has an intuitive interpretation, in particular in the context of operational risk. The Lévy measure of a single operational risk cell measures the expected number of losses per unit time with a loss amount in a prespecified interval. For our compound Poisson model, the Lévy measure Π_i of the cell process S_i is completely determined by the frequency parameter $\lambda_i > 0$ and the distribution function F_i of the cell's severity: $\Pi_i([0, x)) := \lambda_i P(X^i \le x) = \lambda_i F_i(x)$ for $x \in [0, \infty)$. The corresponding one-dimensional tail integral is defined as

$$\overline{\Pi}_i(x) := \Pi_i([x, \infty)) = \lambda_i P(X^i > x) = \lambda_i \overline{F}_i(x). \tag{3}$$

Our goal is modelling multivariate operational risk. Hence, the question is how different one-dimensional compound Poisson processes $S_i(\cdot) = \sum_{k=1}^{N_i(\cdot)} X_k^i$ can be used to construct a d-dimensional compound Poisson process $S = (S_1, S_2, \ldots, S_d)$ with in general dependent components. It is worthwhile to recall the similar situation in the case of the more restrictive setting of static random variables. It is well-known that the dependence structure of a random vector can be disentangled from its marginals by introducing a distributional copula. Similarly, a multivariate tail integral

$$\overline{\Pi}(x_1, \ldots, x_d) = \Pi([x_1, \infty) \times \cdots \times [x_d, \infty)), \quad x \in [0, \infty]^d, \tag{4}$$

can be constructed from the marginal tail integrals (3) by means of a Lévy copula. This representation is the content of Sklar's theorem for Lévy processes with positive jumps, which basically says that every multivariate tail integral $\overline{\Pi}$ can be decomposed into its marginal tail integrals and a Lévy copula $\widehat{C}$ according to

$$\overline{\Pi}(x_1, \ldots, x_d) = \widehat{C}(\overline{\Pi}_1(x_1), \ldots, \overline{\Pi}_d(x_d)), \quad x \in [0, \infty]^d. \tag{5}$$

For a precise formulation of this Theorem we refer to Cont & Tankov [CT04], Theorem 5.6. Now we can define the following prototypical LDA model.

Definition 1 (Multivariate Compound Poisson Model).
(1) All aggregate loss processes S_i for $i = 1, \ldots, d$ are compound Poisson processes with tail integral $\overline{\Pi}_i(\cdot) = \lambda_i F_i(\cdot)$.
(2) The dependence between different cells is modelled by a Lévy copula $\widehat{C} : [0, \infty)^d \to [0, \infty)$, i.e. the tail integral of the d-dimensional compound Poisson process $S = (S_1, \ldots, S_d)$ is defined by

$$\overline{\Pi}(x_1, \ldots, x_d) = \widehat{C}(\overline{\Pi}_1(x_1), \ldots, \overline{\Pi}_d(x_d)).$$

2.3 The Bivariate Clayton Model

A bivariate model is particularly useful to illustrate how dependence modelling via Lévy copulas works. Therefore, we now focus on two operational risk cells as in Definition 1(1). The dependence structure is modelled by a Clayton Lévy copula, which is similar to the well-known Clayton copula for distri-

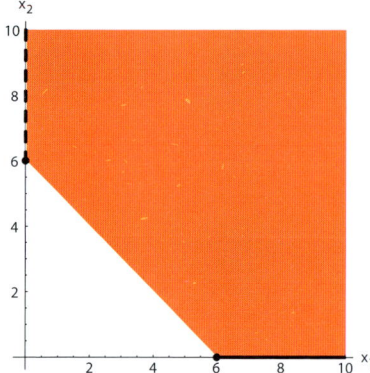

Fig. 1. Decomposition of the domain of the tail integral $\overline{\Pi}^+(z)$ for $z = 6$ into a simultaneous loss part $\overline{\Pi}_{\|}^+(z)$ (*orange area*) and independent parts $\overline{\Pi}_{\perp 1}(z)$ (*solid black line*) and $\overline{\Pi}_{\perp 2}(z)$ (*dashed black line*)

bution functions and parameterized by $\vartheta > 0$ (see Cont & Tankov [CT04], Example 5.5):

$$\widehat{C}_\vartheta(u,v) = (u^{-\vartheta} + v^{-\vartheta})^{-1/\vartheta}, \quad u,v \geq 0.$$

This copula covers the whole range of positive dependence. For $\vartheta \to 0$ we obtain independence and then, as we will see below, losses in different cells never occur at the same time. For $\vartheta \to \infty$ we get the complete positive dependence Lévy copula given by $\widehat{C}_\|(u,v) = \min(u,v)$. We now decompose the two cells' aggregate loss processes into different components (where the time parameter t is dropped for simplicity),

$$\begin{aligned} S_1 &= S_{\perp 1} + S_{\|1} = \sum_{k=1}^{N_{\perp 1}} X^1_{\perp k} + \sum_{l=1}^{N_\|} X^1_{\|l}, \\ S_2 &= S_{\perp 2} + S_{\|2} = \sum_{m=1}^{N_{\perp 2}} X^2_{\perp m} + \sum_{l=1}^{N_\|} X^2_{\|l}, \end{aligned} \quad (6)$$

where $S_{\|1}$ and $S_{\|2}$ describe the aggregate losses of cell 1 and 2 that is generated by "common shocks", and $S_{\perp 1}$ and $S_{\perp 2}$ describe aggregate losses of one cell only. Note that apart from $S_{\|1}$ and $S_{\|2}$, all compound Poisson processes on the right-hand side of (6) are mutually independent. The frequency of simultaneous losses is given by

$$\widehat{C}_\vartheta(\lambda_1, \lambda_2) = \lim_{x \downarrow 0} \overline{\Pi}_{\|2}(x) = \lim_{x \downarrow 0} \overline{\Pi}_{\|1}(x) = \left(\lambda_1^{-\vartheta} + \lambda_2^{-\vartheta}\right)^{-1/\vartheta} =: \lambda_\|,$$

which shows that the number of simultaneous loss events is controlled by the Lévy copula. Obviously, $0 \leq \lambda_\| \leq \min(\lambda_1, \lambda_2)$, where the left and right

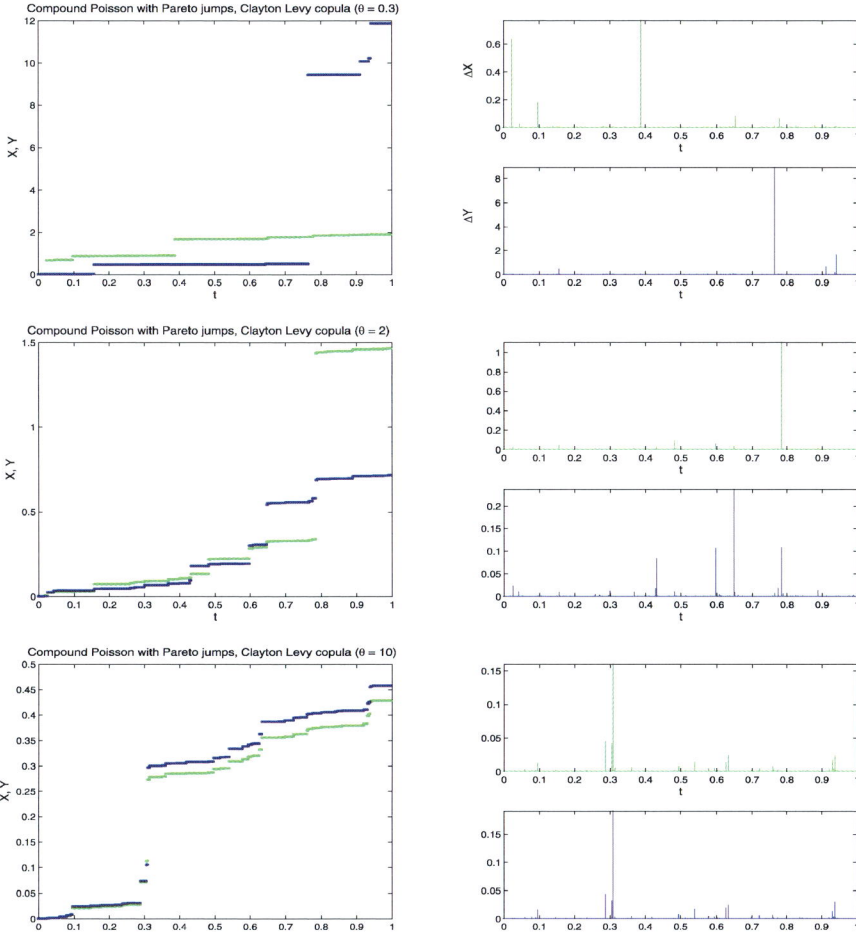

Fig. 2. Two-dimensional LDA Clayton Pareto model (with Pareto tail index $\alpha = 1/2$) for different parameter values. *Left column:* compound processes, *right column:* frequencies and severities. *Upper row:* $\delta = 0.3$ (low dependence), *middle row:* $\delta = 2$ (medium dependence), *lower row:* $\delta = 10$ (high dependence)

bounds refer to $\vartheta \to 0$ and $\vartheta \to \infty$, respectively. Consequently, in the case of independence, losses never happen at the same instant of time.

Also the severity distributions of $X_\parallel^1$ and $X_\parallel^2$ as well as their dependence structure are determined by the Lévy copula. To see this, define the joint survival function as

$$\overline{F}_\parallel(x_1, x_2) := P\left(X_\parallel^1 > x_1, X_\parallel^2 > x_2\right) = \frac{1}{\lambda_\parallel} \widehat{C}_\vartheta \left(\overline{\Pi}_1(x_1), \overline{\Pi}_2(x_2)\right) \qquad (7)$$

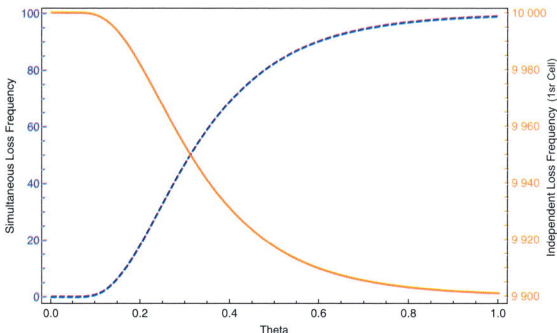

Fig. 3. Visualisation of the cells' loss frequencies controlled by the Clayton Lévy copula for $\lambda_1 = 10\,000$ and $\lambda_2 = 100$. *Left blue axis:* frequency $\lambda_\|$ of the simultaneous loss processes $S_{\|1}$ and $S_{\|2}$ as a function of the Lévy Clayton copula parameter ϑ (*blue, dashed line*). *Right orange axis:* frequency $\lambda_{\perp 1}$ of the independent loss process $S_{\perp 1}$ of the first cell as a function of the Lévy Clayton copula parameter ϑ (*orange, solid line*)

with marginals

$$\overline{F}_{\|1}(x_1) = \lim_{x_2 \downarrow 0} \overline{F}_\|(x_1, x_2) = \frac{1}{\lambda_\|} \widehat{C}_\vartheta(\overline{\Pi}_1(x_1), \lambda_2) \tag{8}$$

$$\overline{F}_{\|2}(x_2) = \lim_{x_1 \downarrow 0} \overline{F}_\|(x_1, x_2) = \frac{1}{\lambda_\|} \widehat{C}_\vartheta(\lambda_1, \overline{\Pi}_2(x_2)) . \tag{9}$$

In particular, it follows that $F_{\|1}$ and $F_{\|2}$ are different from F_1 and F_2, respectively. To explicitly extract the dependence structure between the severities of simultaneous losses $X_\|^1$ and $X_\|^2$ we use the concept of a distributional survival copula. Using (7)–(9) we see that the survival copula S_ϑ for the tail severity distributions $\overline{F}_{\|1}(\cdot)$ and $\overline{F}_{\|2}(\cdot)$ is the well-known distributional Clayton copula; i.e.

$$S_\vartheta(u,v) = (u^{-\vartheta} + v^{-\vartheta} - 1)^{-1/\vartheta}, \quad 0 \leq u, v \leq 1 .$$

For the tail integrals of the independent loss processes $S_{\perp 1}$ and $S_{\perp 2}$. we obtain for $x_1, x_2 \geq 0$

$$\overline{\Pi}_{\perp 1}(x_1) = \overline{\Pi}_1(x_1) - \overline{\Pi}_{\|1}(x_1) = \overline{\Pi}_1(x_1) - \widehat{C}_\vartheta(\overline{\Pi}_1(x_1), \lambda_2) ,$$

$$\overline{\Pi}_{\perp 2}(x_2) = \overline{\Pi}_2(x_2) - \overline{\Pi}_{\|2}(x_2) = \overline{\Pi}_2(x_2) - \widehat{C}_\vartheta(\lambda_1, \overline{\Pi}_2(x_2)) ,$$

so that $\lambda_{\perp 1} = \lambda_1 - \lambda_\|$, $\lambda_{\perp 2} = \lambda_2 - \lambda_\|$.

2.4 Analytical Approximations for Operational VAR

In this section we turn to the quantification of total operational loss encompassing all operational risk cells and, therefore, we focus on the total aggregate loss process S^+ defined in (2). Our goal is to provide some general insight to multivariate operational risk and to find out, how different dependence structures (modelled by Lévy copulas) affect OpVAR, which is the standard metric in operational risk measurement. We need some notation to define it properly.

The tail integral associated with S^+ is given by

$$\overline{\Pi}^+(z) = \Pi\left(\left\{(x_1,\ldots,x_d) \in [0,\infty)^d : \sum_{i=1}^d x_i \geq z\right\}\right), \quad z \geq 0. \quad (10)$$

For $d = 2$ we can write

$$\overline{\Pi}^+(z) = \overline{\Pi}_{\perp 1}(z) + \overline{\Pi}_{\perp 2}(z) + \overline{\Pi}_{\parallel}^+(z), \quad z \geq 0, \quad (11)$$

where $\overline{\Pi}_{\perp 1}(\cdot)$ and $\overline{\Pi}_{\perp 2}(\cdot)$ are the independent jump parts and

$$\overline{\Pi}_{\parallel}^+(z) = \Pi\left(\{(x_1, x_2) \in (0,\infty)^2 : x_1 + x_2 \geq z\}\right), \quad z \geq 0,$$

describes the dependent part due to simultaneous loss events.

Since for every compound Poisson process with intensity $\lambda > 0$ and positive jumps with distribution function F, the tail integral is given by $\overline{\Pi}(\cdot) = \lambda \overline{F}(\cdot)$, it follows from (11) that the total aggregate loss process S^+ is again compound Poisson with frequency parameter and severity distribution

$$\lambda^+ = \lim_{z \downarrow 0} \overline{\Pi}^+(z) \quad \text{and} \quad F^+(z) = 1 - \overline{F}^+(z) = 1 - \frac{\overline{\Pi}^+(z)}{\lambda^+}, \quad z \geq 0. \quad (12)$$

This result proves now useful to determine a bank's total operational risk consisting of several cells. Before doing that, recall the definition of OpVAR for a single operational risk cell (henceforth called stand-alone OpVAR.) For each cell, stand-alone OpVAR at confidence level $\kappa \in (0,1)$ and time horizon t is the κ-quantile of the aggregate loss distribution, i.e.

$$\text{VAR}_t(\kappa) = G_t^{\leftarrow}(\kappa) = \inf\{x \in \mathbb{R} : P(S(t) \leq x) \geq \kappa\}. \quad (13)$$

In Böcker & Klüppelberg [BK05, BK06, BK07a, BK07b] it was shown that OpVAR at high confidence level can be approximated by a closed-form expression, if the loss severity is subexponential, i.e. heavy-tailed. As this is common believe we consider in the sequel this approximation, which can be written as

$$\text{VAR}_t(\kappa) \sim F^{\leftarrow}\left(1 - \frac{1-\kappa}{EN(t)}\right), \quad \kappa \uparrow 1, \quad (14)$$

where the symbol "$\sim$" means that the ratio of left and right hand side converges to 1. Moreover, $EN(t)$ is the cell's expected number of losses in the time interval $[0,t]$. Important examples for subexponential distributions are lognormal, Weibull, and Pareto. We want to emphasize already here that such first order asymptotics work extremely well for heavy-tailed Pareto-like tails, which are realistic in operational risk. Since the loss frequencies only enter as their mean $EN(t)$, any sophisticated modelling of the loss number process is superfluous, see Böcker & Klüppelberg [BK06] for more details. Instead all effort should be directed into a more accurate modelling of the loss severity distribution.

Here, we extend the idea of an asymptotic OpVAR approximation to the multivariate problem. In doing so, we exploit the fact that S^+ is a compound Poisson process with parameters as in (12). In particular, if F^+ is subexponential, we can apply (14) to estimate total OpVAR. Consequently, if we are able to specify the asymptotic behaviour of $\overline{F}^+(x)$ as $x \to \infty$ we have automatically an approximation of $\text{VAR}_t(\kappa)$ as $\kappa \uparrow 1$.

To make more precise statements about OpVAR, we focus our analysis on Pareto distributed severities with distribution function

$$\overline{F}(x) = \left(1 + \frac{x}{\theta}\right)^{-\alpha}, \quad x > 0,$$

with shape parameters $\theta > 0$ and tail parameter $\alpha > 0$. Pareto's law is the prototypical parametric example for a heavy-tailed distribution and suitable for operational risk modelling. As a simple consequence of (14), in the case of a compound Poisson model with Pareto severities (Pareto–Poisson model) analytic OpVAR is given by

$$\text{VAR}_t(\kappa) \sim \theta \left[\left(\frac{\lambda t}{1-\kappa}\right)^{1/\alpha} - 1\right] \sim \theta \left(\frac{\lambda t}{1-\kappa}\right)^{1/\alpha}, \quad \kappa \uparrow 1. \qquad (15)$$

To demonstrate the kind of results we obtain by such approximation methods we consider a Pareto–Poisson model, where the severity distributions F_i of the first (say) $b \leq d$ cells are tail equivalent with tail parameter $\alpha > 0$ and dominant to all other cells, i.e.

$$\lim_{x \to \infty} \frac{\overline{F}_i(x)}{\overline{F}_1(x)} = \left(\frac{\theta_i}{\theta_1}\right)^\alpha, \ i = 1, \ldots, b, \quad \lim_{x \to \infty} \frac{\overline{F}_i(x)}{\overline{F}_1(x)} = 0, \ i = b+1, \ldots, d. \qquad (16)$$

In the important cases of complete positive dependence and independence, closed-form results can be found and may serve as extreme cases concerning the dependence structure of the model.

Theorem 1. *Consider a compound Poisson model with cell processes $S_1, \ldots, S_d$ with Pareto distributed severities satisfying (16). Let $\text{VAR}_t^i(\cdot)$ be the stand-alone OpVAR of cell i.*

(1) *If all cells are completely dependent with the same frequency λ for all cells, then S^+ is compound Poisson with parameters*

$$\lambda^+ = \lambda \quad \text{and} \quad \overline{F}^+(z) \sim \left(\sum_{i=1}^b \theta_i\right)^\alpha z^{-\alpha}, \quad z \to \infty,$$

and total OpVAR *is asymptotically given by*

$$\text{VAR}^+_{\|t}(\kappa) \sim \sum_{i=1}^{b} \text{VAR}^i_t(\kappa), \qquad \kappa \uparrow 1. \tag{17}$$

(2) *If all cells are independent, then* S^+ *is compound Poisson with parameters*

$$\lambda^+ = \lambda_1 + \cdots + \lambda_d \quad \text{and} \quad \overline{F}^+(z) \sim \frac{1}{\lambda^+} \sum_{i=1}^{b} \left(\frac{\theta_i}{z}\right)^\alpha \lambda_i, \quad z \to \infty, \tag{18}$$

and total OpVAR is asymptotically given by

$$\text{VAR}^+_{\perp t}(\kappa) \sim \left[\sum_{i=1}^{b} \left(\text{VAR}^i_t(\kappa)\right)^\alpha\right]^{1/\alpha}, \qquad \kappa \uparrow 1. \tag{19}$$

Theorem 1 states that for the completely dependent Pareto-Poisson model, total asymptotic OpVAR is simply the sum of the dominating cell's asymptotic stand-alone OpVARs. Recall that this is similar to the new proposals of Basel II, where the standard procedure for calculating capital charges for operational risk is just the simple-sum VAR. To put it another way, regulators implicitly assume complete dependence between different cells, meaning that losses within different business lines or risk categories always happen at the same instants of time.

Very often, the simple-sum OpVAR (17) is considered to be the worst case scenario and, hence, as an upper bound for total OpVAR in general, which in the heavy-tailed case can be grossly misleading. To see this, assume the same frequency λ in all cells also for the independent model, and denote by $\text{VAR}^+_{\|}(\kappa)$ and $\text{VAR}^+_{\perp}(\kappa)$ completely dependent and independent total OpVAR, respectively.

Then, as explained in detail in [BK06] for heavy-tailed severity data with $\overline{F}_i(x_i) \sim (x_i/\theta_i)^{-\alpha}$ as $x_i \to \infty$, subadditivity of OpVAR is violated because the sum of stand-alone OpVARs is smaller than independent total OpVAR. The following table, taken from [RK99], illustrates this.

More general dependence structures can be investigated within the framework of multivariate regular variation. For homogeneous models, in particular for the Clayton Lévy copula, precise results have been derived in Klüppelberg and Resnick [KR07] and applied to find OpVAR approximations in Böcker and Klüppelberg [BK06].

3 Inter-Risk Correlation of Market and Credit Risk

3.1 The Necessity for Risk Aggregation

A core element of modern risk control is the calculation of an aggregated group-wide risk figure, which is used to evaluate the capital adequacy of a financial institution. Until now no standard procedure for risk aggregation has

Table 1. Comparison of total OpVaR for two operational risk cells (each with stand alone VaR of 100 million) in the case of complete dependence ($\|$) and independence ($\perp$) for different values of α

α	$\text{VAR}^+_{\|}$	$\text{VAR}^+_{\perp}$
1.2		178.2
1.1		187.8
1.0	200.0	200.0
0.9		216.0
0.8		237.8
0.7		269.2

emerged, but a widespread approach in the banking industry is "aggregation across risk types", where in a first step marginal, institution-wide loss distributions for all relevant risk types are calculated. These marginal risk figures describe the group-wide, pre-aggregated risk of a given risk type encompassing different legal entities, divisions, regions etc. Then, in a second step, the dependence structure between these pre-aggregated risk-type figures is modelled and finally the total risk can be calculated.

The easiest way of aggregating risks is simply to add up all pre-aggregated risk-type figures (cf. Theorem 1(1) in the case of different operational risk figures). Problems with this procedure have been indicated after Theorem 1 and made transparent in Table 1). Consequently, this yields only a very rough estimate of the bank-wide total risk. Furthermore, banks usually try to reduce overall risk by accounting for diversification between different risk types – measured by correlation – because this allows them to reduce expensive equity capital. Hence, advanced approaches for risk aggregation begin with an analysis of the dependence structure between different risk types.

Important measures of dependence in the context of risk-type aggregation are correlation (which models linear dependence); possible non-linear dependence is often modeled by means of copulas. In practise, a widespread approach for aggregating different risk types is the so-called *square-root-formula approach* or *variance-covariance approach*. Though mathematically justified only in the case of elliptically distributed risk types (with the multivariate normal or t distributions as prominent examples), this approach is very often used as a first approximation because total aggregated capital can then be calculated explicitly without expensive simulations. If $X^T = (X_1, \ldots, X_m)$ is the vector of pre-aggregated risk figures (e.g. economic capital X_i for risk-types $i = 1, \ldots, m$), and R the inter-risk correlation matrix, then total aggregated risk X_{tot} is for elliptically distributed X given by

$$X_{tot} = \sqrt{X^T R X}\,. \qquad (20)$$

Hence, a typical problem of risk aggregation is the estimation of the inter-risk correlation matrix R.

In the sequel we concentrate on the two-dimensional problem consisting of credit risk together with another risk type, which henceforth is referred to as market risk. Credit risk can be more than six times as large as the classical market risk associated with trading activities, and it is clear that in this case total risk (20) is mainly dominated by credit risk alone and, in particular, it is only little affected by inter-risk correlation. However, the exposures of other market-like risk types like financial investment risk, real estate risk, or business risk, which are often measured by banks in the context of economic capital and Basel II compliance, are comparable in volume to overall credit risk, and the question regarding correct modelling of inter-risk correlation again becomes important.

We combine a Merton-like factor model for credit risk with a linear factor model for market risk. Both models are driven by a set of (macroeconomic) factors $Y = (Y_1, \ldots, Y_K)$ where the factor weights are allowed to be zero so that a risk type may only depend on a subset of Y. This section is based on [BH07].

3.2 Modelling Credit and Market Risk

Normal Factor Model for Credit Risk

To describe credit portfolio loss, we choose a classical structural model as it can be found e.g. in Bluhm, Overbeck & Wagner [BOW02]. Within these models, a borrower's credit quality is driven by a so-called "ability-to-pay" process. Consider a portfolio of n loans. Then, default of an individual obligor $i \in \{1, \ldots, n\}$ is described by a Bernoulli random variable L_i with $\mathbb{P}(L_i = 1) = p_i = 1 - \mathbb{P}(L_i = 0)$ where p_i is the obligor's probability of default within time period $[0, T]$ for fixed $T > 0$. Following Merton's idea, counterparty i defaults if its asset value log-return A_i falls below some threshold D_i, sometimes referred to as default point, i.e.

$$L_i = \mathbb{1}_{\{A_i < D_i\}}, \qquad i = 1, \ldots, n. \tag{21}$$

If we denote the exposure at default (perhaps enriched by discounting factors and/or net of recovery rates) of an individual obligor by e_i, portfolio loss is given by

$$L^{(n)} = \sum_{i=1}^{n} e_i \, L_i. \tag{22}$$

In a factor-model approach, the asset values A_i are linked to a set of macroeconomic factors $Y_1, \ldots, Y_K$, which are assumed to be normally distributed and the vector $(Y_1, \ldots, Y_K)$ has been transformed to standard normal.

Definition 2 (Normal factor model for credit risk). *Let $Y = (Y_1, \ldots, Y_K)$ be a random vector of (macroeconomic) factors with multivariate standard* normal distribution. We assume that each of the asset value log-returns A_i for $i = 1, \ldots, n$ linearly depends on Y as well as on a standard normally distributed idiosyncratic factor ε_i (which models the performance of firm i) independent of Y, i.e.

$$A_i = \sum_{k=1}^{K} \beta_{ik} Y_k + \sqrt{1 - \sum_{k=1}^{K} \beta_{ik}^2}\, \varepsilon_i\,, \qquad i = 1, \ldots, n\,, \tag{23}$$

with factor loadings β_{ik} satisfying $R_i^2 := \sum_{k=1}^{K} \beta_{ik}^2 \in [0, 1]$, which is that part of the variance of A_i which can be explained by the systematic factor vector Y. Then $L^{(n)}$ as given in (22) is called normal factor model for credit risk.

Equation (23) implies that log-returns $A_1, \ldots, A_n$ are standard normally distributed, but dependent with correlations

$$\rho_{ij} := \operatorname{corr}(A_i, A_j) = \sum_{k=1}^{K} \beta_{ik} \beta_{jk}\,, \qquad i, j = 1, \ldots, n\,. \tag{24}$$

Owing to the normal factor structure of the model, the default point D_i of every obligor is related to its default probability p_i by

$$D_i = \Phi^{-1}(p_i)\,, \qquad i = 1, \ldots, n\,, \tag{25}$$

where Φ is the standard normal distribution function. Moreover, the joint default probability of two obligors is given by

$$p_{ij} := \mathbb{P}(A_i \leq D_i, A_j \leq D_j) = \begin{cases} \Phi_{\rho_{ij}}(D_i, D_j)\,, & i \neq j\,, \\ p_i\,, & i = j\,, \end{cases} \tag{26}$$

where $\Phi_{\rho_{ij}}$ denotes the bivariate normal distribution function with standardized marginals and correlation ρ_{ij} given by (24). Finally, the default correlation between two different obligors is given by

$$\operatorname{corr}(L_i, L_j) = \frac{p_{ij} - p_i p_j}{\sqrt{p_i(1 - p_i)\, p_j(1 - p_j)}}\,, \qquad i, j = 1, \ldots, n\,. \tag{27}$$

Factor Models for Market Risk

We assume that market risk is already pre-aggregated and can be approximated by a one-dimensional random variable Z, representing the aggregated profit and loss (P/L) distribution due to changes in some market variables, such as interest rates or equity prices.

As in the credit risk model of Definition 2, we explain fluctuations of the P/L random variable Z by means of (macroeconomic) factors $Y = (Y_1, \ldots, Y_K)$. We use the same macroeconomic factors for credit and market risk, where independence of risk from such a factor is indicated by a loading factor 0.

As we want to add market and credit risk quantities, we use the convention that losses correspond to positive values of Z. One can think of Y as a vector describing the healthiness of the economy in the sense that positive (negative) values of the Y_k correspond to a good (bad) economy, implying a decreasing (increasing) market risk.

Definition 3 (Normal factor model for market risk). Let $Y = (Y_1, \ldots, Y_K)$ be a random vector of (macroeconomic) factors with multivariate standard normal distribution. Then, the normal factor model for the pre-aggregated market risk P/L is given by

$$Z = -\sigma \left(\sum_{k=1}^{K} \gamma_k Y_k + \sqrt{1 - \sum_{k=1}^{K} \gamma_k^2} \, \eta \right) \qquad (28)$$

with factor loadings satisfying $\sum_{k=1}^{K} \gamma_k^2 \in [0,1]$, which is that part of the variance of Z which can be explained by the systematic factor Y. Furthermore, η is a standard normally distributed idiosyncratic factor, independent of Y. Finally, σ is the standard deviation of Z.

Definition 4 (Normal factor model for credit and market risk). Let $Y = (Y_1, \ldots, Y_K)$ be a random vector of (macroeconomic) factors with multivariate standard normal distribution. Let the credit portfolio loss $L^{(n)}$ be given by (22) and the asset value log-returns A_i for $i = 1, \ldots, n$ are modeled by the normal factor model (23). Let Z be the pre-aggregated market risk P/L modeled by the normal factor model (28). When the credit model's idiosyncratic factors ε_i for $i = 1, \ldots, n$ are independent of η, then we call $(L^{(n)}, Z)$ the normal factor model for credit and market risk.

In order to account for possible heavy tails for Z we introduce the following global shock approach.

Definition 5 (Shock model for market risk). Let $Y = (Y_1, \ldots, Y_K)$ be a random vector of (macroeconomic) factors with multivariate standard normal distribution and let η be the standard normally distributed idiosyncratic factor, independent of Y. Further, let W be a positive random variable, independent of Y and η. Then the shock model for the pre-aggregated market risk P/L is given by the normal mixture model

$$\tilde{Z} = -\sigma W \left(\sum_{k=1}^{K} \gamma_k Y_k + \sqrt{1 - \sum_{k=1}^{K} \gamma_k^2} \, \eta \right), \qquad (29)$$

where σ is a scaling factor. If $W = \sqrt{\nu/S_\nu}$ and S_ν is a χ_ν^2 distributed random variable with ν degrees of freedom, then we call $\tilde{Z}$ a t_ν-model for the pre-aggregated market risk P/L.

The mixing variable W can be interpreted as a "global shock" driving the variance of all factors. Such an overarching shock may occur from political distress, severe economic recession or some natural disaster.

3.3 Inter-Risk Correlation

We now investigate the correlation between credit risk $L^{(n)}$ and market risk Z, which is defined as

$$\operatorname{corr}\left(L^{(n)}, Z\right) = \frac{\operatorname{cov}\left(L^{(n)}, Z\right)}{\sqrt{\operatorname{var}(L^{(n)})}\sqrt{\operatorname{var}(Z)}}. \tag{30}$$

Within our modelling framework, we are able to analytically investigate inter-risk correlation yielding closed-form results.

First we assume that both market and credit risk have a normally distributed factor structure.

Theorem 2 (Inter-risk correlation for the normal factor model). *Suppose that credit portfolio loss $L^{(n)}$ and market risk Z are described by the normal factor model of Definition 4. Then correlation between $L^{(n)}$ and Z is given by*

$$\operatorname{corr}\left(L^{(n)}, Z\right) = \frac{\sum_{i=1}^{n} r_i\, e_i \exp\left(-\tfrac{1}{2} D_i^2\right)}{\sqrt{2\pi\, \operatorname{var}(L^{(n)})}}, \tag{31}$$

where D_i is the default point (25)

$$r_i := \operatorname{corr}(A_i, Z) = \sum_{k=1}^{K} \beta_{ik} \gamma_k, \qquad i = 1, \ldots, n, \tag{32}$$

and

$$\operatorname{var}\left(L^{(n)}\right) = \sum_{i,j=1}^{n} e_i\, e_j\, (p_{ij} - p_i\, p_j), \tag{33}$$

where p_{ij} the joint default probability (26).

Proof. Using $\mathbb{E}(Z) = 0$ and that η in (28) is independent of Y (and thus of L_i), the covariance between $L^{(n)}$ and Z is

$$\operatorname{cov}\left(L^{(n)}, Z\right) = \mathbb{E}\left(Z L^{(n)}\right) = -\sigma \sum_{i=1}^{n} e_i \sum_{k=1}^{K} \gamma_k\, \mathbb{E}(Y_k L_i). \tag{34}$$

Recall the definition of L_i in (21) with A_i as in (23), and define for $k \in \{1,\ldots,K\}$

$$A_i^{(-k)} = \sum_{\substack{l=1 \\ l \neq k}}^{K} \beta_{il} Y_l + \sqrt{1 - \sum_{j=1}^{K} \beta_{ij}^2}\, \varepsilon_i.$$

Conditioning on Y_k yields for the expectation

$$\mathbb{E}(Y_k\, L_i) = \mathbb{E}\left(Y_k\, \mathbb{E}\left(\mathbb{1}_{\{A_i < D_i\}} \mid Y_k\right)\right)$$

$$= \mathbb{E}\left(Y_k\, \mathbb{P}\left(A_i^{(-k)} \leq D_i - \beta_{ik} Y_k\right)\right)$$

$$= \mathbb{E}\left(Y_k\, \Phi\left(\frac{D_i - \beta_{ik} Y_k}{\sqrt{1 - \beta_{ik}^2}}\right)\right),$$

where we have used that $A_i^{(-k)}$ is normally distributed with variance $1 - \beta_{ik}^2$. By partial integration and the fact that for the density φ of the standard normal distribution $y\,\varphi(y)$ has antiderivative $\varphi(y)$, we obtain

$$\mathbb{E}(Y_k\, L_i) = \int_{-\infty}^{\infty} y\, \Phi\left(\frac{D_i - \beta_{ik} y}{\sqrt{1 - \beta_{ik}^2}}\right) \varphi(y)\, dy$$

$$= -\frac{\beta_{ik}}{\sqrt{1 - \beta_{ik}^2}} \int_{-\infty}^{\infty} \varphi\left(\frac{D_i - \beta_{ik} y}{\sqrt{1 - \beta_{ik}^2}}\right) \varphi(y)\, dy.$$

The right-hand side is $-\beta_{ik}$ times the density of a random variable $U = \sqrt{1 - \beta_{ik}^2}\, X + \beta_{ik} Y$ for standard normal iid X, Y at point D_i. Since U is then again standard normal, we obtain

$$\mathbb{E}(Y_k\, L_i) = -\beta_{ik}\, \varphi(D_i) = -\frac{\beta_{ik}}{\sqrt{2\pi}} e^{-\frac{D_i^2}{2}}. \qquad (35)$$

Plugging this into (34) with r_i as in (32) this yields

$$\mathrm{cov}\left(L^{(n)}, Z\right) = \frac{\sigma}{\sqrt{2\pi}} \sum_{i=1}^{n} e_i\, r_i\, e^{-\frac{D_i^2}{2}}.$$

Furthermore, from (22) we calculate

$$\mathrm{var}\left(L^{(n)}\right) = \sum_{i,j=1}^{n} e_i\, e_j\, \left(\mathbb{E}(L_i L_j) - \mathbb{E}(L_i)\mathbb{E}(L_j)\right)$$

$$= \sum_{i,j=1}^{n} e_i\, e_j\, (p_{ij} - p_i\, p_j),$$

where p_{ij} is the joint default probability (26). □

Note that r_i may become negative if (some) factor weights β_{ik} and γ_k have different signs. Therefore, in principal, also negative inter-risk correlations can occur between the credit and market portfolio. Typical values for the inter-risk correlation lie in a range between 10% and 60% and vary significantly within the banking sector. A similar result can be obtained for the shock model of Definition 5.

Theorem 3 (Inter-risk correlation for the t_ν factor model). *Suppose that credit portfolio loss $L^{(n)}$ is described by the normal factor model of Definition 2. Denote by Z and $\widetilde{Z}$ the market risk described by the normal factor and by the shock model of Definition 3 and Definition 5, respectively. If W has finite second moment, then*

$$\operatorname{corr}\left(L^{(n)}, \widetilde{Z}\right) = \frac{\mathbb{E}(W)}{\sqrt{\mathbb{E}(W^2)}} \operatorname{corr}\left(L^{(n)}, Z\right). \tag{36}$$

For the t_ν model with $\nu > 2$ we get

$$\operatorname{corr}\left(L^{(n)}, \widetilde{Z}\right) = f(\nu) \operatorname{corr}\left(L^{(n)}, Z\right) \tag{37}$$

with

$$f(\nu) := \sqrt{\frac{\nu-2}{2}} \frac{\Gamma\left(\frac{\nu-1}{2}\right)}{\Gamma\left(\frac{\nu}{2}\right)}. \tag{38}$$

Proof. Since $\mathbb{E}(Z) = 0$, we obtain with

$$\operatorname{cov}\left(L^{(n)}, \widetilde{Z}\right) = \mathbb{E}(W) \operatorname{cov}\left(L^{(n)}, Z\right) \quad \text{and} \quad \operatorname{var}\left(\widetilde{Z}\right) = \mathbb{E}\left(W^2\right) \operatorname{var}(Z)$$

that

$$\operatorname{corr}\left(L^{(n)}, \widetilde{Z}\right) = \frac{\mathbb{E}(W)}{\sqrt{\mathbb{E}(W^2)}} \operatorname{corr}\left(L^{(n)}, Z\right).$$

For the t_ν model with $\nu > 0$ we have $W = \sqrt{\nu/S}$, where S is χ_ν^2 distributed with density

$$f_\nu(s) = \frac{2^{-\nu/2}}{\Gamma\left(\frac{\nu}{2}\right)} e^{-s/2} s^{\nu/2-1}, \qquad s \geq 0.$$

It follows for $\nu > 1$ that

$$\mathbb{E}\left(\frac{1}{\sqrt{S}}\right) = \frac{2^{-\nu/2}}{\Gamma\left(\frac{\nu}{2}\right)} \int_0^\infty e^{-s/2} s^{\nu/2-3/2}\, ds = \frac{\Gamma\left(\frac{\nu-1}{2}\right)}{\sqrt{2}\, \Gamma\left(\frac{\nu}{2}\right)}.$$

Analogously, for $\nu > 2$ we calculate $\mathbb{E}\left(\frac{1}{S}\right) = \left(\frac{1}{\nu-2}\right)$. Plugging this into (36) gives formula (37). □

Remark 1. Since $\mathbb{E}(W) > 0$, by the Cauchy–Schwarz inequality,

$$0 < \frac{\mathbb{E}(W)}{\sqrt{\mathbb{E}(W^2)}} \leq 1.$$

As a consequence thereof, given a positive inter-risk correlation $\mathrm{corr}(L^{(n)}, Z) \in (0, 1]$ for normally distributed market risk, introducing a shock into the model results in a smaller inter-risk correlation (36). For the t_ν model this situation is depicted in Fig. 4. □

The fact that $\mathrm{corr}(L^{(n)}, Z)$ linearly depends on the correlations r_i and thus on the factor loadings γ_k implies the following Proposition, which can be used to estimate upper bounds for the inter-risk correlation, when no specific information about market risk is available.

Proposition 1 (Inter-risk correlation bounds). *Suppose that credit portfolio loss $L^{(n)}$ and market risk Z are described by the normal factor model of Definition 4. Assume that the market model factor loadings γ_k for $k = 1, \ldots, K$ are unknown. Then correlation between $L^{(n)}$ and Z is bounded by*

$$\left| \mathrm{corr}\left(L^{(n)}, Z \right) \right| \leq \frac{\sum_{i=1}^n e_i \sqrt{\sum_{k=1}^K \beta_{ik}^2} \exp\left(-\tfrac{1}{2} D_i^2\right)}{\sqrt{2\pi \, \mathrm{var}(L^{(n)})}} \leq 1, \qquad (39)$$

where $\mathrm{var}\left(L^{(n)}\right)$ *is given in (33).*

Proof. Since the obligor's exposures e_i are assumed to be positive, it follows from (31) that

$$\left| \mathrm{corr}\left(L^{(n)}, Z \right) \right| \leq \frac{\sum_i e_i |r_i| \exp\left(-\tfrac{1}{2} D_i^2\right)}{\sqrt{2\pi \sum_{ij} e_i e_j (p_{ij} - p_i p_j)}}.$$

From $\sum_{k=1}^K \gamma_k^2 \leq 1$ it follows by the Cauchy–Schwartz inequality that

$$|r_i| = \left| \sum_{k=1}^K \beta_{ik} \gamma_k \right| \leq \left(\sum_{k=1}^K \beta_{ik}^2 \right)^{1/2} \left(\sum_{k=1}^K \gamma_k^2 \right)^{1/2} \leq \left(\sum_{k=1}^K \beta_{ik}^2 \right)^{1/2}.$$

The right-hand side is bounded by one, since $\sum_{k=1}^K \gamma_k^2 = 1$ corresponds to the correlation of the degenerate case of model (28). □

Therefore, solely based on the parametrization of the normal credit factor model and the assumption of a normally distributed, pre-aggregated market risk, bounds for the inter-risk correlation can be derived. Moreover, from the explicit form of (37) in Theorem 3 it is clear that a similar result holds also for the t_ν distributed market risk.

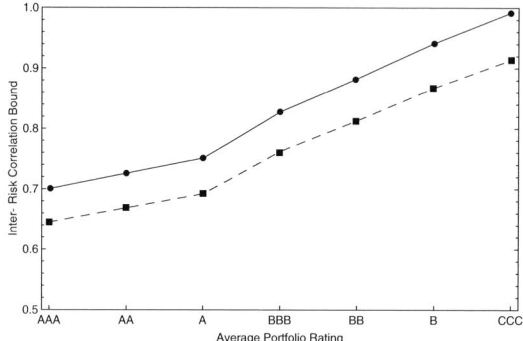

Fig. 4. LHP approximations of the inter-risk correlation bound as a function of the average portfolio rating according to (41). The *solid line* corresponds to the normal factor model (formally $\nu \to \infty$) and the *dashed line* to the shock model with $\nu = 5$. The uniform asset correlation is assumed to be $\rho = 10\,\%$.

One-Factor Approximations

Instructive examples regarding the inter-risk correlation and its bounds can be obtained for one-factor models and they are useful to explain general characteristics of inter-risk correlation. As shown in Böcker & Hillebrand [BH07], Sect. 4.1, such a common one-factor framework for both credit and market risk can be defined consistently, and in the sequel we want to summarize some of their results.

Within the one-factor framework, the credit portfolio is assumed to be homogenous; i.e. for $i = 1, \ldots, n$ exposure $e_i = e$, default probability $p_i = p$, and factor loadings $\beta_{ik} = \beta_k$ for $k = 1, \ldots, K$, i.e. these quantities are the same for all credits of the portfolio. and both market and credit risk are systematically explained only by one single factor $\tilde{Y} := \frac{1}{\sqrt{\rho}} \sum_{k=1}^{K} \beta_k Y_k$, which is a compound of all Y_k for $k = 1, \ldots, K$, where $\rho := \sum_{k=1}^{K} \beta_k^2$ is the uniform asset correlation of the credit portfolio; i.e. for any two asset value log-returns A_i, A_j the correlation is equal to ρ. The situation simplifies further in the case of a sufficiently large portfolio, where we consider $n \to \infty$, resulting in the so-called large homogenous portfolio (LHP) approximation (see also Bluhm, Overbeck & Wagner [BOW02], Sect. 2.5.1.)

$$\frac{L^{(n)}}{n\,e} \stackrel{\text{a.s.}}{\to} \Phi\left(\frac{D - \sqrt{\rho}\,\tilde{Y}}{\sqrt{1-\rho}}\right) =: L, \qquad n \to \infty,$$

where $D = \Phi^{-1}(p)$ and ne is the total exposure of the credit portfolio. The LHP approximation plays an important role in the context of credit portfolio modelling; e.g. it is the underlying assumption in the calculation formula for regulatory capital charges in the *internal-ratings-based* (IRB) approach of Basel II.

Adopting the LHP approximation for the t_ν market model with the normal model as formal limit model with $\lim_{\nu \to \infty} f(\nu) = 1$, inter-risk correlation simplifies considerably. From (26) we get the joint default probability $p_{12} = \Phi_\rho(D, D)$ for two arbitray firms in the portfolio, and from (32) we see that $r = \sum_{k=1}^{K} \beta_k \gamma_k$. Then

$$\mathrm{corr}(L, \widetilde{Z}) = f(\nu) \frac{r\, e^{-D^2/2}}{\sqrt{2\pi(p_{12} - p^2)}}, \tag{40}$$

which is $\mathrm{corr}(L, Z)$ for the normal model with $f(\nu) = 1$. The bound (39) simplifies to

$$|\mathrm{corr}(L, \widetilde{Z})| \leq f(\nu) \frac{\sqrt{\rho}\, e^{-D^2/2}}{\sqrt{2\pi(p_{12} - p^2)}}. \tag{41}$$

According to equations (40) and (41), inter-risk correlation and its bound are functions of the homogeneous asset correlation ρ and the average default probability p and thus on the average rating structure of the credit portfolio. This is depicted in Fig. 4 where LHP approximations of the inter-risk correlation bound are plotted as a function of the average portfolio rating.

A crucial point in the above approximation is the homogeneity of the credit portfolio. Even if actual credit portfolios are rarely exactly homogenous, the derived LHP approximations is a useful approximation in practice for the upper inter-risk correlation bound. Let us consider the normal factor model and so equation (41). For a loss distribution of a general credit portfolio (obtained for instance by Monte Carlo simulation) with expected loss μ, standard deviation ς, and total exposure e_{tot}, estimators $\hat{p}$ and $\hat{\rho}$ for p and ρ, respectively, can be found by moment matching; i.e. by comparing the expected loss and the variance of the simulated portfolio with those of an LHP:

$$\hat{\mu} = e_{\mathrm{tot}}\, \hat{p} \tag{42}$$
$$\hat{\varsigma}^2 = e_{\mathrm{tot}}^2 \left(\hat{p}_{12} - \hat{p}^2\right) = e_{\mathrm{tot}}^2 \left[\Phi_{\hat{\rho}}\left(\Phi^{-1}(\hat{p}), \Phi^{-1}(\hat{p})\right) - \hat{p}^2\right]. \tag{43}$$

From (41) we then obtain the following moment estimator for the upper inter-risk correlation bound

$$\widehat{B}_{\mathrm{LHP}}(\hat{p}, \hat{\rho}) = f(\nu) \frac{e_{\mathrm{tot}}}{\hat{\varsigma}} \frac{\sqrt{\hat{\rho}}\, \exp\left[-\frac{1}{2}\left(\Phi^{-1}(\hat{p})\right)^2\right]}{\sqrt{2\pi}}. \tag{44}$$

4 Conclusion

In this paper we suggested separate models for operational risk, credit risk and market risk, aiming at an integrated model quantifying the overall risk of a financial institution. In doing so, we adopted the common idea that "risk" of a financial position or even an entire bank can be separated into different risk types.

In general, however, such a silo approach often causes problems when risk-type definitions are overlapping, or the classification into risk types is unrealistic or even not possible. We want to present a simple but convincing example.

Consider a fixed-rate corporate bond, where the investor receives fixed, regular interest payments (with a rate set at the time the bond is issued) until the bond matures, called the coupon rate. On one hand, such an investment bears market risk, in particular interest rate risk: If market interest rates rise, then the market price of the bond will fall, because new bonds are expected to be issued with higher coupon rates, making old bonds less attractive. On the other hand, the bond also has credit risk, since the coupon rate of a bond also depends on the financial health of the issuer; i.e. on the credit rating of the company. The higher the company's default probability is, the less likely is that it will be able to pay the interest on the bond and to pay-off the bond at maturity. In this example (and of course also for more complex financial instruments) it does not make sense to distinguish market from credit risk, the only threat for the trader is a decrease in the market value of the bond.

Furthermore, professional trading of financial instruments requires a complex IT-infrastructure, and so also bears a significant fraction of operational risk. However, even in our simple example of the coupon bond, the question regarding its operational VAR remains unsolved. Similar problems arise in the context of other Pillar II risk types such as business and strategic risk, which are currently only poorly considered within a firm's enterprise risk management process. For a novel approach to this particular risk see Böcker [B07].

In accordance with Alan Greenspan we belief that a reliable and functioning risk management system is the basis for success in banking. Therefore, future research has to tackle the problem of how total risk (beyond that of market, credit and operational risk) can be measured and managed properly. To achieve such a "grand unified theory" of risk, a more holistic view on risk instead of the widespread silo approach is called for.

About the Authors/Disclaimer

Klaus Böcker is senior risk controller at UniCredit Group, Munich Branch at HypoVereinsbank AG. Claudia Klüppelberg holds the chair of Mathematical Statistics at the Center for Mathematical Sciences of the Munich University of Technology. The opinions expressed in this article are those of the authors and do not reflect the views of UniCredit Group or HypoVereinsbank AG.

References

[BII04] Basel Committee on Banking Supervision: International Convergence of Capital Measurement and Capital Standards, Basel (2004)

[BS73]   Black, Scholes Black, F., Scholes, M.: The Pricing of Options and Corporate Liabilities, Journal of Political Economy **81**, 637–654 (1973)
[BOW02]   Bluhm, C., Overbeck, L., Wagner, L. An Introduction to Credit Risk Modeling. Chapman & Hall/CRC, Baco Raton (2003)
[B07]   Böcker, K.: Modelling business risk. In Preparation (2007)
[BH07]   Böcker, K., Hillebrand, M.: Interaction of market and credit risk: an analysis of inter-risk correlation and risk aggregation. Submitted for publication (2007)
[BK05]   Böcker, K., Klüppelberg, C.: Operational VaR: a closed-form approximation. RISK, December, 90–93 (2005)
[BK06]   Böcker, K., Klüppelberg, C.: Multivariate models for operational risk. PRMIA "2007 Enterprise Risk Management Symposium Award for Best Paper: New Frontiers in Risk Management Award". Submitted for publication (2006).
[BK07a]   Böcker, K., Klüppelberg, C.: Modelling and measuring multivariate operational risk with Lévy copulas. Submitted for publication (2007)
[BK07b]   Böcker, K., Klüppelberg, C.: Multivariate operational risk: dependence modelling with L copulas. 2007 ERM Symposium Online Monograph, Society of Actuaries, and Joint Risk Management section newsletter. To appear (2007).
[CEBS]   Committee of European Banking Supervisors (CEBS): Application of the Supervisory Review Process under Pillar 2 (CP03 revised), Consultation Paper (2005)
[CT04]   Cont, R., Tankov, P.: Financial Modelling with Jump Processes. Chapman & Hall/CRC, Baco Raton (2004)
[EKM97]   Embrechts, P., Klüppelberg, C., Mikosch, T.: Modelling Extremal Events for Insurance and Finance. Springer, Berlin (1997)
[FK02]   Föllmer, H., Klüppelberg, C.: Finanzmathematik. In: Deutsche Forschungsgemeinschaft (Hrsg.) Perpektiven der Forschung und ihrer Förderung. Aufgaben und Finanzierung 2002–2006. Wiley-VHC, Weinheim (2002)
[H06]   Hillebrand, M.: Modelling and estimating dependent loss given default. RISK, September, 120–125 (2006)
[KR07]   Klüppelberg, C., Resnick, S.I.: The Pareto copula, aggregation of risks and the Emperor's socks. Submitted (2007)
[RK99]   Rootzén, H. and Klüppelberg, C. (1999) A single number can't hedge against economic catastrophes. Ambio **28**, No 6, 550–555. Royal Swedish Academy of Sciences.

Numerical Simulation for Asset-Liability Management in Life Insurance

Thomas Gerstner[1], Michael Griebel[1], Markus Holtz[1], Ralf Goschnick[2], and Marcus Haep[2]

[1] Institut für Numerische Simulation, Universität Bonn, Nussallee 15, 53115 Bonn, Germany, `griebel@ins.uni-bonn.de`
[2] Zürich Gruppe Deutschland, Poppelsdorfer Allee 25–33, 53115 Bonn, Germany `marcus.haep@zurich.com`

Summary. New regulations and stronger competitions have increased the demand for stochastic asset-liability management (ALM) models for insurance companies in recent years. In this article, we propose a discrete time ALM model for the simulation of simplified balance sheets of life insurance products. The model incorporates the most important life insurance product characteristics, the surrender of contracts, a reserve-dependent bonus declaration, a dynamic asset allocation and a two-factor stochastic capital market. All terms arising in the model can be calculated recursively which allows an easy implementation and efficient evaluation of the model equations. The modular design of the model permits straightforward modifications and extensions to handle specific requirements. In practise, the simulation of stochastic ALM models is usually performed by Monte Carlo methods which suffer from relatively low convergence rates and often very long run times, though. As alternatives to Monte Carlo simulation, we here propose deterministic integration schemes, such as quasi-Monte Carlo and sparse grid methods for the numerical simulation of such models. Their efficiency is demonstrated by numerical examples which show that the deterministic methods often perform much better than Monte Carlo simulation as well as by theoretical considerations which show that ALM problems are often of low effective dimension.

1 Introduction

The scope of asset-liability management is the responsible administration of the assets and liabilities of insurance contracts. Here, the insurance company has to attain two goals simultaneously. On the one hand, the available capital has to be invested as profitably as possible (asset management), on the other hand, the obligations against policyholders have to be met (liability management). Depending on the specific insurance policies these obligations are often quite complex and can include a wide range of guarantees and option-like features, like interest rate guarantees, surrender options (with or without surrender fees) and variable reversionary bonus payments. Such bonus payments are typically linked to the investment returns of the company. Thereby, the insurance company has to declare in each year which part of the investment returns is given to the policyholders as reversionary bonus, which part is saved in a reserve account for future bonus payments and which part is kept by the shareholders of the company. These management decisions depend on the financial situation of the company as well as on strategic considerations and legal requirements. A maximisation of the shareholders' benefits has to be balanced with a competitive bonus declaration for the policyholders. Moreover, the exposure of the company to financial, mortality and surrender risks has to be taken into account. These complex problems are investigated with the help of ALM analyses. In this context, it is necessary to estimate the medium- and long-term development of all assets and liabilities as well as the interactions between them and to determine their sensitivity to the different types of risks. This can either be achieved by the computation of particular scenarios (stress tests) which are based on historical data, subjective expectations, and guidelines of regulatory authorities or by a stochastic modelling and simulation. In the latter case, numerical methods are used to simulate a large number of scenarios according to given distribution assumptions which describe the possible future developments of all important variables, e.g. of the interest rates. The results are then analysed using statistical figures which illustrate the expected performance or the risk profile of the company.

In recent years, such stochastic ALM models for life insurance policies are becoming more and more important as they take financial uncertainties more realistically into account than an analysis of a small number of deterministically given scenarios. Additional importance arises due to new regulatory requirements as Solvency II and the International Financial Reporting Standard (IFRS). Consequently, much effort has been spent on the development of these models for life insurance policies in the last years, see, e.g., [2, 4, 7, 13, 19, 24, 33] and the references therein. However, most of the ALM models described in the existing literature are based on very simplifying assumptions in order to focus on special components and effects or to obtain analytical solutions. In this article, we develop a general model framework for the ALM of life insurance products. The complexity of the model is chosen such that most of the models previously proposed in the literature and the most important features of life insurance product management are included. All terms arising in the model can be calculated recursively which allows an straightforward implementation and efficient evaluation of the model equations. Furthermore, the model is designed to have a modular organisation which permits straightforward modifications and extensions to handle specific requirements.

In practise, usually Monte Carlo methods are used for the stochastic simulation of ALM models. These methods are robust and easy to implement but suffer from their relatively low convergence rates. To obtain one more digit accuracy, Monte Carlo methods need the simulation of a hundred times as many scenarios. As the simulation of each scenario requires a run over all time points and all policies in the portfolio of the company, often very long run times are needed to obtain approximations of satisfactory accuracy. As a consequence, a more frequent and more comprehensive risk management, extensive sensitivity investigations or the optimisation of product parameters and management rules are often not possible. In this article we propose deterministic numerical integration schemes, such as quasi-Monte Carlo methods (see e.g. [37, 22, 40]) and sparse grid methods (see, e.g., [9, 15, 16, 23, 38, 42]) for the numerical simulation of ALM models. These methods are alternatives to to Monte Carlo simulation, which have a faster rate of convergence, exploit the smoothness and the anisotropy of the integrand and have deterministic upper bounds on the error. In this way, they often can significantly reduce the number of required scenarios and computing times as we show by numerical experiments. The performance of these numerical methods is closely related to the effective dimension and the smoothness of the problem under consideration. Here, we show that ALM problems are often of very low effective dimension (in the sense that the problem can well be approximated by sums of very low-dimensional functions) which can, to some extent, explain the efficiency of the deterministic methods. Numerical results based on a general ALM model framework for participating life insurance products demonstrate that these deterministic methods in fact often perform much better than Monte Carlo simulation even for complex ALM models with many time steps. Quasi-Monte Carlo methods based on Sobol sequences and dimension-adaptive sparse grids based on one-dimensional Gauss–Hermite quadrature formulae turn out to be the most efficient representatives of several quasi-Monte Carlo and sparse grid variants, respectively. For further details, see [17, 18, 19].

The remainder of this article is as follows: In Sect. 2, we describe the model framework. In Sect. 3, we then discuss how this model can be efficiently simulated by numerical methods for multivariate integration. In Sect. 4, we present numerical results which illustrate possible application of the ALM model and analyse the efficiency of different numerical approaches. The article finally closes in Sect. 5 with concluding remarks.

2 The ALM Model

In this section, we closely follow [19] and describe an ALM model framework for the simulation of the future development of a life insurance company. We first indicate the overall structure of the model and introduce a simplified balance sheet which represents the assets and liabilities of the company. The different modules (capital market model, liability model, management model) and the evolution of the balance sheet items are then specified in the following sections.

2.1 Overall Model Structure

The main focus of our model is to simulate the future development of all assets and liabilities of a life insurance company. To this end, the future develop-

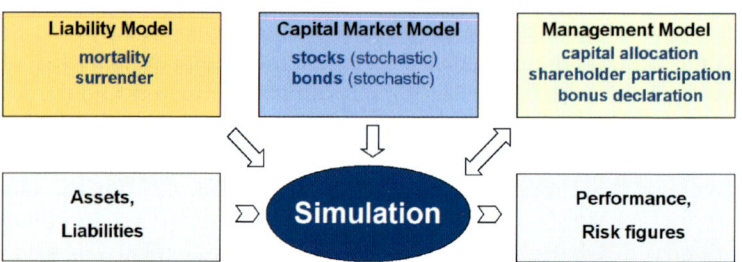

Fig. 1. Overall structure of the ALM model

ment of the capital markets, the policyholder behaviour and the company's management has to be modelled. We use a stochastic capital market model, a deterministic liability model which describes the policyholder behaviour and a deterministic management model which is specified by a set of management rules which may depend on the stochastic capital markets. The results of the simulation are measured by statistical performance and risk figures which are based on the company's most important balance sheet items. They are used by the company to optimise management rules, like the capital allocation, or product parameters, like the surrender fee. The overall structure of the model is illustrated in Fig. 1.

We model all terms in discrete time. Here, we denote the start of the simulation by time $t = 0$ and the end of the simulation by $t = T$ (in years). The interval $[0, T]$ is decomposed into K periods $[t_{k-1}, t_k]$ with $t_k = k\,\Delta t$, $k = 1, \ldots, K$ and a period length $\Delta t = T/K$ of one month.

The asset side consists of the market value C_k of the company's assets at time t_k. On the liability side, the first item is the book value of the actuarial reserve D_k, i.e., the guaranteed savings part of the policyholders after deduction of risk premiums and administrative costs. The second item is the book value of the allocated bonuses B_k which constitute the part of the surpluses that have been credited to the policyholders via the profit participation. The free reserve F_k is a buffer account for future bonus payments. It consists of surpluses which have not yet been credited to the individual policyholder accounts, and is used to smooth capital market oscillations and to achieve a stable and low-volatile return participation of the policyholders. The last item, the equity or company account Q_k, consists of the part of the surpluses which is kept by the shareholders of the company and is defined by

$$Q_k = C_k - D_k - B_k - F_k$$

such that the sum of the assets equals the sum of the liabilities. Similar to the bonus reserve in [24], Q_k is a hybrid determined as the difference between a market value C_k and the three book values D_k, B_k and F_k. It may be interpreted as hidden reserve of the company as discussed in [29]. The balance sheet items at time t_k, $k = 0, \ldots, K$, used in our model are shown in Table 1. In a sensitivity analysis for sample parameters and portfolios it is shown in [19]

Table 1. Simplified balance sheet of the life insurance company

Assets		Liabilities	
Capital	C_k	Actuarial reserve	D_k
		Allocated bonus	B_k
		Free reserve	F_k
		Equity	Q_k

that this model captures the most important behaviour patterns of the balance sheet development of life insurance products. Similar balance sheet models have already been considered in, e.g., [2, 3, 24, 33, 29].

2.2 Capital Market Model

We assume that the insurance company invests its capital either in fixed interest assets, i.e., bonds, or in a variable return asset, i.e., a stock or a basket of stocks. For the modelling of the interest rate environment we use the Cox-Ingersoll-Ross (CIR) model [11]. The CIR model is a one-factor mean-reversion model which specifies the dynamics of the short interest rate $r(t)$ at time t by the stochastic differential equation

$$dr(t) = \kappa(\theta - r(t))dt + \sqrt{r(t)}\sigma_r dW_r(t), \qquad (1)$$

where $W_r(t)$ is a standard Brownian motion, $\theta > 0$ denotes the mean reversion level, $\kappa > 0$ denotes the reversion rate and $\sigma_r \geq 0$ denotes the volatility of the short rate dynamic. In the CIR model, the price $b(t,\tau)$ at time t of a zero coupon bond with a duration of τ periods and with maturity at time $T = t + \tau \Delta t$ can be derived in closed form by

$$b(t,\tau) = A(\tau) e^{-B(\tau) r(t)} \qquad (2)$$

as an exponential affine function of the prevailing short interest rate $r(t)$ with

$$A(\tau) = \left(\frac{2h e^{(\hat{\kappa}+h)\tau \Delta t/2}}{2h + (\hat{\kappa}+h)(e^{h\tau \Delta t} - 1)} \right)^{2\kappa\theta/\sigma_r^2}, \quad B(\tau) = \frac{2(e^{h\tau \Delta t} - 1)}{2h + (\hat{\kappa}+h)(e^{h\tau \Delta t} - 1)},$$

and $h = \sqrt{\hat{\kappa}^2 + 2\sigma_r^2}$. To model the stock price uncertainty, we assume that the stock price $s(t)$ at time t evolves according to a geometric Brownian motion

$$ds(t) = \mu s(t)dt + \sigma_s s(t)dW_s(t), \qquad (3)$$

where $\mu \in \mathbb{R}$ denotes the drift rate and $\sigma_s \geq 0$ denotes the volatility of the stock return. By Itô's lemma, the explicit solution of this stochastic differential equation is given by

$$s(t) = s(0) e^{\left(\mu - \sigma_s^2/2\right)t + \sigma_s W_s(t)}. \qquad (4)$$

Usually, stock and bond returns are correlated. We thus assume that the two Brownian motions satisfy $dW_s(t)dW_r(t) = \rho dt$ with a constant correlation coefficient $\rho \in [-1,1]$. These and other models which can be used to simulate the bond and stock prices are discussed in detail, e.g., in [6, 25, 28].

In the discrete time case, the short interest rate, the stock prices and the bond prices are defined by $r_k = r(t_k)$, $s_k = s(t_k)$ and $b_k(\tau) = b(t_k, \tau)$. For the solution of equation (1), we use an Euler-Maruyama discretization[3] with step size Δt, which yields

$$r_k = r_{k-1} + \kappa(\theta - r_{k-1})\Delta t + \sigma_r \sqrt{|r_{k-1}|}\sqrt{\Delta t}\,\xi_{r,k}, \tag{5}$$

where $\xi_{r,k}$ is a $N(0,1)$-distributed random variable. For the stock prices one obtains

$$s_k = s_{k-1} e^{\left(\mu - \sigma_s^2/2\right)\Delta t + \sigma_s \sqrt{\Delta t}\left(\rho \xi_{r,k} + \sqrt{1-\rho^2}\,\xi_{s,k}\right)}, \tag{6}$$

where $\xi_{s,k}$ is a $N(0,1)$-distributed random variable independent of $\xi_{r,k}$. Since

$$\mathrm{Cov}\left(\rho \xi_{r,k} + \sqrt{1-\rho^2}\,\xi_{s,k}, \xi_{r,k}\right) = \rho,$$

the correlation between the two Wiener processes $W_s(t)$ and $W_r(t)$ is respected. More information on the numerical solution of stochastic differential equations can be found, e.g., in [22, 30].

2.3 Management Model

In this section, we discuss the capital allocation, the bonus declaration mechanism and the shareholder participation.

Capital Allocation

We assume that the company rebalances its assets at the beginning of each period. Thereby, the company aims to have a fixed portion $\beta \in [0,1]$ of its assets invested in stocks, while the remaining capital is invested in zero coupon bonds with a fixed duration of τ periods. We assume that no bonds are sold before their maturity. Let P_k be the premium income at the beginning of period k and let C_{k-1} be the total capital at the end of the previous period. The part N_k of $C_{k-1} + P_k$ which is available for a new investment at the beginning of period k is then given by

$$N_k = C_{k-1} + P_k - \sum_{i=1}^{\tau-1} n_{k-i}\, b_{k-1}(\tau - i),$$

---

[3] An alternative to the Euler-Maruyama scheme, which is more time consuming but avoids time discretization errors, is to sample from a noncentral chi-squared distribution, see [22]. In addition, several newer approaches exist to improve the balancing of time and space discretization errors, see, e.g., [21]. This and the time discretization error are not the focus of this article, though.

where n_j denotes the number of zero coupon bonds which were bought at the beginning of period j. The capital A_k which is invested in stocks at the beginning of period k is then determined by

$$A_k = \max\{\min\{N_k, \beta(C_{k-1} + P_k)\}, 0\} \tag{7}$$

so that the side conditions $0 \leq A_k \leq \beta(C_{k-1}+P_k)$ are satisfied. The remaining money $N_k - A_k$ is used to buy $n_k = (N_k - A_k)/b_{k-1}(\tau)$ zero coupon bonds with duration $\tau \Delta t$.[4] The portfolio return rate p_k in period k resulting from the above allocation procedure is then determined by

$$p_k = \left(\Delta A_k + \sum_{i=0}^{\tau-1} n_{k-i}\, \Delta b_{k,i}\right) / (C_{k-1} + P_k), \tag{8}$$

where $\Delta A_k = A_k(s_k/s_{k-1}-1)$ and $\Delta b_{k,i} = b(t_k, \tau-i-1) - b(t_{k-1}, \tau-i)$ denote the changes of the market values of the stock and of the bond investments from the beginning to the end of period k, respectively.

Bonus Declaration

In addition to the fixed guaranteed interest, a variable reversionary bonus is annually added to the policyholder's account, which allows the policyholder to participate in the investment returns of the company (contribution principle). The bonus is declared by the company at the beginning of each year (principle of advance declaration) with the goal to provide a low-volatile, stable and competitive return participation (average interest principle). Various mathematical models for the declaration mechanism are discussed in the literature. In this article, we follow the approach of [24] where the declaration is based on the current reserve rate γ_{k-1} of the company, which is defined in our framework by the ratio of the free reserve to the allocated liabilities, i.e.,

$$\gamma_{k-1} = \frac{F_{k-1}}{D_{k-1} + B_{k-1}}.$$

The annual interest rate is then defined by

$$\hat{z}_k = \max\{\hat{z}, \omega(\gamma_{k-1} - \gamma)\}.$$

Here, $\hat{z}$ denotes the annual guaranteed interest rate, $\gamma \geq 0$ the target reserve rate of the company and $\omega \in [0,1]$ the distribution ratio or participation coefficient which determines how fast excessive reserves are reduced. This way, a fixed fraction of the excessive reserve is distributed to the policyholders if the reserve rate γ_{k-1} is above the target reserve rate γ while only the guaranteed

---

[4] Note that due to long-term investments in bonds it may happen that $N_k < 0$. This case of insufficient liquidity leads to $n_k < 0$ and thus to a short selling of bonds.

interest is paid in the other case. In our model this annual bonus has to be converted into a monthly interest

$$z_k = \begin{cases} (1+\hat{z}_k)^{1/12} - 1 & \text{if } k \bmod 12 = 1 \\ z_{k-1} & \text{otherwise} \end{cases}$$

which is given to the policyholders in each period k of this year.

Shareholder Participation

Excess returns $p_k - z_k$, conservative biometry and cost assumptions as well as surrender fees lead to a surplus G_k in each period k which has to be divided among the free reserve F_k and the equity Q_k. In case of a positive surplus, we assume that a fixed percentage $\alpha \in [0,1]$ is saved in the free reserve while the remaining part is added to the equity account. Here, a typical assumption is a distribution according to the 90/10-rule which corresponds to the case $\alpha = 0.9$. If the surplus is negative, we assume that the required capital is taken from the free reserve. If the free reserves do not suffice, the company account has to cover the remaining deficit. The free reserve is then defined by

$$F_k = \max\{F_{k-1} + \min\{G_k, \alpha\, G_k\}, 0\}. \qquad (9)$$

The exact specification of the surplus G_k and the development of the equity Q_k is derived in Sect. 2.5.

2.4 Liability Model

In this section, we discuss the modelling of the decrement of policies due to mortality and surrender and the development of the policyholder's accounts.

Decrement Model

For efficiency, the portfolio of all insurance contracts is often represented by a reduced number m of model points. Each model point then represents a group of policyholders which are similar with respect to cash flows and technical reserves, see, e.g., [27]. By pooling, all contracts of a model point expire at the same time which is obtained as the average of the individual maturity times.

We assume that the development of mortality and surrender is given deterministically and modelled using experience-based decrement tables. Let q_k^i and u_k^i denote the probabilities that a policyholder of model point i dies or surrenders in the k-th period, respectively. The probabilities q_k^i typically depend on the age, the year of birth and the gender of the policyholder while u_k^i often depends on the elapsed contract time. Let δ_k^i denote the expected number of contracts in model point i at the end of period k. Then, this number evolves over time according to

$$\delta_k^i = \left(1 - q_k^i - u_k^i\right) \delta_{k-1}^i. \qquad (10)$$

We assume that no new contracts evolve during the simulation.

Insurance Products

In the following, we assume that premiums are paid at the beginning of a period while benefits are paid at the end of the period. Furthermore, we assume that all administrative costs are already included in the premium. For each model point $i = 1, \ldots, m$, the guaranteed part of the insurance product is defined by the specification of the following four characteristics:

- premium characteristic: $(P_1^i, \ldots, P_K^i)$ where P_k^i denotes the premium of an insurance holder in model point i at the beginning of period k if he is still alive at that time.
- survival benefit characteristic: $\left(E_1^{i,G}, \ldots, E_K^{i,G}\right)$ where $E_k^{i,G}$ denotes the guaranteed payments to an insurance holder in model point i at the end of period k if he survives period k.
- death benefit characteristic: $\left(T_1^{i,G}, \ldots, T_K^{i,G}\right)$ where $T_k^{i,G}$ denotes the guaranteed payment to an insurance holder in model point i at the end of period k if he dies in period k.
- surrender characteristic: $\left(S_1^{i,G}, \ldots, S_K^{i,G}\right)$ where $S_k^{i,G}$ denotes the guaranteed payment to an insurance holder in model point i at the end of period k if he surrenders in period k.

The bonus payments of the insurance product to an insurance holder in model point i at the end of period k in case of survival, death and surrender, are denoted by $E_k^{i,B}$, $T_k^{i,B}$ and $S_k^{i,B}$, respectively. The total payments E_k^i, T_k^i and S_k^i to a policyholder of model point i at the end of period k in case of survival, death and surrender are then given by

$$E_k^i = E_k^{i,G} + E_k^{i,B}, \quad T_k^i = T_k^{i,G} + T_k^{i,B} \quad \text{and} \quad S_k^i = S_k^{i,G} + S_k^{i,B}. \tag{11}$$

The capital of a policyholder of model point i at the end of period k is collected in two accounts: the actuarial reserve D_k^i for the guaranteed part and the bonus account B_k^i for the bonus part. Both accounts can efficiently be computed in our framework using the recursions

$$D_k^i = \frac{1+z}{1-q_k^i}(D_{k-1}^i + P_k^i) - E_k^{i,G} - \frac{q_k^i}{1-q_k^i}T_k^{i,G} \tag{12}$$

and

$$B_k^i = \frac{1+z_k}{1-q_k^i}B_{k-1}^i + \frac{z_k - z}{1-q_k^i}(D_{k-1}^i + P_k^i) - E_k^{i,B} - \frac{q_k^i}{1-q_k^i}T_k^{i,B} \tag{13}$$

which results from the deterministic mortality assumptions, see, e.g., [2, 46].

Example 1. As a sample insurance product, an endowment insurance with death benefit, constant premium payments and surrender option is considered. Let P^i denote the constant premium which is paid by each of the policyholders in model point i in every period. If they are still alive, the policyholders receive a guaranteed benefit $E^{i,G}$ and the value of the bonus account at maturity d^i. In case of death prior to maturity, the sum of all premium payments and the value of the bonus account is returned. In case of surrender, the policyholder capital and the bonus is reduced by a surrender factor $\vartheta = 0.9$. The guaranteed components of the four characteristics are then defined by

$$P_k^i = P^i, \quad E_k^{i,G} = \chi_k(d^i) E^{i,G}, \quad T_k^{i,G} = k P^i \text{ and } S_k^{i,G} = \vartheta D_k^i,$$

where $\chi_k(d^i)$ denotes the indicator function which is one if $k = d^i$ and zero otherwise. The bonus payments at the end of period k are given by

$$E_k^{i,B} = \chi_k(d^i) B_k^i, \quad T_k^{i,B} = B_k^i \text{ and } S_k^{i,B} = \vartheta B_k^i.$$

We will return to this example in Sect. 3.

2.5 Balance Sheet Model

In this section, we derive the recursive development of all items in the simplified balance sheet introduced in Sect. 2.1.

Projection of the Assets

In order to define the capital C_k at the end of period k, we first determine the cash flows which are occurring to and from the policyholders in our model framework. The premium P_k, which is obtained by the company at the beginning of period k, and the survival payments E_k, the death payments T_k, and the surrender payments S_k to policyholders, which take place at the end of period k, are obtained by summation of the individual cash flows (11), i.e.,

$$P_k = \sum_{i=1}^m \delta_{k-1}^i P_k^i, \quad E_k = \sum_{i=1}^m \delta_k^i E_k^i, \quad T_k = \sum_{i=1}^m q_k^i \delta_{k-1}^i T_k^i, \quad S_k = \sum_{i=1}^m u_k^i \delta_{k-1}^i S_k^i, \tag{14}$$

where the numbers δ_k^i are given by (10). The capital C_k is then recursively given by

$$C_k = (C_{k-1} + P_k)(1 + p_k) - E_k - T_k - S_k \tag{15}$$

where p_k is the portfolio return rate defined in equation (8).

Projection of the Liabilities

The actuarial reserve D_k and the allocated bonus B_k are derived by summation of the individual policyholder accounts (12) and (13), i.e.,

$$D_k = \sum_{i=1}^m \delta_k^i D_k^i \text{ and } B_k = \sum_{i=1}^m \delta_k^i B_k^i.$$

In order to define the free reserve F_k, we next determine the gross surplus G_k in period k which consists in our model of interest surplus and surrender surplus.

The interest surplus is given by the difference between the total capital market return $p_k (F_{k-1} + D_{k-1} + B_{k-1} + P_k)$ on policyholder capital and the interest payments $z_k (D_{k-1} + B_{k-1} + P_k)$ to policyholders. The surrender surplus is given by $S_k/\vartheta - S_k$. The gross surplus in period k is thus given by

$$G_k = p_k F_{k-1} + (p_k - z_k)(D_{k-1} + B_{k-1} + P_k) + (1/\vartheta - 1)S_k.$$

The free reserve F_k is then derived using equation (9). Altogether, the company account Q_k is determined by

$$Q_k = C_k - D_k - B_k - F_k.$$

Note that the cash flows and all balance sheet items are expected values with respect to our deterministic mortality and surrender assumptions from Sect. 2.4, but random numbers with respect to our stochastic capital market model from Sect. 2.2.

Performance Figures

To analyse the results of a stochastic simulation, statistical measures are considered which result from an averaging over all scenarios. Here, we consider the path-dependent cumulative probability of default

$$\mathrm{PD}_k = \mathbb{P}\left(\min_{j=1,\ldots,k} Q_j < 0\right)$$

as a measure for the risk while we use the expected future value $\mathbb{E}[Q_k]$ of the equity as a measure for the investment returns of the shareholders in the time interval $[0, t_k]$. Due to the wide range of path-dependencies, guarantees and option-like features of the insurance products and management rules, closed-form representations for these statistical measures are in general not available so that one has to resort to numerical methods. It is straightforward to include the computation of further performance and risk measures like the variance, the value-at-risk, the expected shortfall or the return on risk capital. To determine the sensitivity $f'(v) = \partial f(v)/\partial v$ of a given performance figure f to one of the model parameters v, finite difference approximations or more recent approaches, like, e.g., smoking adjoints [20], can be employed.

3 Numerical Simulation

In this section, we discuss the efficient numerical simulation of the ALM model described in Sect. 2. The number of operations for the simulation of a single scenario of the model is of order $O(m \cdot K)$ and takes about 0.04 seconds on a dual Intel(R) Xeon(TM) CPU 3.06GH workstation for a representative portfolio with $m = 500$ model points and a time horizon of $K = 120$ periods.

The number of scenarios which have to be generated depends on the accuracy requirements, on the model parameters[5] and on the employed numerical method. In the following, we first rewrite the performance figures of the model as high-dimensional integrals. Then, we survey numerical methods which can be applied to their computation, discuss their dependence on the effective dimension and review techniques which can reduce the effective dimension in certain cases.

3.1 Representation as High-Dimensional Integrals

It is helpful to represent the performance figures of the ALM simulation as high-dimensional integrals to see how more sophisticated methods than Monte Carlo simulation can be used for their numerical computation. To derive such a representation, recall that the simulation of one scenario of the ALM model is based on $2K$ independent normally distributed random numbers $\mathbf{y} = (y_1, \ldots, y_{2K}) = (\xi_{s,1}, \ldots, \xi_{s,K}, \xi_{r,1}, \ldots, \xi_{r,K}) \sim N(\mathbf{0}, \mathbf{1})$. These numbers specify the stock price process (6) and the short rate process (5). Then, the term structure, the asset allocation, the bonus declaration, the shareholder participation and the development of all involved accounts can be derived using the recursive equations of the previous sections. Altogether, the balance sheet items C_K, B_K, F_K and Q_K at the end of period K can be regarded as (usually very complicated) deterministic functions $C_K(\mathbf{y})$, $B_K(\mathbf{y})$, $F_K(\mathbf{y})$, $Q_K(\mathbf{y})$ depending on the normally distributed vector $\mathbf{y} \in \mathbb{R}^{2K}$. As a consequence, the expected values of the balance sheet items at the end of period K can be represented as $2K$-dimensional integrals, e.g.,

$$E[Q_K] = \int_{\mathbb{R}^{2K}} Q_K(\mathbf{y}) \frac{e^{-\mathbf{y}^T \mathbf{y}/2}}{(2\pi)^K} \, d\mathbf{y} \tag{16}$$

for the equity account. Often, monthly discretizations of the capital market processes are used. Then, typical values for the dimension $2K$ range from 60–600 depending on the time horizon of the simulation.

Transformation

The integral (16) can be transformed into an integral over the $2K$-dimensional unit cube which is often necessary to apply numerical integration methods. By the substitution $y_i = \Phi^{-1}(x_i)$ for $i = 1, \ldots, 2K$, where Φ^{-1} denotes the inverse cumulative normal distribution function, we obtain

$$E[Q_K] = \int_{\mathbb{R}^{2K}} Q_K(\mathbf{y}) \frac{e^{-\mathbf{y}^T \mathbf{y}/2}}{(2\pi)^K} \, d\mathbf{y} = \int_{[0,1]^d} f(\mathbf{x}) \, d\mathbf{x} \tag{17}$$

---

[5] The model parameters affect important numerical properties of the model, e.g. the effective dimension (see Sect. 3.3) or the smoothness.

with $d = 2K$ and $f(\mathbf{x}) = Q_k(\Phi^{-1}(\mathbf{x}))$. For the fast computation of $\Phi^{-1}(x_i)$, we use Moro's method [35]. Note that the integrand (17) is unbounded on the boundary of the unit cube, which is undesirable from a numerical as well as theoretical point of view. Note further that different transformations to the unit cube exist (e.g. using the logistic distribution or polar coordinates) and that also numerical methods exist which can directly be applied to the untransformed integral (16) (e.g. Gauss–Hermite rules).

3.2 Numerical Methods for High-Dimensional Integrals

There is a wide range of methods (see, e.g., [12]) available for numerical multivariate integration. Mostly, the integral (17) is approximated by a weighted sum of n function evaluations

$$\int_{[0,1]^d} f(\mathbf{x})\,d\mathbf{x} \approx \sum_{i=1}^{n} w_i f(\mathbf{x}_i) \qquad (18)$$

with weights $w_i \in \mathbb{R}$ and nodes $\mathbf{x}_i \in \mathbb{R}^d$. The number n of nodes corresponds to the number of simulation runs. Depending on the choice of the weights and nodes, different methods with varying properties are obtained. Here, the dimension as well as the smoothness class of the function f should be taken into account.

Monte Carlo

In practise, the model is usually simulated by the Monte Carlo (MC) method. Here, all weights equal $w_i = 1/n$ and uniformly distributed sequences of pseudo-random numbers $\mathbf{x}_i \in (0,1)^{2K}$ are used as nodes. This method is independent of the dimension, robust and easy to implement but suffers from a relative low probabilistic convergence rate of order $O(n^{-1/2})$. This often leads to very long simulation times in order to obtain approximations of satisfactory accuracy. Extensive sensitivity investigations or the optimisation of product or management parameters, which require a large number of simulation runs, are therefore often not possible.

Quasi-Monte Carlo

Quasi-Monte Carlo (QMC) methods are equal-weight rules like Monte Carlo. Instead of pseudo-random numbers, however, deterministic low-discrepancy sequences (see, e.g., [37, 22]) or lattices (see, e.g., [40]) are used as point sets which are chosen to yield better uniformity than random samples. Some popular choices are Halton, Faure, Sobol and Niederreiter–Xing sequences and extensible shifted rank-1 lattice rules based on Korobov or fast component-by-component constructions. From the Koksma–Hlawka inequality it follows that convergence rate of QMC methods is of order $O(n^{-1}(\log n)^d)$ for integrands of bounded variation which is asymptotically better than the $O(n^{-1/2})$ rate of

MC. For periodic integrands, lattice rules can achieve convergence of higher order depending on the decay of the Fourier coefficients of f, see [40]. Using novel digital net constructions (see [14]), QMC methods can also be obtained for non-periodic integrands which exhibit convergence rates larger than one if the integrands are sufficiently smooth.

Product Methods

Product methods for the computation of (17) are easily obtained by using the tensor products of the weights and nodes of one-dimensional quadrature rules, like, e.g., Gauss rules (see, e.g., [12]). These methods can exploit the smoothness of the function f and converge with order $O(n^{-s/d})$ for $f \in C^s([0,1]^d)$. This shows, however, that product methods suffer from the curse of dimension, meaning that the computing cost grows exponentially with the dimension d of the problem, which prevents their efficient applications for high-dimensional ($d > 5$) applications like ALM simulations.

Sparse Grids

Sparse grid (SG) quadrature formulas are constructed using certain combinations of tensor products of one-dimensional quadrature rules, see, e.g., [9, 15, 23, 38, 42]. In this way, sparse grids can, like product methods, exploit the smoothness of f and also obtain convergence rates larger than one. In contrast to product methods, they can, however, also overcome the curse of dimension like QMC methods to a certain extent. They converge with order $O(n^{-s}(\log n)^{(d-1)(s-1)})$ if the integrand belongs to the space of functions which have bounded mixed derivatives of order s. Sparse grid quadrature formula come in various types depending on the one-dimensional basis integration routine, like the trapezoidal, the Clenshaw-Curtis, the Patterson, the Gauss-Legendre or the Gauss–Hermite rule. In many cases, the performance of sparse grids can be enhanced by local adaptivity, see [5, 8], or by a dimension-adaptive grid refinement, see [16].

3.3 Impact of the Dimension

In this section, we discuss the dependence of MC, QMC and SG methods on the nominal and the effective dimension of the integral (17).

Tractability

In contrast to MC, the convergence rate of QMC and SG methods still exhibit a logarithmic dependence on the dimension. Furthermore, also the constants in the O-notation depend on the dimension of the integral. In many cases (particularly within the SG method) these constants increase exponentially with the dimension. Therefore, for problems with high nominal dimension d, such as the ALM of life insurance products, the classical error bounds of the previous section are no longer of any practical use to control the numerical error of the approximation. For instance, even for a moderate dimension of $d = 20$ and for a computationally unfeasibly high number $n = 10^{90}$ of function evaluations, $n^{-1}(\log n)^d > n^{-1/2}$ still holds in the QMC and the MC error bounds. For classical Sobolov spaces with bounded derivatives up to a certain order, it can even be proved (see [39, 41]) that integration is intractable, meaning that for these function classes deterministic methods of the form (18) can never completely avoid the curse of dimension. For weighted Sobolov spaces, however, it is shown in [39, 41] that integration is tractable if the weights decay sufficiently fast. In the next paragraph and in Sect. 4.3 we will give some indications that ALM problems indeed belong to such weighted function spaces.

ANOVA Decomposition and Effective Dimension

Numerical experiments show that QMC and SG methods often produce much more precise results than MC methods for certain integrands even in hundreds of dimensions. One explanation of this success is that QMC and SG methods can, in contrast to MC, take advantage of low effective dimensions. QMC methods profit from low effective dimensions by the fact that their nodes are usually more uniformly distributed in smaller dimensions than in higher ones. SG methods can exploit different weightings of different dimensions by a dimension-adaptive grid refinement, see [16]. The effective dimension of the integral (17) is defined by the ANOVA decomposition, see, e.g., [10]. Here, a function $f : \mathbb{R}^d \to \mathbb{R}$ is decomposed by

$$f(x) = \sum_{u \subseteq \{1,\ldots,d\}} f_u(x_u) \text{ with } f_u(x_u) = \int_{[0,1]^{d-|u|}} f(x) dx_{\{1,\ldots,d\} \setminus u} - \sum_{v \subset u} f_v(x_v)$$

into 2^d sub-terms f_u with $u \subseteq \{1,\ldots,d\}$ which only depend on variables x_j with $j \in u$. Thereby, the sub-terms f_u describe the dependence of the function f on the dimensions $j \in u$. The effective dimension in the truncation sense of a function $f : \mathbb{R}^d \to \mathbb{R}$ with variance $\sigma^2(f)$ is then defined as the smallest integer d_t, such that $\sum_{v \subseteq \{1,\ldots,d_t\}} \sigma_v^2(f) \geq 0.99 \, \sigma^2(f)$ where $\sigma_u^2(f)$ denotes the variances of f_u. The effective dimension d_t roughly describes the number of important variables of the function f. The effective dimension in the superposition sense is defined as the smallest integer d_s, such that $\sum_{|v| \leq d_s} \sigma_v^2(f) \geq 0.99 \, \sigma^2(f)$ where $|v|$ denotes the cardinality of the index set v. It roughly describes the highest order of important interactions between variables in the ANOVA decomposition. For the simple function $f(x_1, x_2, x_3) = x_1 e^{x_2} + x_2$ with $d = 3$, we obtain $d_t = 2$ and $d_s = 2$ for instance. For large d, it is no longer possible to compute all 2^d ANOVA sub-terms. The effective dimensions can still be computed in many cases, though. For details and an efficient algorithm for the computation of the effective dimension in the truncation sense we refer to [45]. For the more difficult problem to compute the effective dimension in the superposition sense, we use the recursive method described in [44].

Dimension Reduction

Typically, the underlying multivariate Gaussian process is approximated by a random walk discretization. In many cases, a substantial reduction of the effective dimension in the truncation sense and an improved performance of the deterministic integration schemes can be achieved if the Brownian bridge or the principal component (PCA) decompositions of the covariance matrix of the underlying Brownian motion is used instead as it was proposed in [1, 36] for option pricing problems. The Brownian bridge construction differs from the standard random walk construction in that rather than constructing the increments sequentially, the path of the Gaussian process is constructed in a hierarchical way which has the effect that more importance is placed on the earlier variables than on the later ones. The PCA decomposition, which is based on the eigenvalues and -vectors of the covariance matrix of the Brownian motion, maximises the concentration of the total variance of the Brownian motion in the first few dimensions.[6] Its construction requires, however, $O(d^2)$ operations instead of $O(d)$ operations which are needed for the random walk or for the Brownian bridge discretization. For large d, this often increases the run times of the simulation and limits the practical use of the PCA construction.

4 Numerical Results

We now describe the basic setting for our numerical experiments and investigate the sensitivities of the performance figures from Sect. 2.5 to the input parameters of the model. Then, the risks and returns of two different asset allocation strategies are compared. Finally, we compute the effective dimensions of the integral (17) in the truncation and superposition sense and compare the efficiency of different numerical approaches for its computation.

4.1 Setting

We consider a representative model portfolio with 50,000 contracts which have been condensed into 500 equal-sized model points. The data of each model

---

[6] Note that without further assumptions on f it is not clear which construction leads to the minimal effective dimension due to possibly non-linear dependencies of f on the underlying Brownian motion. As a remedy, also more complicated covariance matrix decompositions can be employed which take into account the function f as explained in [26].

Table 2. Capital market parameters p used in the simulation and their partial derivatives $f'(p)/f(p)$ for $f \in \{\text{PD}_K, \mathbb{E}[Q_K], \mathbb{E}[F_K]\}$

	stock price model		interest rate model					correlation
	$\mu = 8\%$	$\sigma_s = 20\%$	$\kappa = 0.1$	$\theta = 4\%$	$\sigma_r = 5\%$	$r_0 = 3\%$	$\lambda_0 = -5\%$	$\rho = -0.1$
$\mathbb{E}[Q_K]$	0.028	0.035	0.007	0.085	−0.001	0.156	−0.001	−0.0008
$\mathbb{E}[F_K]$	0.039	−0.008	0.009	0.136	−0.0014	0.212	−0.0014	−0.0002
PD_K	−0.431	0.219	−0.172	−0.884	0.729	−2.122	0.005	0.04

Table 3. Solvency rate, management and product parameters p used in the simulation and their partial derivatives $f'(p)/f(p)$ for $f \in \{\text{PD}_K, \mathbb{E}[Q_K], \mathbb{E}[F_K]\}$

	asset allocation		bonus declaration		shareholder	product parameters		solv. rate
	$\beta = 10\%$	$\tau = 3$	$\omega = 25\%$	$\gamma = 15\%$	$\alpha = 90\%$	$\vartheta = 90\%$	$z = 3\%$	$\gamma_0 = 10\%$
$\mathbb{E}[Q_K]$	0.083	0.004	−0.002	0.009	−0.101	−0.006	−0.086	0.011
$\mathbb{E}[F_K]$	0.002	0.002	−0.009	0.03	0.013	−0.01	−0.22	0.034
PD_K	0.265	−0.054	0	−0.002	0.001	0.08	2.706	−0.504

point i is generated according to the following distribution assumptions: entry age $\underline{x}^i \sim N(36, 10)$, exit age $\overline{x}^i \sim N(62, 4)$, current age $x_0^i \sim U(\underline{x}^i, \overline{x}^i)$ and monthly premium $P^i \sim U(50, 500)$ where $N(\mu, \sigma)$ denotes the normal distribution with mean μ and variance σ, and $U(a, b)$ denotes a uniform distribution in the interval $[a, b]$. In addition, the side conditions $15 \leq \underline{x}^i \leq 55$ and $55 \leq \overline{x}^i \leq 70$ are respected. The probability that the contracts of a model point belong to female policyholders is assumed to be 55%. From the difference of exit age and current age the maturity time $d^i = \overline{x}^i - x^i$ of the contracts is computed. As sample insurance product, an endowment insurance with death benefit, constant premium payments and surrender option is considered as described in Example 1. For simplicity, we assume that the policies have not received any bonus payments before the start of the simulation, i.e., $B_0^i = 0$ for all $i = 1, \ldots, m$. We take the probabilities q_k^i of death from the DAV 2004R mortality table and choose exponential distributed surrender probabilities $u_k^i = 1 - e^{-0.03\Delta t}$. At time t_0, we assume a uniform bond allocation, i.e., $n_j = (1 - \beta)C_0 / \sum_{i=0}^{\tau-1} b_0(i)$ for $j = 1 - \tau, \ldots, 0$. We assume $Q_0 = 0$ which means that the shareholders will not make additional payments to the company to avoid a ruin. This way, $\mathbb{E}[Q_k]$ serves as a direct measure for the investment returns of the shareholders in the time interval $[0, t_k]$. The total initial reserves of the company are then given by $F_0 = \gamma_0 D_0$. In the following, we choose a simulation horizon of $T = 10$ years and a period length of $\Delta t = 1/12$ years, i.e., $K = 120$. In our numerical tests we use the capital market, product and management parameters as displayed in the second rows of Table 2 and 3 unless stated otherwise. In Table 2 and 3 also the sensitivities $f'(v)/f(v)$ (see Sect. 2.5) are displayed for different functions $f \in \{\text{PD}_K, \mathbb{E}[Q_K], \mathbb{E}[F_K]\}$ and different model input parameter v, e.g., $\partial \text{PD}_K / (\partial \mu \, \text{PD}_K) = -0.431$.

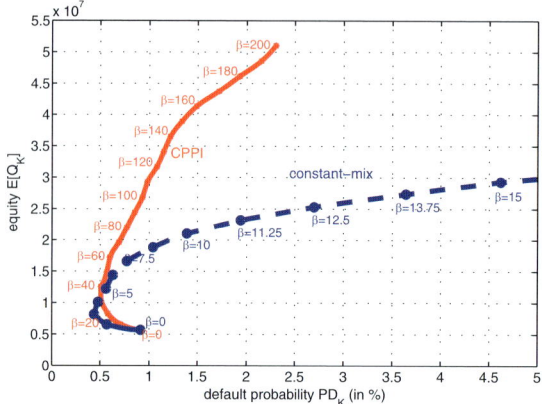

Fig. 2. Risk-return profiles of the different capital allocation strategies

4.2 Capital Allocation

To illustrate possible applications of the ALM model, we compare the constant-mix capital allocation strategy of Sect. 2.3 with an CPPI (constant proportion portfolio insurance) capital allocation strategy (see, e.g., [34]) with respect to the resulting default risk PD_K and returns $\mathbb{E}[Q_K]$. Within the CPPI strategy, the proportion of funds invested in (risky) stocks is linked to the current amount of reserves. The strategy is realised in our model framework by replacing $\beta(C_{k-1} + P_k)$ in equation (7) by βF_{k-1} with $\beta \in \mathbb{R}^+$. The resulting risk-return profiles of the constant-mix strategy and of the CPPI strategy are displayed in Fig. 2 for different choices of β.

We see that the slightly negative correlation $\rho = -0.1$ results in a diversification effect such that the lowest default risk is not attained at $\beta = 0$ but at about $\beta = 2.5\%$ in the constant-mix case and at about $\beta = 40\%$ in the CPPI case. Higher values of β lead to higher returns but also to higher risks. As an interesting result we further see that the CPPI strategy almost always leads to portfolios with much higher returns at the same risk and is therefore clearly superior to the constant-mix strategy almost independently of the risk aversion of the company. The only exception is a constant-mix portfolio with a stock ratio β of 2.5–4%, which could be an interesting option for a very risk averse company.

4.3 Effective Dimension

For the setting of Sect. 4.1, we determine in this section the effective dimensions d_t and d_s of the integral (17) in the truncation and superposition sense, respectively, see Sect. 3.3. The effective dimensions depend on the nominal

Table 4. Truncation dimensions d_t of the ALM integrand (17) for different nominal dimensions d and different covariance matrix decompositions

d	Random walk	Brownian bridge	Principal comp.
32	32	7	12
64	64	7	14
128	124	13	12
256	248	15	8
512	496	16	8

dimension d, on the discretization of the underlying Gaussian process and on all other model parameters. In Table 4, the effective dimensions d_t are displayed which arise by the methods described in [45] for different nominal dimensions d if the random walk, the Brownian bridge and the principal component (PCA) path construction is employed, respectively. One can see that the Brownian bridge and PCA path construction lead to a large reduction of the effective dimension d_t compared to the random walk discretization. In the latter case, the effective dimension d_t is almost as large as the nominal dimension d while in the former cases the effective dimensions are almost insensitive to the nominal dimensions and are bounded by only $d_t = 16$ even for very large dimensions as $d = 512$. In case of the PCA construction, d_t is even slightly decreasing for large d which is related to the so-called concentration of measure phenomenon, see [31]. Further numerical computations using the method described in [44] show that the ALM problem is also of very low effective dimension d_s in the superposition sense. Here, we only consider moderately high nominal dimensions due to the computational costs which increase with d. For $d \leq 32$, we obtain that the integral (17) is 'nearly' additive, i.e. $d_s = 1$, independent of d and independent of the covariance matrix decomposition. Note that the effective dimensions are affected by several parameters of the ALM model. More results which illustrate how the effective dimensions in the truncation sense vary in dependence of the capital market model and of other parameters can be found in [18].

4.4 Convergence Rates

In this section, we compare the following methods for the computation of the expected value (17) with the model parameters specified in Sect. 4.1:

- MC Simulation,
- QMC integration based on Sobol point sets (see [32, 43]),
- dimension-adaptive SG based on the Gauss–Hermite rule (see [16]).

In various numerical experiments, the Sobol QMC method and the dimension-adaptive Gauss–Hermite SG method turned out to be the most efficient representatives of several QMC variants (we compared Halton, Faure, Sobol low discrepancy point sets and three different lattice rules with and without ran-

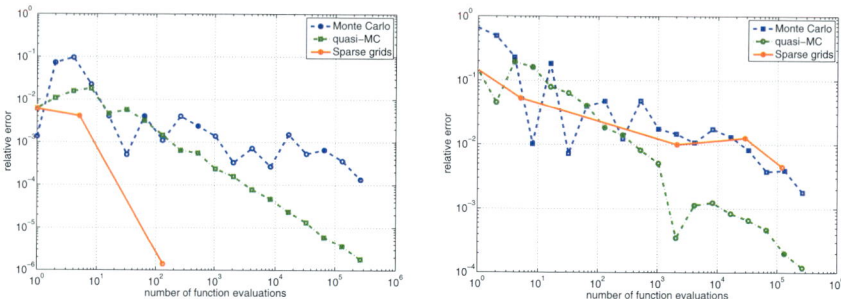

Fig. 3. Errors and required number of function evaluations of the different numerical approaches to compute the expected value (17) with $d = 32$ (*left*) and with $d = 512$ (*right*) for the model parameters specified in Sect. 4.1

domisation) and of several SG variants (we compared trapezoidal, Clenshaw–Curtis, Patterson, Gauss–Legendre and Gauss–Hermite rule and different grid refinement strategies), respectively. The results for $d = 32$ and $d = 512$ are summarised in Fig. 3 where the number n of function evaluations is displayed which is needed to obtain a given accuracy. In both cases we used the Brownian bridge path construction for the stock prices and short interest rates. One can see that the QMC method clearly outperforms MC simulation in both examples. The QMC convergence rate is close to one and nearly independently of the dimension. Moderate accuracy requirements of about 10^{-3}–10^{-4} are obtained by the QMC method about 100-times faster as by MC simulation. For higher accuracy requirements, the advantage of the QMC method is even more pronounced. Recall that these results can not be explained by the Koksma–Hlawka inequality but by the very low effective dimension of the ALM problem, see Sect. 4.3. The performance of the SG method deteriorates for very high dimensions. In the high dimensional case $d = 512$, the SG method is not competitive to QMC. For the moderately high dimension $d = 32$, sparse grids are the most efficient method with a very high convergence rate of almost three. With 129 function evaluation already an accuracy of 10^{-6} is achieved. Further numerical experiments indicate that the performance of the SG method is more sensitive than (Q)MC to different choices of model parameters which affect the smoothness of the integrand, like more aggressive bonus declaration schemes and more volatile financial markets.

5 Concluding Remarks

In this article, we first described a discrete time model framework for the asset-liability management of life insurance products. The model incorporates fairly general product characteristics, a surrender option, a reserve-dependent bonus declaration, a dynamic capital allocation and a two-factor stochastic capital market model. The recursive formulation of the model allows for an efficient computation of the model equations. Furthermore, the model structure is modular and allows to be extended easily. Numerical experiments illustrate that the model captures the main behaviour patterns of the balance sheet development of life insurance products. In the second part of this article, we investigated the application of deterministic integration schemes, such as quasi-Monte Carlo and sparse grid methods for the numerical simulation of ALM models in life insurance. Numerical results demonstrate that quasi-Monte Carlo and sparse grid methods can often outperform Monte Carlo simulation for the ALM of participating life insurance products. Furthermore, quasi-Monte Carlo methods converge nearly independently of the dimension and produce even for high dimensions $d = 512$ more precise results than MC. Sparse grids are the most efficient method for moderately high dimensions, but their performance is more sensitive to different choices of model parameters which affect the smoothness of the integrand and deteriorates for very high dimensions. In these cases additional transformations are required to improve the smoothness of the integrand. To explain the efficiency of the deterministic methods we computed the effective dimension of the ALM problem with and without dimension reduction techniques and showed that ALM problem are often of very low effective dimension in the truncation and also in the superposition sense.

Acknowledgement

This research was supported by the program "Mathematics for innovations in industry and service" of the German Ministry for Education and Research (BMBF).

References

1. P. Acworth, M. Broadie, and P. Glasserman. A comparison of some Monte Carlo and quasi-Monte Carlo methods for option pricing. In *Monte Carlo and Quasi-Monte Carlo Methods in Scientific Computing*, P. Hellekallek, H. Niederreiter, (eds.), pages 1–18. Springer, 1998.
2. A. Bacinello. Fair pricing of life insurance participating contracts with a minimum interest rate guaranteed. *Astin Bulletin*, 31(2):275–297, 2001.
3. A. Bacinello. Pricing guaranteed life insurance participating policies with annual premiums and surrender option. *Astin Bulletin*, 7(3):1–17, 2003.
4. L. Ballotta, S. Haberman, and N. Wang. Guarantees in with-profit and unitized with-profit life insurance contracts: Fair valuation problem in presence of the default option. *Insurance: Math. Ecomonics*, 73(1):97–121, 2006.
5. T. Bonk. A new algorithm for multi-dimensional adaptive numerical quadrature. In W. Hackbusch and G. Wittum, editors, *Adaptive Methods – Algorithms, Theory, and Applications, NNFM 46*, pages 54–68. Vieweg, 1994.
6. D. Brigo and F. Mercurio. *Interest Rate Models – Theory and Practice*. Springer, 2001.

7. E. Briys and F. Varenne. On the risk of life insurance liabilities: Debunking some common pitfalls. *J. Risk and Insurance*, 64:37–57, 1997.
8. Hans-Joachim Bungartz and Stefan Dirnstorfer. Multivariate quadrature on adaptive sparse grids. *Computing*, 71(1):89–114, 2003.
9. Hans-Joachim Bungartz and Michael Griebel. Sparse grids. *Acta Numerica*, 13:1–123, 2004.
10. R. E. Caflisch, W. J. Morokoff, and A. B. Owen. Valuation of mortgage backed securities using Brownian bridges to reduce effective dimension. *J. Comp. Finance*, 1(1):27–46, 1997.
11. J. Cox, J. Ingersoll, and S. Ross. A theory of the term structure of interest rates. *Econometrica*, 53:385–407, 1985.
12. P. Davis and P. Rabinowitz. *Methods of Numerical Integration*. Academic Press, 1984.
13. M. De Felice and F. Moriconi. Market based tools for managing the life insurance company. *Astin Bulletin*, 1(35):79–111, 2005.
14. J. Dick. Walsh spaces containing smooth functions and quasi-monte carlo rules of arbitrary high order. Availabe at http://www2.maths.unsw.edu.au/Contacts/profile.php?logname=josi, 2007.
15. T. Gerstner and M. Griebel. Numerical integration using sparse grids. *Numerical Algorithms*, 18:209–232, 1998.
16. T. Gerstner and M. Griebel. Dimension–adaptive tensor–product quadrature. *Computing*, 71(1):65–87, 2003.
17. T. Gerstner, M. Griebel, and M. Holtz. Numerical simulation methods for asset-liability management problems in life insurance. In preparation, 2007.
18. T. Gerstner, M. Griebel, and M. Holtz. The Effective Dimension of Asset-Liability Management Problems in Life Insurance. In C. Fernandes, H. Schmidli, and N. Kolev, editors, *Proc. Third Brazilian Conference on Statistical Modelling in Insurance and Finance*, pages 148–153, 2007.
19. T. Gerstner, M. Griebel, M. Holtz, R. Goschnick, and M. Haep. A general asset-liability management model for the efficient simulation of portfolios of life insurance policies. *Insurance: Math. Ecomonics*, 2007. In press.
20. M. Giles and P. Glasserman. Smoking adjoints: Fast Monte Carlo greeks. *Risk*, 19:88–92, 2006.
21. M.B. Giles. Multi-level Monte Carlo path simulation. To appear in Operations Research, 2007.
22. Paul Glasserman. *Monte Carlo Methods in Financial Engineering*. Springer, 2003.
23. M. Griebel. Sparse grids and related approximation schemes for higher dimensional problems. In *Proceedings of the conference on Foundations of Computational Mathematics (FoCM05), Santander*, 2005.
24. A. Grosen and P. Jorgensen. Fair valuation of life insurance liabilities: The impact of interest rate guarantees, surrender options and bonus policies. *Insurance: Math. Economics*, 26(1):37–57, 2000.
25. J Hull. *Options, Futures and other Derivative Securities*. Prentice Hall, Upper Saddle River, 2000.
26. Junichi Imai and Ken Seng Tan. Minimizing effective dimension using linear transformation. In H. Niederreiter, editor, *Monte Carlo and Quasi-Monte Carlo Methods 2002*, pages 275–292. Springer, 2004.

27. Reinhold Jaquemod. *Abschlussbericht der DAV-Arbeitsgruppe Stochastisches Unternehmensmodell für deutsche Lebensversicherungen.* Schriftenreihe angewandte Versicherungsmathematik. Versicherungswirtschaft, 2005.
28. Ioannis Karatzas and Steven E. Shreve. *Methods of Mathematical Finance.* Springer, New York, 1998.
29. A. Kling, A. Richter, and J. Russ. The interaction of guarantees, surplus distribution, and asset allocation in with-profit life insurance policies. *Insurance: Math. Economics*, 40(1):164–178, 2007.
30. Peter E. Kloeden and Eckhard Platen. *Numerical Solution of Stochastic Differential Equations.* Springer, 1992.
31. Michel Ledoux. *The Concentration of Measure Phenomenon.* Mathematical Surveys and Monographs. American Mathematical Society, 2001.
32. C. Lemieux, M. Cieslak, and K. Luttmer. RandQMC user's guide: A package for randomized quasi-Monte Carlo methods in C. Available at http://www.math.ucalgary.ca/~lemieux, 2004.
33. K. Miltersen and S. Persson. Guaranteed investment contracts: Distributed and undistributed excess returns. *Scand. Actuarial J.*, 23:257–279, 2003.
34. Thomas Moller and Mogens Steffensen. *Market-Valuation Methods in Life and Pension Insurance.* International Series on Actual Science. Cambridge University Press, 2007.
35. B. Moro. The full Monte. *RISK*, 8(2), 1995.
36. B. Moskowitz and R. Calfisch. Smoothness and dimension reduction in quasi-Monte Carlo methods. *J. Math. Comp. Modeling*, 23:37–54, 1996.
37. H. Niederreiter. *Random Number Generation and Quasi-Monte Carlo Methods.* SIAM, Philadelphia, 1992.
38. E. Novak and K. Ritter. High dimensional integration of smooth functions over cubes. *Numer. Math.*, 75:79–97, 1996.
39. E. Novak and H. Wozniakowski. Intractability results for integration and discrepancy. *J. Complexity*, 17:388–441, 2001.
40. H. Sloan and S. Joe. *Lattice Methods for Multiple Integration.* Oxford University Press, New York, 1994.
41. I.H. Sloan and H. Wozniakowski. When are quasi-Monte Carlo algorithms efficient for high-dimensional integrals? *J. Complexity*, 14:1–33, 1998.
42. S.A. Smolyak. Quadrature and interpolation formulas for tensor products of certain classes of functions. *Dokl. Akad. Nauk SSSR*, 4:240–243, 1963.
43. I. M. Sobol. The distribution of points in a cube and the approximate evaluation of integrals. *J. Comp. Mathematics and Math. Physics*, 7:86–112, 1967.
44. X. Wang and I. Sloan. Why are high-dimensional finance problems often of low effective dimension? *SIAM J. Sci. Comput.*, 27(1):159–183, 2005.
45. Xiaoqun Wang and Kai-Tai Fang. The effective dimension and quasi-Monte Carlo integration. *J. Complexity*, 19(2):101–124, 2003.
46. Kurt Wolfsdorf. *Versicherungsmathematik Teil 1 Personenversicherung.* Teubner, 1986.

On the Dynamics of the Forward Interest Rate Curve and the Evaluation of Interest Rate Derivatives and Their Sensitivities

Cristian Croitoru[1], Christian Fries[2,3], Willi Jäger[1], Jörg Kampen[1,4], and Dirk-Jens Nonnenmacher[2,5]

[1] Interdisciplinary Center for Scientific Computing (IWR) Heidelberg, INF 368, 69120 Heidelberg, Germany
{cristian.croitoru,jaeger}@iwr.uni-heidelberg.de
[2] Dresdner Bank AG, Jürgen-Ponto-Platz 1, 60301 Frankfurt am Main, Germany
[3] now at DZBank AG, Platz der Republik, 60265 Frankfurt am Main, Germany
mail@christian-fries.de
[4] now at Weierstrass Institute for Applied Analysis and Stochastics, Mohrenstr. 39, 10117 Berlin, Germany, kampen@wias-berlin.de
[5] now at HSH Nordbank AG, Gerhart-Hauptmann-Platz 50, 20095 Hamburg, Germany

Summary. We present an overview of in project developed new techniques in computing key quantities of financial markets. Our approach is generic in the sense that the techniques apply essentially in the general frame work of markets which are described by systems of stochastic differential equations. We exemplify our methods in the LIBOR market model which is a standard interest rate market model widely used in practice and has its name from the daily quoted London interbank offered rates. The LIBOR market model has been developed in recent years beyond the classical framework in the direction of incomplete market models (with stochastic volatility and with jumps). Particular challenges are the high dimensionality (up to 20–40 factors), the calibration, and related problems of derivative prices evaluation and computation of sensitivities. We show how advanced Monte-Carlo techniques can be combined with analytic results about transition densities in order to obtain highly efficient and accurate numerical schemes for computing some of the key quantities in financial markets, especially hedging parameters.

1 Introduction

Evaluation of derivatives and their sensitivities leads to mathematical problems of determining expected values of initial value systems of stochastic differential equations and stochastic derivatives described in the framework of

---

[1] Project supported by BMBF Program 'Mathematics for innovations in industry and services' and in cooperation with Dresdner Bank AG

Malliavin calculus or, equivalently, to Cauchy problems of (systems) of partial differential equations. Evaluation of high-dimensional interest rate derivatives and their sensitivities are based typically on Monte Carlo simulation methods. An alternative (and usually incompatible) method are sparse grids which were investigated in a preceding project. In praxis one uses Monte-Carlo techniques because they are well-tested for higher dimensionial advanced models, and because they are practically the most robust with respect to the curse of dimension and the curse of low regularity. The latter problem occurs mainly for two types of reasons in finance. The first are inherent low regularity of solutions of mathematical models such as free boundary problems describing exotic products, e.g. American derivatives, where the holder of the option has the right to exercise at any time where the contract is valid. Lower regularity of derivatives may also occur apart from exotics in market models based on systems of Feller processes where jump measures depend on the state of underlyings. The second type of reason is low regularity which may not appear on the mathematical level but appears at least on the numerical or computational level. It may often be the case that degenerate diffusions can be avoided on the mathematical level, but come back on the numerical level when we have to deal with low volatilities. It may even be that we are forced to deal with degenerate equations for purely computational reasons of dimension reduction. Indeed in the case of the analytical WKB-representation of transition densities, which we use here, it turns out that higher order approximations (orders beyond quadratic precision with respect to time) are computable in a reasonbale amount of time within the current technical possibilities if some dimension reduction techniques are applied. There are, however, some subtleties which have to be taken into account when applying the techniques of analytic approximation of transition densities in order to obtain numerical schemes based on Monte-Carlo techniques. The first one is that we cannot simply sample from the analytical WKB-approximation. Therefore a simple prior scheme is introduced from which we sample. The analytic approximation is used then to obtain the weights in the Monte-Carlo scheme. Mathematically, derivatives can simply be obtained by derivation of the analytic WKB-approximation. The resulting scheme (both for evaluation of derivatives and sensitivities) was called the proxy scheme in [FrKa]. However, it turned out that the resulting Monte-Carlo scheme for sensitivities has unbounded variance if the product of the squared volatility and time is small. Therefore, a refined Monte-Carlo scheme was proposed in [KKS] which ensures that the variance is small.

In this paper we shall give an overview about the techniques involved starting with the description of the basic framework in Sect. 2 and of the general results for analytic WKB-expansion obtained in [Ka] in Sect. 3. In Sect. 4 we recall and improve the basic proxy scheme of [FrKa] for evaluation of financial derivatives using higher order WKB-approximations for the Monte-Carlo schemes. We also point out how dimension reduction can be put on a solid mathematical basis (while the details are to complex to describe it here and will be dealt within an upcoming paper). In Sect. 5 we describe the niceties of exploding variance and how this affects the construction of robust schemes for sensitivities. In Sect. 6 we apply the method to the LIBOR market model and discuss some numerical results.

2 The Mathematical Framework of Market Models, Derivatives and Sensitivities

We consider a Markovian system (S, B) of underlying asset process in $\mathbb{R}_+^n \times \mathbb{R}_+$ ($\mathbb{R}_+ := \{x : x > 0\}$) on a filtered probability space $(\Omega, \mathcal{F}, (\mathcal{F}_t)_{t \in [t_0,T]}, P)$, consisting of n risky assets $S = (S^1, ..., S^n)$, and a numeraire B, where the filtration $(\mathcal{F}_t)$ satisfies the usual conditions and that the system (t, S_t) is Markovian with respect to this filtration. Moreover we assume that S has an absolute continuous transition kernel with density $p(t, x, s, y)$, which has derivatives of any order in $0 \le t < s$, $x, y \in \mathbb{R}_+^n$. Further we assume that B_t, $t > 0$, is adapted to $(S_t, 0 \le t \le s)$ and is of finite variation. The dynamics of the system (S, B) is given by

$$\frac{dS^i}{S^i} = r(t, S)dt + \sum_{j=1}^n \sigma^{ij}(t, S)dW^j, \quad \frac{dB}{B} = r(t, S)dt, \quad 1 \le i, j \le n, \qquad (1)$$

in the (risk-neutral) measure P. In (1) $W = (W^1, ..., W^n)^\top$ is an adapted n-dimensional standard Wiener process. Here, $(\mathcal{F}_t)$ is the P-augmentation of the filtration generated by W. We could include additional jump terms but since the theory of WKB-expansions of these extended models has not been published yet we stick to the case of systems which have models with continuous paths. We assume that the process has a transition density $p(t, x, s, y)$ which is differentiable with respect to $x, y \in \mathbb{R}_+^n$, $s, t \in [t_0, T]$, $t > s$, up to any order. For an $(\mathcal{F}_\cdot)$-stopping time τ and payoff $f(S_\tau)B_\tau$ contingent claims are priced at time t_0 (assuming $B(0) = b_0 = 1$) by a formula of type

$$v(t_0, x_0) = E_Q \ f(S_\tau^{t_0, x_0})$$

where Q is some equivalent measure (e.g. [Du]). In case of complete markets this so-called risk-neutral measure is unique. In general, however, markets are incomplete, and additional criteria and methods have to be introduced in order to select a risk-neutral measure. Finding the criteria of selection (modeling) as well as analyzing, computing, and estimating parameters of the related equations (typically Hamilton–Jacoby equations) is a topic of ongoing research which will be also considered in the next BMBF-project. Here we assume that a measure has been chosen and computed or is given (as in the case of a complete market). For deterministic τ, say $\tau \equiv T$, the value process of the European claim for $t_0 \le t \le T$ is

$$v_t := v(t, S_t, B_t) := B_t E^{\mathcal{F}_t} f(S_T) = e^{\int_{t_0}^t r(s, S_s)ds} E_t \ f(S_T),$$

where E_t denotes conditional expectation with respect to $\mathcal{F}_t$, and the discounted price process $u_t := v_t/B_t$ can be expressed in terms of the transition density via

$$u_t := u(t, S_t) := E^{\mathcal{F}_t} f(S_T) = \int_{\mathbb{R}^n} p(t, S_t, T, y) f(y) dy,$$

and where

$$u(t, s) = \int p(t, s, T, y) f(y) dy \tag{2}$$

is the unique solution of the Cauchy problem

$$\begin{cases} \dfrac{\partial u}{\partial t} + \dfrac{1}{2} \sum_{i,j=1}^n s^i s^j \left(\sigma\sigma^\top\right)^{ij}(t,s) \dfrac{\partial^2 u}{\partial s^i \partial s^j} + \sum_{i=1}^n s^i r(t,s) \dfrac{\partial u}{\partial s^i} = 0, \\ u(T, s) = f(s). \end{cases} \tag{3}$$

It is clear then that $p(\cdot, \cdot, T, y)$ is the fundamental solution of the latter equation (3) with $p(T, s, T, y) = \delta(s - y)$, where δ the Dirac distribution. The optimal stopping problems for the pricing of American and Bermudean options lead to related Cauchy free boundary problems.

3 WKB-Expansions

Next we turn to some general results on analytic expansions of the fundamental solution (i.e. the transition density) p (cf. [Ka] for details). If we write the Cauchy problem (3) in logarithmic coordinates $x_i = \ln(s_i)$, then it becomes a Cauchy problem on the domain $[0, T] \times \mathbb{R}^n \to \mathbb{R}$ where we denote the corresponding diffusion coefficients by $a_{ij}(t, x)$ and $b_i(t, x)$. In order to simplify the notation we consider the time-homogenous case and drop the dependence on time t for the moment (we come back to the time-dependent case later in the context of the LIBOR market model). Pointwise valid analytic expansions of the fundamental solution p exist if

(A) the matrix norm of $(a_{ij}(x))$ is bounded below and above by $0 < \lambda < \Lambda < \infty$ uniformly in x,
(B) the smooth functions $x \to a_{ij}(x)$ and $x \to b_i(x)$ and all their derivatives are bounded.

For more subtle (and partially weaker conditions) we refer to [Ka]. Further, we denote the additional condition

(C) there exists a constant c such that for each multiindex α and for all $1 \le i, j, k \le n$,

$$\left|\dfrac{\partial a_{jk}}{\partial x^\alpha}\right|, \left|\dfrac{\partial b_i}{\partial x^\alpha}\right| \le c \exp\left(c|x|^2\right). \tag{4}$$

Then

Theorem 1. *If the hypotheses (A), (B) are satisfied, then the fundamental solution p has the representation*

$$p(\delta t, x, y) = \frac{1}{\sqrt{2\pi\delta t}^n} \exp\left(-\frac{d^2(x,y)}{2\delta t} + \sum_{k \geq 0} c_k(x,y)\delta t^k\right), \quad (5)$$

where d and c_k are smooth functions, which are unique global solutions of the first order differential equations (6), (7) and (9) below. Especially,

$$(\delta t, x, y) \to \delta t \ln p(\delta t, x, y) = -\frac{n}{2}\delta t \ln 2\pi\delta t - \frac{d^2}{2} + \sum_{k \geq 0} c_k(x,y)\delta t^{k+1}$$

is a smooth function which converges to $-\frac{d^2}{2}$ as $\delta t \searrow 0$, where d is the Riemannian distance induced by the line element $ds^2 = \sum_{ij} a_{ij}^{-1} dx_i dx_j$, where with a slight abuse of notation (a_{ij}^{-1}) denotes the inverse matrix of (a_{ij}). If the hypotheses (A), (B) and (C) are satisfied, then, in addition, the functions d and c_k, $k \geq 0$, equal their Taylor expansion around y globally.

The recursion formulas for d and c_k, $k \geq 0$, are

$$d^2 = \frac{1}{4}\sum_{ij} d^2_{x_i} a_{ij} d^2_{x_j}, \quad (6)$$

where $d^2_{x_k}$ denotes the derivative of the function d^2 with respect to the variable x_k, with the boundary condition $d(x,y) = 0$ for $x = y$,

$$-\frac{n}{2} + \frac{1}{2}Ld^2 + \frac{1}{2}\sum_i \left(\sum_j (a_{ij}(x) + a_{ji}(x)) \frac{d^2_{x_j}}{2}\right) \frac{\partial c_0}{\partial x_i}(x,y) = 0, \quad (7)$$

where

$$c_0(x,y) = -\frac{1}{2}\ln\sqrt{\det(a_{ij}(y))}, \quad (8)$$

and for $k + 1 \geq 1$ we obtain

$$(k+1)c_{k+1}(x,y) + \frac{1}{2}\sum_{ij} a_{ij}(x) \left(\frac{d^2_{x_i}}{2}\frac{\partial c_{k+1}}{\partial x_j} + \frac{d^2_{x_j}}{2}\frac{\partial c_{k+1}}{\partial x_i}\right)$$

$$= \frac{1}{2}\sum_{ij} a_{ij}(x) \sum_{l=0}^{k} \frac{\partial c_l}{\partial x_i}\frac{\partial c_{k-l}}{\partial x_j} + \frac{1}{2}\sum_{ij} a_{ij}(x)\frac{\partial^2 c_k}{\partial x_i \partial x_j} + \sum_i b_i(x)\frac{\partial c_k}{\partial x_i},$$

with boundary conditions

$$c_{k+1}(x,y) = R_k(y,y) \text{ if } x = y, \quad (9)$$

R_k being the right side of (9). We see that the Riemannian distance d has to be approximated in regular norm in order to get an accurate WKB-expansion in general. How this can be accomplished is shown in [Ka2]. Designing numerical schemes we work with approximations both with respect to time and with respect to spatial variables, of course. In order to analyze the time truncation error we consider WKB-approximations of the fundamental solution p of the form

$$p_l(t,x,T,y) = \frac{1}{\sqrt{2\pi\delta t}^n} \exp\left(-\frac{d^2(x,y)}{2\delta t} + \sum_{k=0}^{l} c_k(x,y)\delta t^k\right), \quad (10)$$

i.e. we assume that the coefficients d^2 and c_k, $0 \le k \le l$ have been computed up to order l. Let $(A)_t, (B)_t, (C)_t$ denote the analogies to assumptions (A), (B), (C) for time- and space-dependent coefficients, and let us denote the domain of the Cauchy problem by $D = (0,T) \times \mathbb{R}^n$. For integers $n \ge 0$ and real numbers $\delta \in (0,1)$ let $C^{m+\delta/2, n+\delta}(D)$ be the space of m (n) times differentiable functions such that the m-th (n-th) derivative with respect to time (space) is Hölder continuous with exponent $\frac{\delta}{2}$ (δ). Furthermore, $|\cdot|_{m+\delta/2, n+\delta}$ denote the natural norms associated with these function spaces. Then a consequence of Safanof's theorem (cf. [Kr]) is

Theorem 2. *Assume that $(A)_t, (B)_t$ and $(C)_t$ are satisfied and let $g \in C^{2+\delta}(\mathbb{R}^n)$ and $f \in C^{\delta/2, \delta}(D)$. If*

$$c \le -\lambda \text{ for some } \lambda > 0, \quad (11)$$

then the Cauchy problem

$$\begin{cases} \dfrac{\partial w}{\partial t} + \dfrac{1}{2}\sum_{ij} a_{ij}(t,x) \dfrac{\partial^2 w}{\partial x_i \partial x_j} + \sum_i b_i(t,x) \dfrac{\partial w}{\partial x_i} + c(t,x)w = f(\delta t, x) \text{ in } D \\ w(T,x) = g(x) \text{ for } x \in \mathbb{R}^n \end{cases} \quad (12)$$

has a unique solution w, and there exists a constant c depending only on δ, n, λ, Λ and $K = \max\{|a|_\delta, |b|_\delta, |c|_\delta\}$ such that

$$|w|_{1+\delta/2, 2+\delta} \le c\left[|f|_{\delta/2, \delta} + |g|_{2+\delta}\right]. \quad (13)$$

It can be shown (cf. [Ka, KKS]) that the truncation error

$$u^\Delta(t,x) = u(t,x) - u_l(t,x), \quad (14)$$

where

$$u(t,x) = \int_{\mathbb{R}^n} g(y) p(t,x,T,y) dy, \quad (15)$$

and

$$u_l(t,x) = \int_{\mathbb{R}^n} g(y) p_l(t,x,T,y) dy, \quad (16)$$

satisfies the following

Theorem 3. *Assume that conditions (A), (B), and (C) hold and that $g \in C_0^\delta(\mathbb{R}^n)$. Then*

$$|u(t,x,y) - u_l(t,x,y)|_{1+\delta/2, 2+\delta} = O\left(\delta t^{l-\frac{\delta}{2}}\right) \tag{17}$$

Finer error estimates based on Taylor expansions of WKB-functions c_k, $k \geq 0$, will be presented in [FrKa2]. A refined analysis in the special case of reducible diffusion will be presented in [CrKa]. For the classical LIBOR market model some simplifications are possible by a global transform to the Laplacian. We recall (cf. [Ka])

Proposition 1. *There is a global coordinate transformation for the operator in (3) such that the second order part of the transformed operator equals the Laplacian, if $a_{ij} = (\sigma\sigma^\top)_{ij}$ for a (square) matrix function σ which satisfies*

$$\sum_{l=1}^n \frac{\partial \sigma_{ik}(x)}{\partial x_l} \sigma_{lj}(x) = \sum_{l=1}^n \frac{\partial \sigma_{ij}(x)}{\partial x_l} \sigma_{lk}(x), \quad x \in \mathbb{R}^n. \tag{18}$$

If the condition of Proposition 1 is satisfied, then coordinate transformation leads to second order coefficients of the form $a_{ij} \equiv \delta_{ij}$, so that the solution of (6) becomes

$$d^2(x,y) = \sum_i (x_i - y_i)^2. \tag{19}$$

If conditions (A), (B), (C) and (18) hold, then in the transformed coordinates, explicit formulas for the coefficient functions $c_k, k \geq 0$ can be computed via the formulas

$$c_0(x,y) = \sum_i (y_i - x_i) \int_0^1 b_i(y + s(x-y))ds, \tag{20}$$

and

$$c_{k+1}(x,y) = \int_0^1 R_k^L(y + s(x-y), y) s^k ds, \tag{21}$$

where

$$R_k^L(x,y) = \frac{1}{2}\sum_i \sum_{l=0}^k \frac{\partial c_l}{\partial x_i} \frac{\partial c_{k-l}}{\partial x_i} + \frac{1}{2}\Delta c_k + \sum_i b_i(x) \frac{\partial c_k}{\partial x_i}. \tag{22}$$

Assuming that b_i has a power series representation (here $\Delta x := (x-y)$)

$$b_i^y(x) = \sum_\gamma b_i^{y\gamma} \Delta x^\gamma \tag{23}$$

we obtain the recursion formulas

$$c_0(x,y) = -\sum_i \Delta x_i \int_0^1 b_i(y + s\Delta x)ds = -\sum_i \sum_\gamma b_i^{y\gamma} \Delta x^{\gamma+1_i} \frac{1}{1+|\gamma|} \tag{24}$$

and

$$c_{k+1}(x,y) = \sum_i \sum_{\delta \leq \alpha} \left\{ \frac{1}{2} \sum_{l=0}^{k} \sum_{\beta+\gamma=\delta} (\beta_i+1)(\gamma_i+1) c^y_{l(\beta+1_i)} c^y_{(k-l)(\gamma+1_i)} \right.$$
$$\left. + \frac{1}{2}(\delta_i+2)(\delta_i+1) c_{k(\delta+2_i)} + b_i^{y\delta}(\delta_i+1) c_{k(\delta+1_i)} \right\} p_{k\delta}^{y\alpha} \Delta x^\delta \quad (25)$$

where we use, with $\delta_\Sigma := \sum_{i=1}^{n} \delta_i$ (for more details consider [Ka] and [CrKa]),

$$\int_0^1 (y+s(x-y))^\alpha s^{k-1} ds = \sum_{\delta=0}^{\alpha} \frac{1}{\delta_\Sigma + k} \left[\prod_{i=1}^{n} \left(\frac{\alpha_i!}{\delta_i!(\alpha_i - \delta_i)!} \right) y^{(\alpha-\delta)} \right] \Delta x^\delta$$
$$=: \sum_{\delta=0}^{\alpha} p_{k\delta}^{y\alpha} \Delta x^\delta.$$

4 The Proxy Scheme for Evaluation of Financial Derivatives

We reconsider the proxy scheme of [FrKa]. There exist several refinements of this scheme (cf. [Fr]), but the basic idea is the following. An option pricing formula with target scheme S^* is reinterpreted in the framework of weighted Monte-Carlo schemes by an auxiliary proxy scheme S^p with weights w:

$$E_\mathbb{Q}\left(f(S^*(T))|\mathcal{F}_t\right) = E_\mathbb{Q}\left(f(S^p(T))w|\mathcal{F}_t\right) \quad (26)$$

We consider (conditioned on $\mathcal{F}_0$ and at $S(0) = 0$, for simplification of notation)

$$E_\mathbb{Q}(f(S^*)) = \int f(S') p_T^*(S') dS' = \int f(S') \frac{p_T^*(S')}{p_T^p(S')} p_T^p(S') dS' \quad (27)$$

$$\approx \frac{1}{n} \sum_{i=1}^{n} f(S^p(T, \omega_i)) \underbrace{\frac{p_T^*(S^p(T,\omega_i))}{p_T^p(S^p(T,\omega_i))}}_{w_i \text{ weights}} \quad (28)$$

where we put the time parameter of the fundamental solution as a subscript (since time dependence is not essential here), and drop the dependence on $S(0) = 0$ (since this reference is always the same and can be understood implicitly). In principle this idea can be applied for sensitivities. We have

$$\frac{\partial}{\partial X} E_\mathbb{Q}(f(S^*)) = \int f(S') \frac{\partial}{\partial X}\left(p_T^*(S')\right) dS' = \int f(S') \frac{\frac{\partial}{\partial X} p_T^*(S')}{p_T^p(S')} p_T^p(S') dS'$$

$$\approx \frac{\partial}{\partial X}\left(\frac{1}{n}\sum_{i=1}^{n} f(S^p(T,\omega_i)) \underbrace{\frac{p_T^*(S^p(T,\omega_i))}{p_T^p(S^p(T,\omega_i))}}_{w_i \text{ weights}}\right) \qquad (29)$$

with a parameter or an underlying X (higher derivatives work analogously, of course). The weighted Monte-Carlo scheme gives us the opportunity to sample from a simple prior scheme (e.g. an Euler scheme) and then use a more accurate scheme to get the correct weights. This is to avoid sampling from a higher order WKB-approximation which is practically impossible. This was the idea in [FrKa], where numerical schemes of slightly higher order than the Euler scheme were used as target scheme. In an upcoming paper [FrKa] we shall use higher order approximation of the LIBOR kernel. However, it turns out higher order WKB-approximation are computationally expensive for the full factor model. It is therefore useful to perform a dimension reduction by principal component analysis at the reference starting point of the scheme. But there are additional problems here in order to put everything on a solid mathematical basis. First of all, the transition densities are transition densities of degenerate diffusions, and the WKB-theory is derived for non-degenerate diffusions. However, degenerate diffusions behave differently with respect to uniqueness, existence and regularity. It can be shown that there are solutions to the degenerate equations which are closest to the non-dengenerate transition densities (cf. [Ka] for the basic ideas and [FrKa2] for a detailed study). The second problem is the adaptation of the error estimate. This will be considereed in [FrKa2]. A third problem was already mentioned in [FrKa]. The change of measure in the reinterpretation requires that

$$p^p(T_i, S, T_{i+1}, S') = 0 \Rightarrow p^*(T_i, S, T_{i+1}, S') = 0.$$

This can be ensured by an Euler subdiscretization (cf. [Fr]). In equation (29) we put the derivative 'outside' which indicates that the sensitivities can be computed by numerical differentiation of the whole scheme. In case of the WKB-approximation of the target scheme we can do the derivative directly and can avoid numerical differentiation. But it turned out that estimators have to be refined in order to have bounded variance estimators if the product of the squared volatility and time is small as was discovered in [KKS], and we shall describe it in the next section.

5 Refinement of the Proxy Scheme for Computation of Sensitivities

A detailed analysis (which were originally discovered by numerical observations of exploding variance for the estimator in (29) for small time) shows that a simple choice of the prior can lead to exploding variance for small time or small volatilities. Let us consider this phenomenon (cf [KKS] for more details): consider a smooth function $f : \mathbb{R}^n_+ \to \mathbb{R}_+$ and the transition function $p : \mathbb{R}^n_+ \times \mathbb{R}^n_+ \to \mathbb{R}_+$, where we drop the parameter T as a subscript of the transition function because it is of no importance for the following discussion. We want to estimate probabilistic representations for the integral

$$I(x) := \int p(x,y) f(y) dy,$$

and derivatives such as its gradient

$$\nabla_x I(x) = \int \nabla_x p(x,y) f(y) dy.$$

Let Y be some random variable with density ϕ (our prior) on $\mathbb{R}^n_+$, $\phi > 0$. with samples $_mY$ for $m = 1, ..., M$, where M is some large integer. Then,

$$I(x) = E\, p(x, \zeta) \frac{f(Y)}{\phi(Y)} \tag{30}$$

may be estimated by the unbiased Monte Carlo estimator

$$\widehat{I}(x) := \frac{1}{M} \sum_{m=1}^{M} p(x, {_mY}) \frac{f({_mY})}{\phi({_mY})}. \tag{31}$$

Hence, in accordance with the previous section, the estimator corresponding to

$$\nabla_x I(x) = E\, \frac{\partial}{\partial x} p(x, Y) \frac{f(Y)}{\phi(Y)} \tag{32}$$

is

$$\widehat{\nabla_x I}(x) := \frac{1}{M} \sum_{m=1}^{M} \frac{\partial}{\partial x} p(x, {_mY}) \frac{f({_mY})}{\phi({_mY})}. \tag{33}$$

However even the natural choice of a prior $\phi(\cdot) := p(s, \sigma; x_0, \cdot)$, where p is the lognormal distribution, i.e.

$$p(s, \sigma; x_0, y) := \frac{1}{(2\pi\sigma^2 s)^{n/2}} \prod_{i=1}^{n} \frac{\exp\left[-\frac{1}{2\sigma^2 s} \ln^2 \frac{y^i}{x_0^i}\right]}{y^i} \tag{34}$$

such that $p(s, \sigma; x_0, \cdot)$ is the density of the random variable $(x_0^1 e^{\sigma\sqrt{s}\xi^1}, ..., x_0^n e^{\sigma\sqrt{s}\xi^n})$, with ξ^i, $i = 1, ..., d$, are i.i.d. standard normal random variables, leads for the simple choice of $f \equiv ||x_0||$ (a constant of order x_0 in magnitude) to an variance

$$\text{Var}\left[\frac{\widehat{\partial I}}{\partial x^j}(x_0)\right] = \frac{||x_0/x_0^j||^2}{M} \frac{1}{\sigma^2 s} \tag{35}$$

which explodes when $\sigma^2 s$ goes to zero. Therefore we choose estimators in accordance to the following theorem (cf. [KKS] for a proof).

Theorem 4. Let λ be a reference density on $\mathbb{R}^n$ with $\lambda(z) \neq 0$ for all z (for example, the standard normal density). Let ξ be an $\mathbb{R}^n$-valued random variable with density λ and $g : \mathbb{R}_+^n \times \mathbb{R}^n \to \mathbb{R}_+^n$ be a smooth map with $\nabla_z g(x,z) \neq 0$, such that for each $x \in \mathbb{R}_+^n$ the random variable $Y^x := g(x,\xi)$ has a density $\phi(x,\cdot)$ on $\mathbb{R}_+^n$. Then, we have the probabilistic representation

$$\nabla_x I(x) = E\, \nabla_x \frac{p(x, Y^x) f(Y^x)}{\phi(x, Y^x)} = E\, \nabla_x \frac{p(x, g(x,\xi)) f(g(x,\xi))}{\phi(x, g(x,\xi))}, \qquad (36)$$

with corresponding Monte Carlo estimator

$$\widehat{\nabla_x I}(x) = \frac{1}{M} \sum_{m=1}^{M} \nabla_x \frac{p(x, g(x,_m\xi)) f(g(x,_m\xi))}{\phi(x, g(x,_m\xi))}. \qquad (37)$$

Let $|\cdot|$ denote either a vector norm or a compatible matrix norm. Then it holds

$$E\left|\nabla_x \frac{p(x, g(x,\xi)) f(g(x,\xi))}{\phi(x, g(x,\xi))}\right|^2 \leq 2M_4^2 \left(M_2^2 M_3^2 + 4M_1^2 M_5^2 + 4M_1^2 M_3^2 M_6^2\right), \qquad (38)$$

hence the second moments of the Monte Carlo samplers for the components of $\partial I/\partial x$ are bounded by the right-hand-side of (38), if for fixed $x \in \mathbb{R}_+^n$, there are $\alpha_1, ..., \alpha_6 > 1$ with

$$\frac{1}{\alpha_4} + \frac{1}{\alpha_1} + \frac{1}{\alpha_5} = 1, \quad \frac{1}{\alpha_4} + \frac{1}{\alpha_2} + \frac{1}{\alpha_3} = 1, \quad \frac{1}{\alpha_4} + \frac{1}{\alpha_1} + \frac{1}{\alpha_6} + \frac{1}{\alpha_3} = 1,$$

such that,

$$E\, f^{2\alpha_1}(g(x,\xi)) = \int f^{2\alpha_1}(y) \phi(x,y) dy \leq M_1^{2\alpha_1},$$

$$E\, |\nabla_y f(g(x,\xi))|^{2\alpha_2} = \int |\nabla_y f(y)|^{2\alpha_2} \phi(x,y) dy \leq M_2^{2\alpha_2},$$

$$E\left|\frac{\partial g}{\partial x}(x,\xi)\right|^{2\alpha_3} \leq M_3^{2\alpha_3},$$

$$E\left(\frac{p(x, g(x,\xi))}{\phi(x, g(x,\xi))}\right)^{2\alpha_4} = \int \left(\frac{p(x,y)}{\phi(x,y)}\right)^{2\alpha_4} \phi(x,y) dy \leq M_4^{2\alpha_4},$$

$$E\left|\frac{p_x(x, g(x,\xi))}{p(x, g(x,\xi))} - \frac{\phi_x(x, g(x,\xi))}{\phi(x, g(x,\xi))}\right|^{2\alpha_5} =$$

$$\int \left|\frac{p_x(x,y)}{p(x,y)} - \frac{\phi_x(x,y)}{\phi(x,y)}\right|^{2\alpha_5} \phi(x,y) dy \leq M_5^{2\alpha_5},$$

and

$$E\left|\frac{p_y(x,g(x,\xi))}{p(x,g(x,\xi))} - \frac{\phi_y(x,g(x,\xi))}{\phi(x,g(x,\xi))}\right|^{2\alpha_6} =$$
$$\int\left|\frac{p_y(x,y)}{p(x,y)} - \frac{\phi_y(x,y)}{\phi(x,y)}\right|^{2\alpha_6}\phi(x,y)dy \leq M_6^{2\alpha_6},$$

with shorthands $p_x := \nabla_x p$, etc.

Similar results can be obtained for higher order derivatives and for derivatives with respect to parameters.

6 Applications to the LIBOR Market Model

We recall a LIBOR market model with respect to a tenor structure $0 < T_1 \ldots < T_{n+1}$ in the terminal measure P_{n+1} (induced by the terminal zero coupon bond $B_{n+1}(t)$). The dynamics of the forward LIBORs $L_i(t)$, defined in the interval $[0,T_i]$ for $1 \leq i \leq n$, are governed by the following system of SDE's (e.g., see [Sc]),

$$dL_i = -\sum_{j=i+1}^n \frac{\delta_j L_i L_j \gamma_i^\top \gamma_j}{1+\delta_j L_j}\,dt + L_i\,\gamma_i^\top dW_{n+1} =: \mu_i(t,L) + L_i\,\gamma_i^\top dW_{n+1}, \quad (39)$$

where $\delta_i = T_{i+1} - T_i$ are day count fractions and $t \to \gamma_i(t) = (\gamma_{i,1}(t),\ldots,\gamma_{i,d}(t))$, $(\gamma_i^\top \gamma_j)_{i,j=1}^n =: \rho$ are deterministic volatility vector functions defined in $[0,T_i]$. We denote the matrix with rows $\gamma_i^\top$ by Γ and assume that Γ is invertible. In (39), $(W_{n+1}(t) \mid 0 \leq t \leq T_n)$ is a standard d-dimensional Wiener process under the measure P_{n+1} with d, $1 \leq d \leq n$, being the number of driving factors. In what follows, we consider the full-factor LIBOR model with $d = n$ in the time interval $[0; T_1)$.

6.1 WKB Approximations for the LIBOR Kernel

In order to apply the recursion formulas for the coefficient functions c_0 (cf. (24)) and c_{k+1} (cf. (25)) we first observe that the transition density $p^L(s,u,t,v)$ of the classical LIBOR process (39) can be transformed to the equation of form

$$dY_i = \mu_i^Y(t,Y)dt + dW_{n+1}^i, \quad 1 \leq i \leq n. \tag{40}$$

For the transitional density $p^Y(s,x,t,y)$, we can compute c_k, $k = 0, 1, \ldots$ in (5) by (20)-(21) with $b_i(y) = \mu_i^Y(t,y)$. After that, we find $p^L(s,u,t,v)$ by density transformation formula. Note that the LIBOR-drift is an analytic function. Nevertheless, it can be written in form of a power series only locally. In [Ka3] techniques of regular polynomial interpolation are developed as a byproduct. Current research in [CrKa] experiments with higher order approximations of the transition density based on the formulas (24) and (25) and regular polynomial interpolation. Currently WKB-approximations have been tested numerically in a full factor LIBOR market model up to c_1 (cf. [KKS]). We finally report on that numerical experiments.

6.2 Case Study: European Swaptions

The estimators (31) and (37) are applied to pricing European swaptions and computing Deltas in the classical LIBOR market model. For more details we refer to [KKS]. Simulations for more complex products are currently done and show an excellent accuraccy of the WKB-expansion. A (payer) swaption contract payoff with maturity T_i and strike θ with principal \$1 is given by

$$u(L) = \frac{B_{n+1}(0)}{B_{n+1}(T_1)} \left(\sum_{j=i}^{n} B_{j+1}(T_1) \left(\delta_j L_j(T_1) \right) - \theta \right)^+ . \quad (41)$$

In the LIBOR market model (39) we take $\delta_i \equiv 0.5$ when $i \geq 1$, flat 3.5 % initial LIBOR curve and constant volatility loadings

$$\gamma_i(t) \equiv 0.2 e_i, \quad (e_i \text{ unit vector of standard basis in } \mathbb{R}^n)$$

and with input correlation matrix ρ with components,

$$\rho_{ij} = \exp\left[\frac{|j-i|}{n-1} \ln \rho_\infty\right], \quad 1 \leq i, j \leq n \quad (42)$$

with $n > 2$ and $\rho_\infty = 0.3$ (for more general correlation structures we refer to [Sc]). We consider at-the-money ($\theta = 3.5\,\%$) swaption over a period $[T_1, T_{19}]$. In our experiments, we take as φ a canonical lognormal approximation of transitional kernel $p_{ln}^L(s,x,t,y)$

$$\frac{1}{\sqrt{2\pi(t-s)}^n} \prod_{i=1}^{n} \frac{\Gamma_{ii}^{-1}}{v_i} \exp\left(-\frac{(\Gamma^{-1}((\log\frac{v_1}{u_1}\ldots\log\frac{v_n}{u_n}) - \mu^{ln}(s,t,x))^T)^2}{2(t-s)}\right)$$

with

$$\mu_i^{ln}(s,t,x) = (t-s)\left(\frac{|\gamma_i|^2}{2} - \sum_{j=i+1}^{n} \frac{|\gamma_i||\gamma_j|\rho_{ij}\delta_j x_j}{1+\delta_j x_j}\right), \quad 1 \leq i \leq n.$$

The bias achieves 5 % for European swaptions and 3 % for Deltas, see Table 1 and Table 2. Next we consider the estimators (31) and (37) with payoff (41)

Table 1. (the values are in basis points)

T_1	$\widehat{I}_{ex}$ (SD)	$\widehat{I}_{ln}$ (SD)	$\widehat{I}_0$ (SD)	$\widehat{I}_1$ (SD)
0.5	129.6(0.4)	128.9(0.4)	129.1(0.4)	128.4(0.4)
1.0	179.1(0.5)	179.4(0.5)	180.6(0.6)	178.7(0.5)
2.0	243.8(0.8)	246.0(0.8)	251.4(0.8)	245.1(0.8)
5.0	351.2(1.3)	357.8(1.3)	376.3(1.4)	349.4(1.3)
10.0	430.3(2.0)	453.3(2.2)	499.4(2.1)	430.6(1.8)

Table 2. (the values are in basis points)

T_1	$\widehat{\frac{\partial I_{ex}}{\partial x_1}}^{(h)}$ (SD)	$\widehat{\frac{\partial I_{ln}}{\partial x_1}}^{(h)}$ (SD)	$\widehat{\frac{\partial I_0}{\partial x_1}}^{(h)}$ (SD)	$\widehat{\frac{\partial I_1}{\partial x_1}}^{(h)}$ (SD)
0.5	2475.3(5.6)	2470.3(6.0)	2485.4(6.0)	2470.5(6.0)
1.0	2450.6(6.2)	2451.7(6.2)	2480.0(6.6)	2450.1(6.1)
2.0	2401.4(6.4)	2405.2(6.4)	2460.3(6.6)	2400.4(6.4)
5.0	2257.2(7.1)	2261.2(7.2)	2386.7(7.4)	2239.1(6.9)
10.0	2017.9(8.3)	2077.3(8.8)	2299.8(9.0)	2010.2(7.7)

at $x = L(0)$, where

$$\varphi(x, \cdot) = p_{ln}^L(0, x, T_1, \cdot),$$
$$p(x, \cdot) = p_0^L(0, x, T_1, \cdot) \quad \text{and} \quad p(x, \cdot) = p_1^L(0, x, T_1, \cdot).$$

denoted by $\widehat{I}_0$ and $\widehat{I}_1$, correspondingly. A comparison is done with "exact" values which are obtained by simulating M LIBOR trajectories (39) by log-Euler scheme with very small time step, $\Delta t = \delta_i/10$, and take

$$\widehat{I}_{ex} = \frac{1}{M}\sum_{m=1}^{M} u\left({}_m L_{T_1}^{0,x}\right), \tag{43}$$

$$\widehat{\frac{\partial I_{ex}}{\partial x_i}} = \frac{1}{M}\sum_{m=1}^{M} \frac{u\left({}_m L_{T_1}^{0,x+\Delta_i x}\right) - u\left({}_m L_{T_1}^{0,x+\Delta_i x}\right)}{2\Delta_i x}, \quad 1 \le i \le n,$$

where $\Delta_i x = (\Delta \delta_{ij})_{j=0}^n$. Analogously to (43), we compute $\widehat{I}_{ln}$ and $\widehat{\frac{\partial I_{ln}}{\partial x_i}}$ on LIBORs simulated according to lognormal approximation of transition kernel $p_{ln}^L(0, x, T_1, \cdot)$. Table 1 and Table 2, we show 0-values of European swaptions and the Deltas, computed via estimators $\widehat{I}_{ex}$, $\widehat{I}_{ln}$, $\widehat{I}_0$, $\widehat{I}_1$ and $\widehat{\frac{\partial I_{ex}}{\partial x_1}}$, $\widehat{\frac{\partial I_{ln}}{\partial x_1}}$, $\widehat{\frac{\partial I_0}{\partial x_1}}$, $\widehat{\frac{\partial I_1}{\partial x_1}}$, correspondingly, for different maturities T_1. To compute the values in the tables, M is taken equal to 3×10^6 and 2×10^6 correspondingly, to keep standard deviations within 0.5 % relative to the values. The WKB approximation is computed up to first order, i.e. we have computed c_0 and c_1. This leads to a very close estimate of the European swaptions and Deltas, even for large maturities. This experiments have been confirmed recently in the case of more complex products, even Bermudean options. This will be published in a revised form of [KKS]. In Table 1 and Table 2 T_1 denotes the maturity.

References

[CrKa] Croitoru, C., Kampen, J.: Accurate numerical schemes and computation of a class of linear parabolic initial value problems. (in preparation)

[Du] Duffie, D.: Dynamic Asset Pricing Theory. Princeton, Princeton University Press (2001)

[FrKa] Fries, C., Kampen, J.: Proxy Simulation Schemes for generic robust Monte-Carlo sensitivities, process oriented importance sampling and high accuracy drift approximation (with applications to the LIBOR market model). Journal of Computational Finance, Vol. 10, Nr. 2

[FrKa2] Fries, C., Kampen, J.: Proxy Simulation Schemes for generic robust Monte-Carlo sensitivities based on dimension reduced higher order analytic expansions of transition densities. (in preparation)

[Fr] Fries, C.: Mathematical Finance. Theory, Modeling, Implementation. Wiley, Hoboken (2007), http://www.christian-fries.de/finmath/book

[Ka] Kampen, J.: The WKB-Expansion of the fundamental solution of linear parabolic equations and its applications. Book, submitted to Memoirs of the American Mathematical Society (electronically published at SSRN 2006)

[Ka2] Kampen, J.: How to compute the length of a geodesic on a Riemannian manifold with small error in arbitrarily regular norms. WIAS preprint (to appear)

[Ka3] Kampen, J.: Regular polynomial interpolation and global approximation of global solutions of linear partial differential equations. WIAS preprint (2007)

[KKS] Kampen, J., Kolodko, A., Schoenmakers, J.: Monte Carlo Greeks for callable products via approximative Greenian Kernels. WIAS preprint, revised version to appear in SIAM Journal of computation (2007)

[Kr] Krylov, N.V.: Lectures on Elliptic and Parabolic Equations in Hölder Spaces. Graduate Studies in Mathematics, Vol. 12, American Mathematical Society (1996)

[Sc] Schoenmakers, J.: Robust Libor Modelling and Pricing of Derivative Products. Financial Mathematics. Chapman & Hall/CRC (2005)

Printing: Krips bv, Meppel, The Netherlands
Binding: Stürtz, Würzburg, Germany